JOINING FIBRE-REINFC

JOINING FIBRE-REINFORCED PLASTICS

Edited by

F. L. MATTHEWS

Director, Centre for Composite Materials,
Imperial College of Science and Technology,
London, UK

ELSEVIER APPLIED SCIENCE
LONDON and NEW YORK

ELSEVIER APPLIED SCIENCE PUBLISHERS LTD
Crown House, Linton Road, Barking, Essex IG11 8JU, England

Sole Distributor in the USA and Canada
ELSEVIER SCIENCE PUBLISHING CO., INC.
52 Vanderbilt Avenue, New York, NY 10017, USA

WITH 17 TABLES AND 189 ILLUSTRATIONS

British Library Cataloguing in Publication Data

Joining fibre-reinforced plastics.
1. Fiber reinforced plastics—Joints
I. Matthews, F. L.
668.4'94 TP1177.5.F5

Library of Congress Cataloging in Publication Data

Joining fibre-reinforced plastics.

Bibliography: p.
Includes index.
1. Fiber reinforced plastics—Joints. I. Matthews, F. L.
TA455.P55J65 1986 620.1'923 86-11648

ISBN 1-85166-019-4

Printed in Great Britain by Galliard (Printers) Ltd, Great Yarmouth

Preface

The performance of a structure, or component, is critically dependent upon the behaviour of any joints it contains. All too often a weight or strength advantage, brought about by clever design or use of materials, is lost because the characteristics of the associated joints were not properly understood.

As fibre-reinforced composite materials have become more widely used, so the need for reliable, efficient, load-carrying joints has become apparent. In recent years, a very large number of articles on most aspects of joints in composites has appeared in the literature.

It seemed timely, therefore, to gather together the major features of joints' behaviour in one publication. I wish to thank my colleagues (Terry Collings, Ken Liechti, Steve Johnson, Dave Dillard, Bob Adams, and John Hart-Smith) for allowing themselves to be persuaded to participate in this endeavour; without their collaboration there would be no book. A special word of thanks is also due to Hal Brinson, who, although himself unable to contribute, found the authors for Chapter 4.

We believe that this text will be useful to research workers in universities and other establishments, and to engineers and designers in all branches of industry using fibre-reinforced plastics for load-bearing structures.

F. L. MATTHEWS

Preface

Contents

List of Contributors

R. D. ADAMS

Professor, Department of Mechanical Engineering, University of Bristol, Queens Building, University Walk, Bristol BS8 1TR, UK

T. A. COLLINGS

Principal Scientific Officer, Royal Aircraft Establishment, Structures Department, X-34 Building, Farnborough, Hampshire GU14 6TD, UK

D. A. DILLARD

Assistant Professor, Department of Engineering Science and Mechanics, Virginia Polytechnic Institute and State University, Blacksburg, Virginia 24061, USA

L. J. HART-SMITH

Senior Engineer Scientist, Department CI 253, Mail Stop 36-90, Douglas Aircraft Company, McDonnell Douglas Corporation, Long Beach, California 90846, USA

W. S. JOHNSON

Senior Research Engineer, Materials Division, NASA—Langley Research Center, Hampton, Virginia 23665, USA

K. M. LIECHTI

Assistant Professor, Department of Aerospace Engineering and Engineering Mechanics, University of Texas, Austin, Texas 78712-1085, USA

F. L. MATTHEWS

Director, Centre for Composite Materials, Imperial College of Science and Technology, Prince Consort Road, London SW7 2BY, UK

Chapter 1

Introduction

F. L. MATTHEWS

Centre for Composite Materials,
Imperial College of Science and Technology,
London, UK

1.1. INTRODUCTION

Ideally, a structure would be designed without joints, the latter being a source of weakness and/or excess weight. In practice limitations on component size imposed by manufacturing processes, and the requirements of inspection, accessibility, repair and transportation/assembly, mean that some load-carrying joints are inevitable in all large structures.

The designer has essentially two basic techniques at his disposal for joining components of fibre-reinforced plastics, mechanical fastening and adhesive bonding. The advantages and disadvantages of these approaches may be summarised as follows:

Mechanically Fastened Joints

Advantages:

(i) No surface preparation of components required.
(ii) Disassembly possible without component damage.
(iii) No abnormal inspection problems.

Disadvantages:

(i) Holes cause unavoidable stress concentrations.
(ii) Can incur a large weight penalty.

Bonded Joints

Advantages:

(i) Stress concentration can be minimised.
(ii) Incur a small weight penalty.

Disadvantages:

(i) Disassembly impossible without component damage.
(ii) Can be severely weakened by environmental effects.
(iii) Require surface preparation.
(iv) Integrity difficult to confirm by inspection.

It is well known that, for metals, fatigue is the major cause of structural failure and many fatigue failures are associated with some type of joint.[1] Although the fatigue-to-static strength ratio is on the whole better for composites than metals, it is still necessary to have a detailed understanding of the behaviour of joints in composites. The reasons for this are associated with the basic characteristics of metals and composites.

Because of their ductile nature an accurate stress analysis is usually not required for metals, and design procedures for joints are reasonably straightforward and well-established. The essential features of lap joints, the usual type of load-carrying joint and the subject of this book, mean that the structural elements being joined will be subjected to high in-plane direct and shear stresses and relatively high through-thickness direct and shear stresses. Fibre-reinforced plastics (FRP) composites have a relatively low in-plane shear strength, and are particularly weak in through-thickness shear and tensile modes. Hence a comprehensive stress analysis is required to provide an understanding of the factors controlling joint strength, to give guidance on the strengths to be expected from a particular design, and to assist in producing improved joint designs with consequent gains in efficiency.

Experimentally determined behaviour of both small and full-size joints is also required to complement the theoretical work. Knowledge of failure modes, at macroscopic and microscopic levels, is essential if theoretical models are to be refined. One of the problems with composites is the very large number of materials that can be evolved by changing fibre, matrix, fibre orientation, lay-up, stacking sequence, etc. In view of the prohibitive cost of testing all possible combinations a general predictive method is clearly needed. Although, at present, no universal technique exists, reliable methods are being used in a number of (restricted) cases.

In the present text, lap joints are emphasised only to limit the scope and not because other types of joint are considered to be unimportant. For example, flange joints play a vital role in chemical plant where pipes are joined to vessels, and in turbojets where casing segments join. When such components are manufactured from FRP, difficulties will be encountered which are again associated with the anisotropic nature of the material.[2] Likewise, special joints which are neither lap nor flange type may be needed for particular situations. A good example of this is the root attachment of helicopter rotor blades which can be effected by wrapping continuous fibres around a pin.[3] Another example of composites in joining is the use of FRP, with either thermosetting or thermoplastics matrices, to manufacture rivets.[4] The object here is to reduce galvanic corrosion problems associated with aluminium alloy, steel and stainless steel fasteners in carbon fibre composites.

In the current book, then, single, double, scarf and stepped lap joints will be discussed, although the emphasis will change from chapter to chapter. For bonded joints all types will be examined, although most consideration will be given to single and double laps, whereas for mechanical joints attention will be focused almost exclusively on double lap configurations.

1.2. MECHANICALLY FASTENED JOINTS

1.2.1. Experimentally determined strength

The general behaviour of mechanically fastened joints can be conveniently discussed, as in Chapter 2, under three broad headings: fastener type; materials; geometric factors.

Although mechanically fastened joints in FRP exhibit the same failure modes as in metals, the mechanisms by which damage initiates and propagates is fundamentally different and extremely complex.[5] The dependence of damage development and associated strength on through-thickness restraint means that torque-tightened bolts are superior to rivets which, in turn, are superior to other forms of mechanical fastening.

The influence on failure of the relative values of fibre and matrix failure strain explains, in part, the strength differences between CFRP (carbon or graphite fibre-reinforced epoxy resin), GFRP (glass fibre-reinforced epoxy), KFRP ('Kevlar' fibre-reinforced epoxy), and GRP (glass fibre-reinforced polyester resin). Lay-up and stacking sequence are also important since they determine the stress distribution around the fastener hole.

Joint behaviour, and particularly the failure mode, is also dependent on joint geometry, i.e. width in single-hole joints or pitch in multi-hole joints; end (or edge) distance; hole diameter; laminate thickness.

In addition to the above, joint performance is also determined by several other factors including: hole quality; the fit of the fastener in the hole (i.e. the relative diameters of fastener and hole); and the fit of the fastener in the washer (for bolted joints). The latter effect can be particularly important in fatigue.[5]

1.2.2. Stress analysis and strength prediction

Until quite recently far more theoretical work had been undertaken for bonded than for mechanically fastened joints.[6,7] This was largely for historical reasons. Bonded joints in metals had received fairly extensive analytical treatment in the 1940s and 1950s, and this foundation was rapidly extended when composites came into large-scale use in the late 1960s. Mechanically fastened joints in metals, on the other hand, had received very little analytical consideration and this, coupled with an initial reluctance to make such joints in composites, meant that the topic got off to a slow start. Developments have, however, been rapid in the past few years as both continuum methods and finite element techniques have been brought to bear on the problem. Both these approaches, together with methods of strength prediction, are discussed in Chapter 3.

Many of the classical continuum analyses are based on the fundamental work of Bickley[8] for isotropic materials. A rigorous approach demands the inclusion of anisotropic elasticity, friction at the fastener/hole interface, finite plate dimensions, fastener flexibility and fastener/hole tolerance. Such analyses are two-dimensional and hence strictly apply only to pin-loaded holes. It is clear that the length of contact arc on the pin surface, and the regions of slip and non-slip within this arc, should be determined from the analysis, not assumed at the outset. Also, the commonly made assumption that pin bearing pressure is distributed in a cosinusoidal fashion is seen to be seriously in error in certain circumstances.

If the failure process is to be accurately modelled a three-dimensional method, which includes the two-phase nature of composites, is required. Currently this means using finite element techniques. To include all the effects mentioned above, together with through-thickness stresses and the influence of bolt tightening, is a formidable and expensive task which has not apparently been attempted to date. However, the simplified three-dimensional analyses that have been undertaken show encouraging correlation between stress distributions and observed failure modes.

No universally applicable method of strength prediction is available. Although for some failure modes two-dimensional methods give good results, their success appears to be specific to the particular material under investigation.

1.3. BONDED JOINTS

1.3.1. Experimentally determined strength

The vital difference between bonded and mechanically fastened joints in FRP is in the respective dimensions of the load transfer elements: fractions of a millimetre for the adhesive layer, several millimetres for the fasteners. The very thin adhesive layer gives rise to special difficulties, as indicated by Brinson.[9]

As with mechanically fastened joints, a thorough understanding of the initiation and progress of the failure process is vital if analytical models and widely applicable predictive methods are to be derived. Because they allow a detailed description of the failure behaviour, the methods of fracture mechanics are emphasised in Chapter 4.

Because of the mixed-mode nature of debonding, it is suggested that three different specimen geometries be used to determine fracture properties. The double cantilever beam is used when Mode I effects are relevant, the cracked lap shear specimen for mixed Mode I and Mode II effects, and the end-notched flexure specimen for purely Mode II. Single lap shear specimens, commonly used for testing, are useful only for purposes of ranking, but not of rational design.

Detailed information of failure can only be obtained from refined methods of measuring displacements and strains. No single technique is suitable and a combination of pointwise methods, using transducers, and whole field methods, using Moiré interferometry, is needed.

Since the matrix of FRP laminates may be much weaker than current structural adhesives, failure may occur within the composite by delamination or interply fracture, rather than debonding. The exact nature of failure will therefore depend on the surface ply orientation, stacking sequence, joint geometry and loading, in addition to the matrix and adhesive.

1.3.2. Stress analysis and strength prediction

As stated in Section 1.2.2, continuum methods of stress analysis of bonded lap joints in composites are derived from investigations of similar joints between metal adherends. In these analyses a longitudinal section of the joint is considered, giving therefore a two-dimensional, or in the simplest cases one-dimensional, stress distribution. The latter approach, which ignores through-thickness stresses, produces a physically infeasible, i.e. non-zero, adhesive shear stress at the ends of the overlap. Nevertheless, such an approach is useful for parametric studies and can provide valuable general guidance for design purposes.[7]

The inclusion of through-thickness stresses, adherend anisotropy and stacking sequence, non-linear behaviour and geometric refinements (double lap, scarf and stepped lap configurations) introduce additional complexity into the analysis. In some instances it may be difficult to obtain a solution to the governing equations.

When real effects, such as the spew-fillet at the edges of the adhesive, are to be considered, finite element methods must be used if an accurate stress distribution is required. The thin adhesive layer causes particular difficulties for finite element modelling, especially if through-thickness effects are to be included. A satisfactory element mesh within the adhesive can lead to a very large number of elements in the whole joint; there is a clear need for special elements to represent the adhesive layer. The number of elements will probably become unacceptably large when the necessary modelling of the adherends on a ply-by-ply basis is also included.

Comparison with measured strengths shows that it is essential to include the non-linear shear stress–strain behaviour of the adhesive, if meaningful predictions of strength are to be achieved. One difficulty, still to be satisfactorily resolved, is the accurate measurement of such properties.

The above, and other related factors, are discussed in Chapter 5.

1.4. DESIGN

The ultimate objective of the experimental and theoretical investigations described above is to produce information that can be used for design purposes. At any moment in time, design methods will rarely include the most advanced state of current knowledge. This is inevitable since industry has to meet deadlines and to achieve this may have to accept a slightly less efficient product. Such a situation need not worry us provided the results of the latest research work are incorporated into the design process at the earliest opportunity.

The final chapters of the current text address the design of mechanically fastened (Chapter 6) and bonded (Chapter 7) joints. The approach described reflects present-day practice in the aerospace industry which, of necessity, designs to higher safety standards than industry in general. The methods should, therefore, be applicable to all structural joints, no matter what the field of application.

Because there is no universally applicable failure theory for composites, the design of mechanically fastened joints is best treated from a general point of view which does not need to be changed as different materials are considered. One benefit of this approach is to allow definition of a minimum programme of tests to characterise adequately joints in any composite. In describing the behaviour of single fastener joints, and the influence of material variables, there will be some overlap with Chapter 2. However, this is necessary in order to develop a generalised approach to multi-row joints and obtain a measure for structural efficiency.

A major problem with bonded joints is the sensitivity of the adhesive to moisture. The design of joints should therefore ensure that such degradation is precluded by correct choice of adhesive. Joint geometry can be related to applied load intensity and undesirable effects such as 'peel' (through-thickness tensile) stresses can be minimised by suitable tapering the adherends. Adequate understanding of joint behaviour allows us to specify acceptable repair methods, a topic which is assuming ever-increasing importance.

REFERENCES

1. Kirkby, W. T. Report No. AGARD-AG-176 (Ed. H. Liebowitz), 1974, Advisory Group for Aerospace Research & Development, Paris, France.
2. Matthews, F. L., Foulkes, D. M., Godwin, E. W. and Kilty, P. F., *Proc. 14th Reinforced Plastics Congress*, 1984, The British Plastics Federation, London, UK.
3. Stratton, W. K., Scarpati, T. S. and Class, C. A., *Proc. 4th Army Materials Technology Conference, Advances in Joining Technology* (Eds J. J. Burke, A. E. Gorum and A. Tarpinian), 1975, Brook Hill Publishers, Chestnut Hill, Mass., USA.
4. Tanis, C. and Poullos, M., Jr, *Fibrous Composites in Structural Design* (Eds E. M. Lenoe, D. W. Oplinger and J. J. Burke), 1980, Plenum Press, New York, USA.
5. Smith, P. A., PhD Thesis, 1985, Cambridge University, Cambridge, UK.
6. Godwin, E. W. and Matthews, F. L., *Composites*, 1980, **11**, 155.
7. Matthews, F. L., Kilty, P. F. and Godwin, E. W., *Composites*, 1982, **13**, 29.
8. Bickley, W. G., *Phil. Trans. Roy. Soc.*, 1928, **A227**, 399.
9. Brinson, H. F., *Appl. Mech. Rev.*, 1985, **38**, iii.

Chapter 2

Experimentally Determined Strength of Mechanically Fastened Joints

T. A. COLLINGS

Royal Aircraft Establishment, Farnborough, Hampshire, UK

2.1. INTRODUCTION

The use of mechanically fastened joints in fibre-reinforced plastics (FRP) is a logical carry-over from the fastening technique used for structures made from isotropic materials where a wealth of experience and understanding already exists. The enthusiasm of designers to use mechanical joints for composites has in the past been lukewarm due partly to a lack of confidence in the ability of composites to suffer holes and cut outs, and partly to designers extending joining techniques used for isotropic materials to composites without understanding the anisotropic nature and failure mechanisms of composites. This is not the case today because of a new approach in composite thinking and the development of new design philosophies based on an understanding of the mechanisms of failure of FRP.

It is true that FRP components can be considerably weakened by the introduction of holes; this is attributable in part to the large stress concentrations that occur in the region of such discontinuities, and partly to a lack of plasticity. This is evident if we look at the value of the tensile elastic stress concentration due to a circular hole in a unidirectional infinite sheet, which can be as large as 8 in contrast to the value of 3 normally associated with isotropic materials. Furthermore, because most isotropic materials exhibit some degree of plasticity, yielding can occur at regions of high stress and the effect of stress concentrations on the final net failing stress is small. This is not the case for unidirectional FRP, which is essentially elastic to failure, so the effect of stress concentration is to give rise to a low net tensile stress. It is not surprising therefore that the efficiency of mechanical joints in unidirectional FRP is very low indeed. If, however, the degree of anisotropy is reduced in the vicinity of a hole some 'softening' or pseudo-plastic behaviour can be introduced, and an increase in efficiency brought about. Indeed such joint softening can be readily achieved (see ref. 1) by the incorporation of fibres oriented in different directions in the vicinity around a hole.

It is clear from the experimental evidence available that mechanical joints in FRP are indeed promising; however, because of the large number of variables involved, complete characterisation of all joint materials, joint types and failure mechanisms is difficult. The approach set out in this chapter is to demonstrate the behaviour of some types of joints, materials and fasteners, and where possible to infer the influence of important parameters and to highlight design philosophies.

2.2. FAILURE MODES

Mechanically fastened joints in composites display the same failure modes as do metals; i.e. failure can take place in tension or shear or bearing. Because of the nature of FRP two other modes of failure are possible, namely cleavage and pull-out. Figure 1 shows the location of each of the modes. The failure mechanisms of FRP, unlike those of metals, are complex and varied, and will be dependent upon many factors such as fibre type, orientation, surface treatment and matrix, etc. It follows that a knowledge of a wide range of variables is needed if favourable joint conditions are to be achieved and unwanted failure modes avoided.

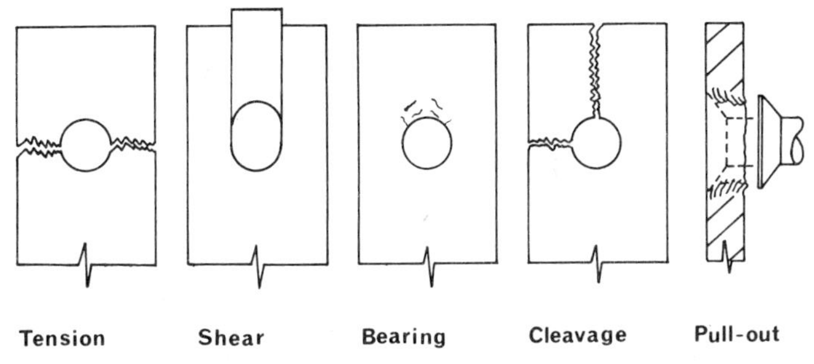

FIG. 1. Modes of failure for mechanical joints in FRP.

Failure of laminates made from unidirectional rovings, in which at least 10% of the fibres are aligned in the direction of the load, will now be considered since this fibre arrangement is usually required to perform at high stresses and at near-optimum strength.

2.2.1. Tension failure

As with conventional materials, the tensile load required to fail a laminate through a section at which holes occur (net section) is less than at a section at which there are no holes (gross section). The stresses at these sections, at failure, are given respectively by

$$\sigma_N = \frac{P}{(w - nd)t} \tag{2.1}$$

and

$$\sigma_G = \frac{P}{wt} \tag{2.2}$$

where P is the failing load of the member, w the joint width at the net section, n the number of holes of diameter d occurring at that section, and t the joint thickness. The tensile strength efficiency achieved at these sections, expressed in the form of average net and gross stress concentrations, is given by

$$k_N = \frac{\sigma_\infty}{\sigma_N} \tag{2.3}$$

and

$$k_G = \frac{\sigma_\infty}{\sigma_G} \tag{2.4}$$

where σ_∞ is the theoretical ultimate tensile strength of a plain laminate.

Since FRP materials are essentially elastic to failure, they cannot take advantage of plastic behaviour and yielding at the hole edge; therefore the effect of stress concentration is to give rise to a correspondingly low net failing stress. Because the value of the net failing stress will depend upon the degree of anisotropy in the region of a hole, fibre orientation will feature strongly in the way failure occurs and in the magnitude of the ultimate failing stress. Figures 2 and 3 show typical tensile failures for different lay-ups of carbon fibre-reinforced plastic (CFRP)[1] and glass fibre-reinforced plastic (GFRP)[2] using unidirectional rovings.

When a CFRP laminate is subjected to a tensile load the fibres parallel to the loading axis will carry most of the load; tensile fracture of a laminate will be governed by failure of these fibres. The process by which tensile failure of the net section can occur for loaded holes can be explained by a mechanism postulated by Potter[3] for a holed laminate being loaded remotely from the hole. Failure occurs at the net section, being initiated by and propagated from the stress concentration at the edge of the hole at a location 90° to the loading axis. This means that all load-carrying axial fibres must have failed sequentially in such a way that failure of one fibre inevitably led to the failure of its immediate neighbours. Since the tensile strength of a loaded hole is almost completely controlled by axial fibres, it is this sequential fibre failure process which will also govern laminate failure. Acoustic emission work on CFRP by Collings and Mead[4] has shown that for a $0/\pm 45°$ laminate, geometrically constrained to fail in tension, at sufficiently low shear and bearing stresses to reduce other possible acoustic emission sources, fibre failure first occurs at approximately 85% of the net strength, after which a stable situation is reached. No further fibre failure occurs until about 95% of net strength

FIG. 2. Tensile failure on $0/\pm 45°$ CFRP.[1]

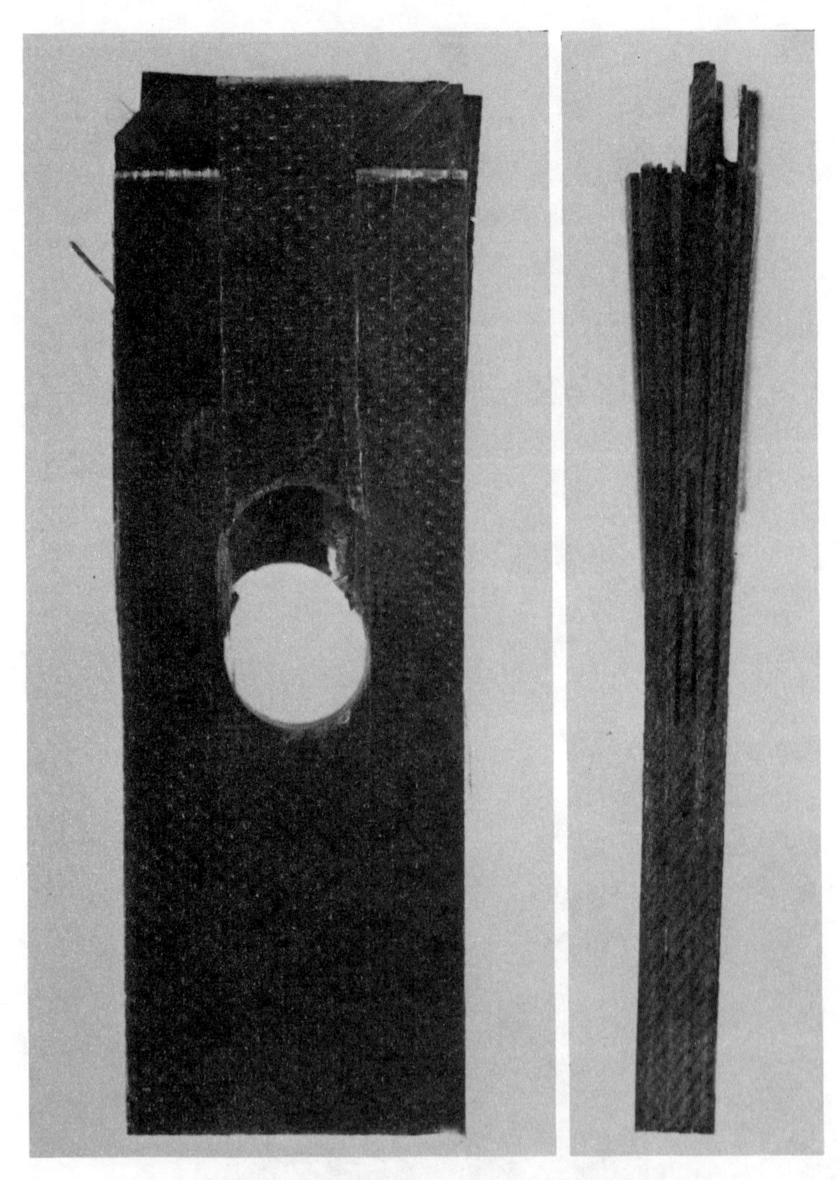

FIG. 3. Tensile failure in $0/\pm45°$ GFRP.[2]

is attained. Failure then proceeds in a catastrophic manner until ultimate failure. The acoustic emission recorded during such a test is given in Fig. 4.

GFRP failures are more complex and experimental evidence (Fig. 3 and refs 2, 5 and 6) has demonstrated, for GFRP made from preimpregnated warp sheet, that failure starts in a forced extrusion of the 0= fibres, over the projected area of the hole, out through the joint end. Failure propagates by in-plane shearing between the 0° and 45° plies across the full width of the laminate. As a consequence only the 45° fibres remain capable of sustaining any axial load at the net section, and failure will occur in tension immediately after in-plane shearing. Because the failure process is shear-limited, conventional tensile failures are difficult to achieve for laminates containing more than 50% 0° fibres.

'Kevlar', an aramid fibre, exhibits a high tensile strength in the fibre direction; however, because of the relatively low lateral cohesion of the fibre, fibrillation (fibre splitting along the fibre axis), rather than fibre fracture, can occur. This gives a stress concentration relief mechanism at the edge of a loaded hole and contributes to the load-bearing capability of 'Kevlar' fibre-reinforced plastic (KFRP) laminates.

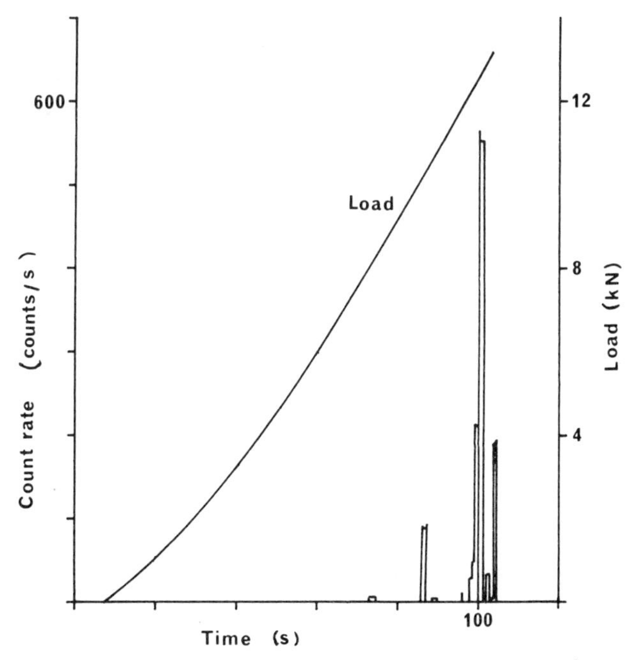

FIG. 4. Acoustic emission of a bolted joint in CFRP.[4]

2.2.2. Shear failure

The shear strength normally quoted for FRP is the interlaminar shear strength obtained using the well-known short beam shear test technique.[7] Unfortunately this is of little value in the estimation of joint shear strength where failure is due to in-plane shear. The in-plane shear strength can be measured using the rail shear test technique[8] but, with stress concentrations at the hole edge, it is unlikely to be a representative test. Strengths of joint shear are therefore best measured using loaded hole shear specimens.

The shear strength, as for conventional isotropic materials, is given as

$$\tau = \frac{P}{2et} \tag{2.5}$$

where e is the distance (parallel to the load) between the hole centre and the free edge, usually known as the end distance. Typical shear-type failures for CFRP laminates are given in Figs 5 and 6.

The joint shear strength of unidirectional FRP, when loaded in the direction of the fibres, is low compared with the in-plane shear strength. The difference between the two indicates the presence of significant shear stress concentrations around the loaded hole. In contrast, the shear strengths of $0/\pm 45°$ laminates are high and since they are known to be insensitive to end distance it can be concluded that the shear stress concentration is small. Therefore, as with tensile strength, fibre orientation will be a major consideration in designing against shear failures.

2.2.3. Bearing failure

Bearing of a pin in a hole gives rise to compressive stresses around the loaded half of the circumference of a hole and as might be expected, as shown by Collings,[1] the compressive strength of a 0° laminate is an important parameter in the determination of bearing strength. Similarly the effect of clamping due to the lateral constraint provided by a tightened bolt will also feature in deciding the magnitude of the bearing strength at which failure occurs. This has been demonstrated by Collings[1,9] for CFRP laminates and by Kretsis and Matthews,[5] Stockdale and Matthews[10] for GFRP laminates, and Matthews and Kalkanis[11] for KFRP laminates.

For practical purposes the bearing strength of a composite material, like that of a conventional material, is usually expressed as an average

Fig. 5. Shear failure in 0/90° CFRP.[1]

stress acting uniformly over the cross-sectional area of the hole, so that

$$\sigma_b = \frac{P}{ndt} \tag{2.6}$$

The mechanism of failure for CFRP has been shown[9] to occur in shear, through both fibres and matrix, initiating at the hole edge where the load intensity is a maximum. The point of failure will depend upon the fibre orientations and the proportion of each orientation contained in the laminate; typical failures are given in Fig. 7. It has also been shown that laminates containing some $\pm 45°$ and/or 90° plies perform well under bolt bearing conditions although this is contrary to what might be expected from a knowledge of the compressive strength behaviour of $\pm 45°$ and 90° laminates. Clearly the mechanism of failure in bearing, for these

Fig. 6. Shear failure in $0/\pm 45°$ CFRP.[1]

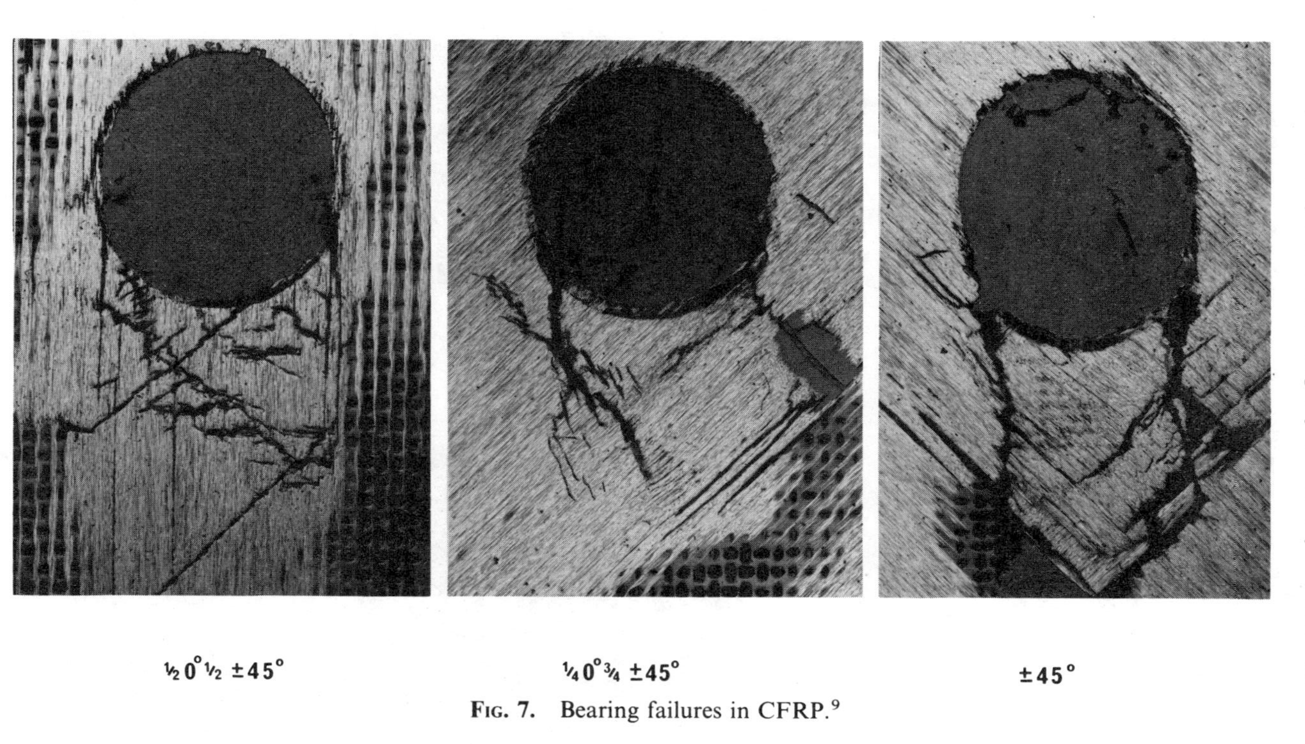

FIG. 7. Bearing failures in CFRP.[9]

orientations, is different from that in compression. The reason for this change is due to the lateral constraint provided by the clamping action of the bolt. This clamping action provides a constraint against the Poisson's expansion normal to the plane of the laminate so that fibres acting transversely to the load will undergo a different mode of failure, i.e. in constrained transverse compression.

The mechanism responsible for the failure of GFRP laminates has, as yet, not been identified. However, it is suggested from the test results reported in ref. 5 that the low modulus of glass fibre favours an instability mechanism that is not fully prevented by lateral constraint. This effect becomes more pronounced as the d/t ratio increases to 3.

For KFRP, using woven fabric, bearing strength test results reported in ref. 11 showed that the compressive properties of 'Kevlar' fibre, in combination with the woven nature of the fibre material used, was

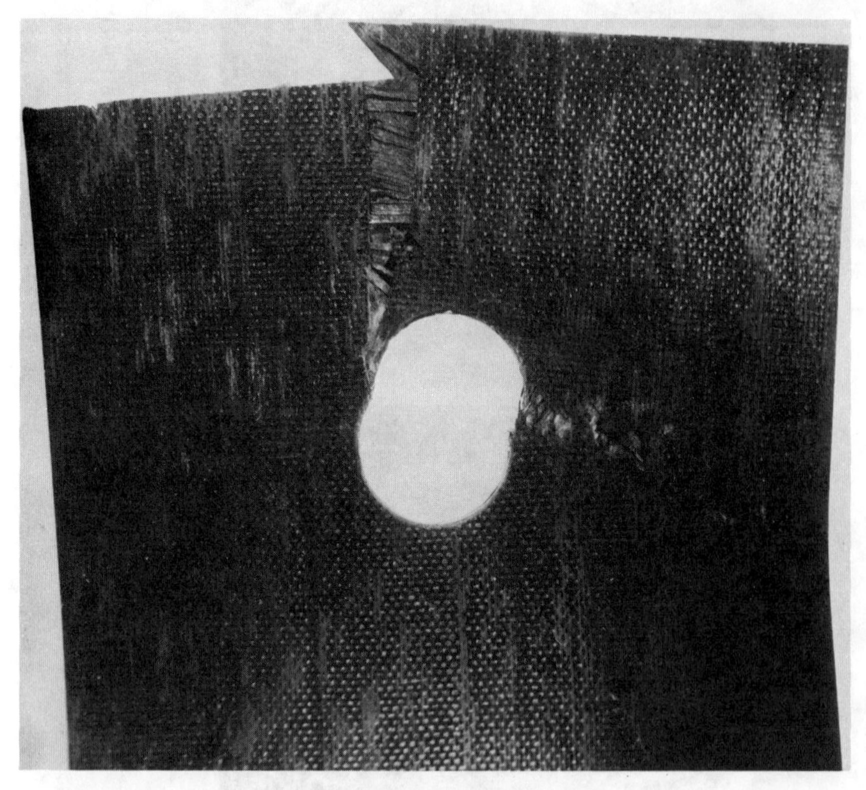

Fig. 8. Cleavage failure in CFRP ($\frac{2}{3}$ 0°, $\frac{1}{3}$ ±45°).[1]

responsible for a failure mechanism that was different from that of either CFRP or GFRP. Due to the fibrillar structure of 'Kevlar' fibre, weak planes of failure exist in the fibre and therefore KFRP laminates are less dominated by the strength of the interface between the fibre and matrix. The relatively weak bonding of these fibrils to each other may therefore feature in the bearing failure process.

2.2.4. Cleavage failure

Cleavage failure only occurs in $0 \pm \theta^\circ$ lay-ups which contain low proportions of non-axial fibres. Failure initiates in a single shear mode; this is followed by the failure of the net section on one side of the laminate. For most applications, lay-ups containing fibre orientations that are susceptible to this type of failure would not be used; if such a lay-up is needed, further reinforcement in the vicinity of the hole would be required. Figure 8 shows a typical cleavage failure in a 2/3 0°, 1/3 $\pm 45^\circ$ CFRP laminate.

2.2.5. Pull-out failure

Failure of the pull-out type is most frequently associated with rivets, in particular countersunk rivets. Since rivets are mostly used in single shear, axial in-plane loading of a joint will impose out-of-plane bending and peeling of the joint. This peeling introduces out-of-plane transverse loads that try to pull the joint apart. Consequently, rivet bending will be severe and through-thickness shearing of the rivet head through the laminate thickness will occur.

2.3. CHOICE OF FASTENERS

Although the choice of fasteners available to designers is quite wide, the inherent weaknesses of composite materials can somewhat reduce this choice if reasonably efficient joints, in terms of strength-to-weight, are to be achieved. Examples of the types of fasteners that have been applied with varied success are screws, rivets and bolts, each of which will be considered in turn.

2.3.1. Self-tapping screws

Perhaps the simplest form of fastener is the self-tapping screw. This provides both a simple and inexpensive connection where access to the reverse side of a joint is impossible. The efficiency of this type of fastener is low since there is a likelihood of thread stripping and for this reason it is not recommended for use where frequent demounting is required. An improvement in screw efficiency can be achieved by the use of heli-coil inserts in the tapped hole in the composite, or by moulding in metal blocks that can subsequently be drilled and tapped by conventional means.

2.3.2. Rivets

Rivets have been shown, in general, to be suitable for joining laminates up to 3 mm thick. The rivet types and forms available are numerous with a choice between solid and hollow, each of which can be obtained in a range of head types and sizes. Since the process of installing them involves a closing pressure that is not always readily controllable, the degree of lateral clamping will vary considerably. In some cases the riveting operation can cause damage to a laminate; this will have implications on the ability of the joint to sustain efficient bearing loads. Matthews *et al*,[12] have related joint strength to the degree of clamping imposed on the laminate by a rivet and suggest that an optimum level of constraint exists. On external surfaces, where surface protrusions cannot be suffered, the use of countersunk rivets must be considered. The use of countersunks will of course impose a restriction on the minimum allowable laminate thickness and the choice of countersink angle will have to be considered. Work by Matthews and Leong[13] shows, for CFRP, that increasing the countersink angle increases joint strength in terms of improved rivet pull-through strength. This is not altogether surprising since a larger countersink angle provides a larger area for resisting rivet pull-through, although some restriction on the tensile strength capacity of the composite will result because of the reduced available net section. A conservative design for the net tensile area across a countersunk hole is given by Lubin[14] as the width of the joint minus an effective hole diameter times the adherend thickness.

It has been shown[13] that non-countersunk rivets are to be preferred to countersunk and that solid rivets give, in general, stronger joints than do hollow rivets.

2.3.3. Bolts

As composites become more committed for use in heavily loaded structures, so the need for strong mechanical fasteners becomes more pressing. In areas where parts are removed for inspection or maintenance bolted connections will be required. The strength behaviour of bolted joints, for CFRP laminates, has been covered extensively by Collings,[1] who investigated a wide range of variables such as lay-up, fibre orientation and bolt diameter, and their influence on the three main failure modes. This investigation showed bolted joints to be the most efficient form of mechanical fastening for CFRP and (perhaps surprisingly) they were superior, on a specific strength basis, to conventional structural materials for all of the joint failure modes.

2.4. MATERIALS

FRP composite materials embrace a wide range of reinforcing fibres, such as glass, carbon, boron and 'Kevlar'; of these, glass and carbon have seen most use in the UK.

Glass fibre is a comparatively cheap fibre with a high specific strength[15] (see Table 1) and has seen extensive commercial utilisation in the form of chopped strand mattings (CSM) and woven rovings (WR). Joint strengths for chopped strand mattings are not strongly direction sensitive since the in-plane properties can be considered isotropic due to the random nature of the fibres. Nevertheless fibre volume fraction will need to be considered in strength calculations because of differing material qualities. Woven fabric material is normally used with the warp direction parallel to the load direction and joint data collection exercises have usually been restricted to this lay-up configuration. More recently woven fabrics have been used in conjunction with unidirectional rovings to add in-plane shear stiffness by orienting the warp at 45° to the major loading direction. Strength data for this configuration are unfortunately not available at present.

The joint strength of CSM and WR and 50/50 mixtures of the two, CSM/WR, has been determined by Matthews *et al.*[17] This work covers a range of bolt diameters from 6 to 12 mm and a range of material thicknesses from 3 to 25 mm. Fibre volume fractions of the materials were typically 0·19–0·44 for CSM, 0·24 for CSM/WR and 0·44 for WR.

In more recent years glass in the form of unidirectional rovings has been used for application in the more highly stressed structural field, such as helicopter blades where high stiffness is not always a major design requirement. Because of its high specific strength combined with its cheapness, glass can be used with full effect to achieve a cost-effective performance.

Carbon fibre is used mostly in the form of preimpregnated sheets which consist of a number of aligned unidirectional rovings (in tows of 6000 or

TABLE 1
Mechanical Properties of GRP, CFRP and KFRP[2,15,16]

Material	*Fibre content (vol. %)*	*Density (g/cm³)*	*Tensile strength (MN/m²)*	*Tensile modulus (GN/m²)*	*Compressive strength (MN/m²)*	*Compressive modulus (GN/m²)*
Glass						
Unidirectional						
Wound epoxide	40–75	1·7–2·2	530–1 730	28–62	310–480	
Unidirectional polyester	32–54	1·6–2·0	410–1 180	21–41	210–480	
Bidirectional						
Satin weave polyester	32–50	1·6–1·9	250–400	14–25	210–280	9–17
Woven roving polyester	28–40	1·5–1·8	230–340	13–17	98–140	8–17
Random						
Preform polyester	14–32	1·4–1·6	70–170	6–12	130–160	
Hand and spray-up polyester	14–24	1·4–1·5	63–140	6–12	130–170	6–9
Carbon						
Unidirectional						
HTS 914	60	1·54	1 800	131	1 200	124
XAS 914	60	1·54	1 900	128	1 350	124
Kevlar 49						
Unidirectional	60	1·38	1 380	75	276	
Woven (fabric)	60	1·33	520	31	172	

10 000 fibres) impregnated with a resin matrix. In this form a variety of structural materials can be made by stacking plies in different directions in much the same way as plywood. With these variations in ply orientation, stiffness and strength properties change and, as a consequence, a large number of laminate configurations need evaluation. Studies on the effect of different laminate orientations and lay-ups have reduced somewhat the number of configurations that are likely to be used for structural members, and this has simplified the collection of design data. Some strength and stiffness properties for unidirectional material have been included in Table 1.

'Kevlar' fibres in composite structures offer a means of improved damage tolerance compared with that achieved with glass or carbon fibres. Whilst 'Kevlar' fibres have high strength along the fibre axis, the transverse strength is low. As already mentioned, the reason for this lies in the fibrillar nature of the fibre, which breaks down under transverse or shear stresses. It is this weak bonding of the fibrils that imparts the damage tolerance property to KFRP. Unlike carbon, 'Kevlar' fibres have a low compressive strength. The reason, it is suggested,[18] is that there is a cooperative buckling of the molecular chains which is permitted because of an allowable bond rotation. Some strength and stiffness properties[16] of 'Kevlar' fibres are included in Table 1.

As an alternative material, unidirectional tows of 'Kevlar' fibres have been woven to give a fabric where most of the fibre is concentrated in heavy warp threads to give plain and twill weaves. These are used in much the same way as glass fabrics, and are to be preferred where the high stiffness property is required in one direction.

2.5. INFLUENCE OF FIBRE ORIENTATION ON FAILURE MODE

The influence of fibre orientation and the proportion of fibres at the various orientations can best be demonstrated by looking at a single bolted joint in a particular family of laminate orientations. It has been shown by Collings[1] that the best overall joint performance for CFRP structures is achieved by using a combination of 0° and $\pm 45^\circ$ plies, where the 0° axis is the principal load axis. As might be expected, the ratio of 0° to $\pm 45^\circ$ plies strongly influences both the material property and the joint behaviour. The strength behaviour of the $0/\pm 45^\circ$ family of laminates, for a 6·35 mm bolt diameter and specific joint geometry, is described in

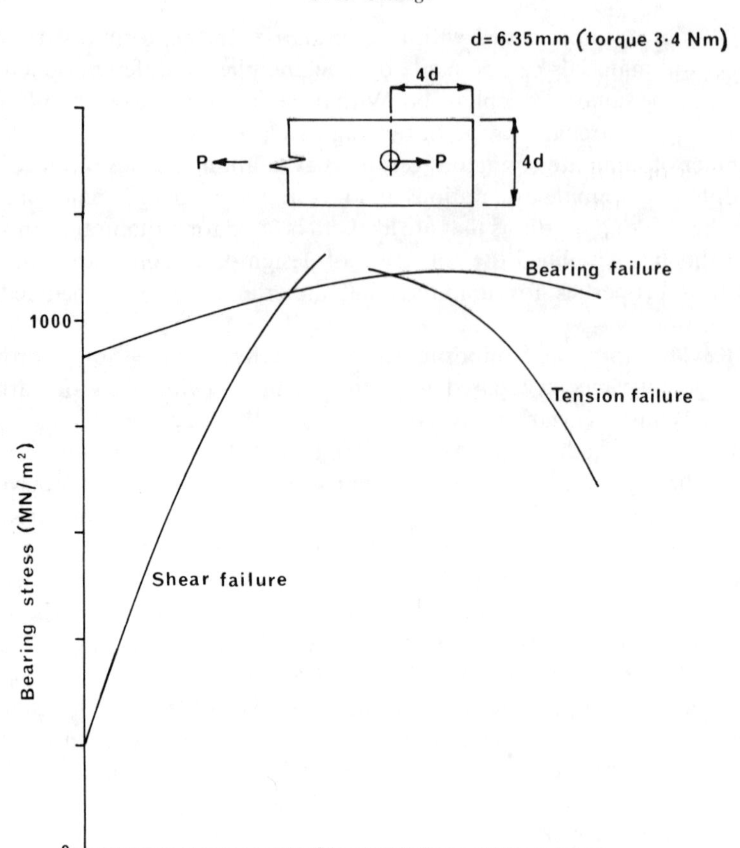

FIG. 9. Influence of fibre orientation on failure mode (0/±45° CFRP).[2]

Fig. 9 for the three main failure modes. As might be predicted, at low levels of 0° material shear failure will predominate until sufficient ±45° material is added to increase the shear strength to the point where bearing becomes the critical mode of failure. Since ±45° material is weak in tension, any continued increase in its contribution to a laminate's total thickness will eventually change the mode of failure to one of tension. Although Fig. 9 demonstrates a design curve for a particular joint geometry and fibre/resin type, many such curves can be derived from a knowledge of the strength behaviour of different lay-ups and fibre/resin types as a function of width or end distance. Such data will be presented later in the chapter (see Sections 2.8.1 and 2.8.2).

2.6. INFLUENCE OF STACKING SEQUENCE ON STRENGTH

The bearing strength of a pin-loaded hole is different from that of a bolt-loaded hole (see Section 2.7) since the lateral constraint provided by the clamping effect of the bolt is known to influence the mechanism of failure and hence the value of the ultimate failing stress. The effect of stacking sequence must therefore be considered separately for the unconstrained and the constrained cases.

The plies of a laminate can be stacked in more than one sequence through the thickness and as a consequence some laminate strength properties are affected. Collings[1] has shown for single bolted joints in two different 0/±45° CFRP laminates, each containing 2/3 |0° |and 1/3 ±45° plies, laid-up using two different stacking sequences, that there

TABLE 2
Lay-Up Number and Stacking Sequence[19]

Ply number to centreline[b]	*Ply orientation (degrees)*[a]				
	Lay-up number				
	1	*2*	*3*	*4*	*5*
1	+	+	+	+	+
2	0	−	−	−	−
3	−	0	+	0	0
4	0	0	−	0	0
5	90	+	90	90	0
6	0	90	0	0	0
7	+	−	0	+	0
8	0	0	0	−	+
9	−	0	0	0	−
10	0	0	0	0	90
Number of plies	20	20	20	20	20
Percentage of 0/±45/90°	50/40/10	50/40/10	50/40/10	50/40/10	50/40/10

[a] + and − refer to ±45° ply orientation.
[b] Lay-ups are all symmetric about the centreline.

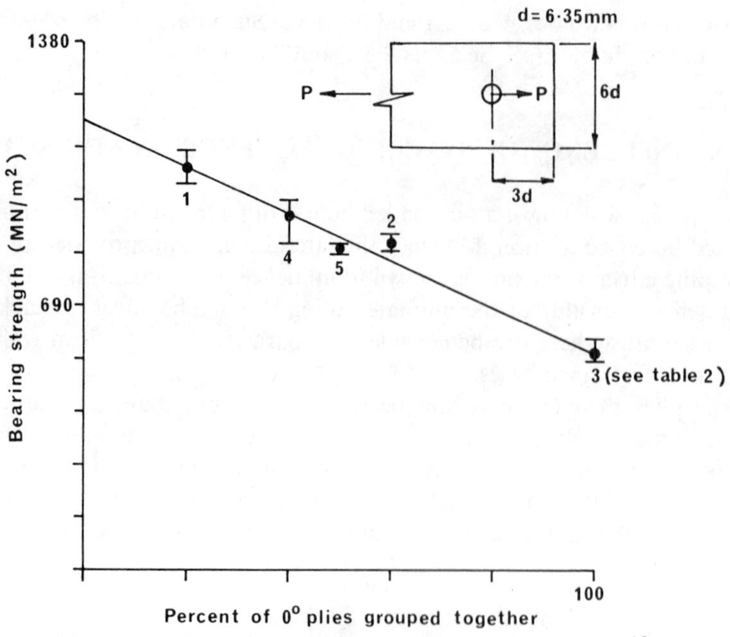

Fig. 10. Effect of grouped 0° plies on bearing strength.[19]

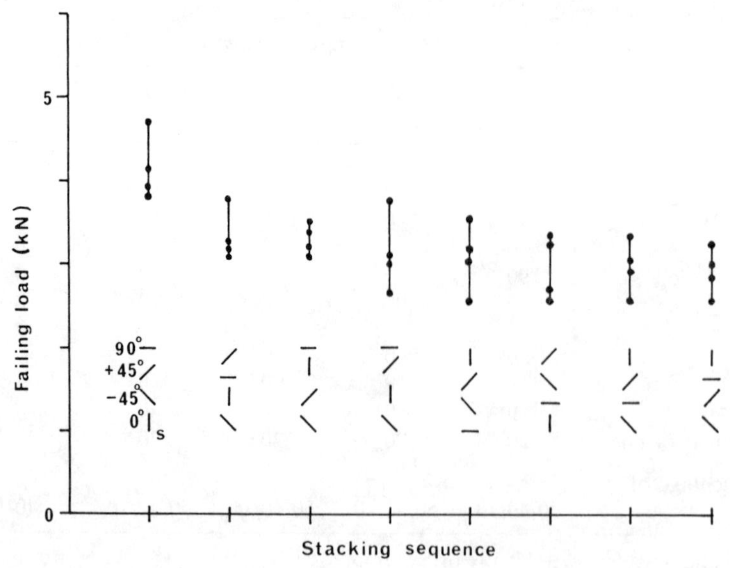

Fig. 11. Effect of stacking sequence on failing load for GFRP.[20]

was no difference between the measured shear strengths, but a difference of 6% between the tensile strengths. Bearing strengths, however, were significantly different, showing a drop of 16% for the more 'blocked' laminate. Garbo and Ogonowski[19] have shown for a 1/2 0°, 2/5 ±45°, 1/10 90° CFRP laminate that the joint bearing strength dropped by 50% as the 0° plies became more 'blocked', as shown in Fig. 10 and Table 2.

Pin-loaded holes (with no lateral constraint) in (0, 90, ±45°) GFRP laminates were investigated by Quinn and Matthews,[20] who showed a definite relationship between bearing strength and stacking sequence. Here it was shown that a $(90/0/\pm45°)_s$ laminate gave the highest bearing strength performance and a $(0/90/\pm45°)_s$ laminate gave the weakest bearing strength, with a strength reduction of 30% compared with that of the best laminate. The results of the investigation are given in Fig. 11.

2.7. EFFECT OF LATERAL CONSTRAINT

Early forms of compression tests carried out on unidirectional CFRP have demonstrated that a substantial degree of lateral constraint is necessary to prevent premature specimen end failure due to a breakdown in the fibre/resin interface resulting in a brush-like failure. Bearing of a pin in a hole gives rise to similar compressive behaviour around the loaded half of the hole, and indeed compression strength has been shown to be a major consideration in the prediction of bearing strength.[9] It is therefore not unexpected that lateral constraint will influence the magnitude of the bearing strength. Recent studies by Collings[9] and Garbo and Ogonowski[19] for CFRP laminates, and Matthews and Kretsis[5] and Stockdale and Matthews[10] for GFRP laminates, have shown that increasing the bolt-tightening torque, and hence the lateral constraint, increases the bearing strength. However, excessive overtightening of bolts may cause damage to the surface of the laminate and therefore the degree of lateral constraint must be kept to the minimum necessary to develop adequate bearing strength. Investigations into the degree of bolt clamping necessary to provide optimum bearing strengths for CFRP have been reported separately by Collings[1] and Garbo and Ogonowski[19] to be 22 MN/m^2.

Recent studies by Shivakumar and Crews[21] indicate that bolt clamping may decrease with time due to the viscoelastic behaviour of the resin. This behaviour is more pronounced with moisture uptake and high temperature. Shivakumar and Crews suggest that the reduction can be

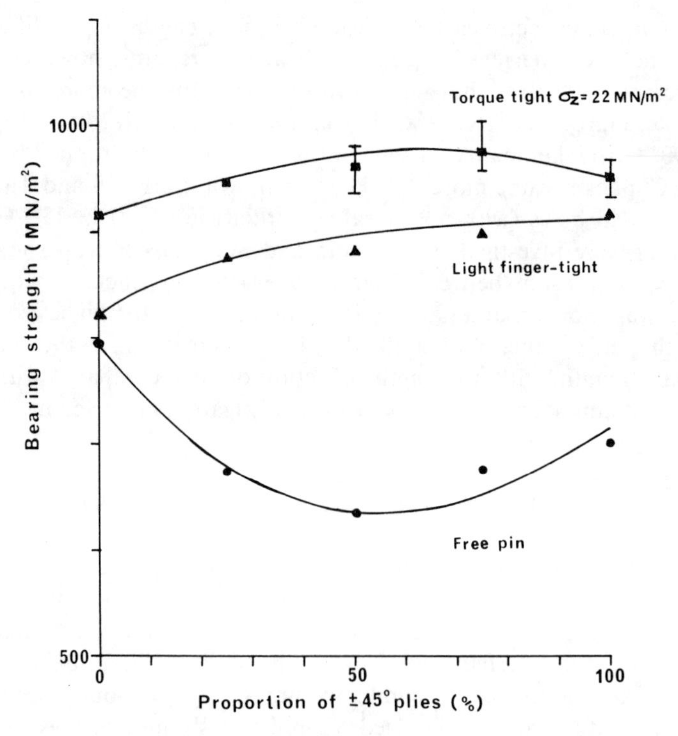

FIG. 12. Effect of lateral constraint on the bearing strength of HTS/914 $0/\pm45°$ laminates.[9]

as much as 60% at moisture levels of 1% at 60°C. On the other hand the loss of strength due to reduced lateral constraint has been demonstrated by Collings[9] (see Fig. 12), for a light finger-tight constraint and for dry $0/\pm45°$ laminates at room temperature, to be only 12% of that of the fully constrained case.

2.8. INFLUENCE OF JOINT GEOMETRY ON FAILURE

2.8.1. Width

Unlike isotropic materials, FRP cannot take advantage of plasticity and yielding at the hole, and therefore the effect of stress concentration on the net tensile failing stress will be strongly dependent upon width. However, it is evident from experimental data that the ultimate net and gross tensile strengths of a laminate are dependent upon ply orientation and that gross tensile strength is dependent upon hole size. Consideration of width effects cannot therefore be discussed without reference to both these parameters. The effect of width is most marked for 0/90° laminates and least for ±45° laminates. Such sensitivity to width has also been demonstrated by Waddoups *et al.*[22] in tests carried out on holed laminates, loaded in tension at the ends, where the value of the gross stress concentration was found to increase with hole diameter.

In general the joint width or fastener pitch is chosen so that a tension failure occurs at a mean stress as close as possible to the bearing strength of the material. For this reason it is more convenient to express the effect of width in terms of the bearing strength that can be sustained by the joint at failure. Typical failure curves for carbon, glass and 'Kevlar' FRP, for bolts loaded in double shear, are given in Figs 13–18.

2.8.2. End distance

The effect of end distance on shear strength, like that of width on tensile strength, depends upon the lay-up under consideration. For most lay-ups shear stress has been shown to decrease with increased end distance and account must be taken of this when designing joints against this mode of

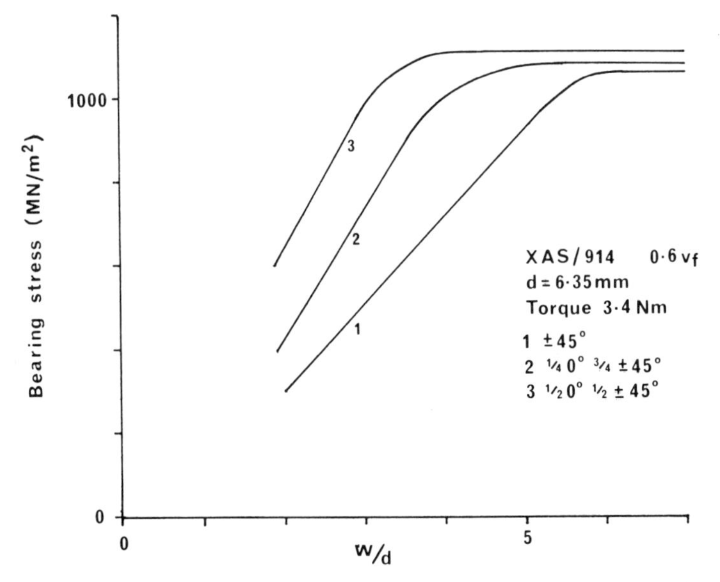

Fig. 13. Variation of bearing stress with w/d for 0/±45° CFRP.[2]

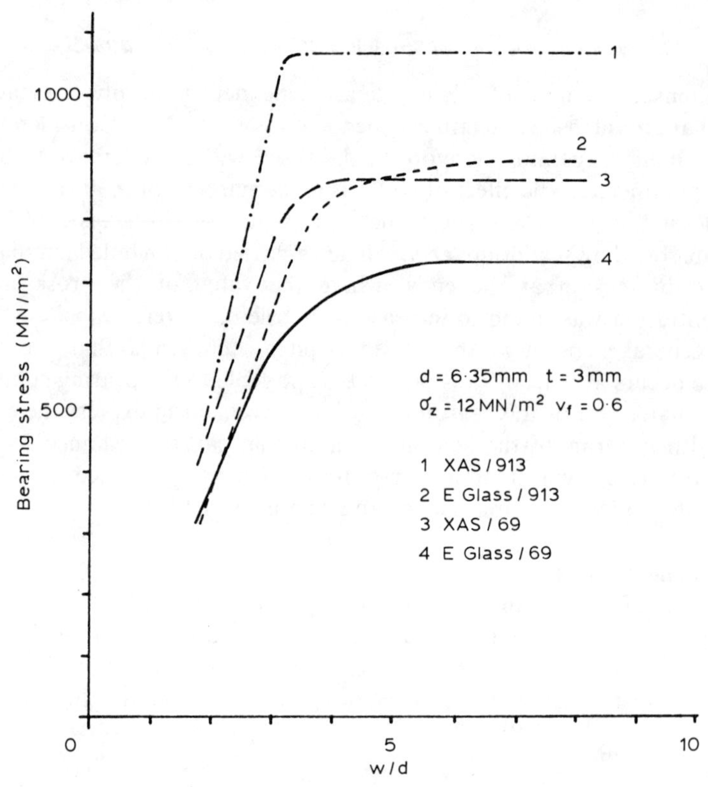

Fig. 14. Variation of bearing stress with w/d for $\frac{1}{3}$ 0°, $\frac{2}{3}$ ±45° laminates.[5]

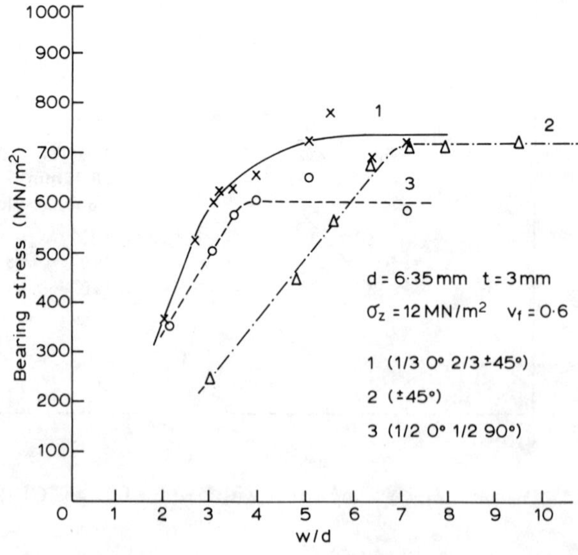

Fig. 15. Variation of bearing stress with w/d for various E Glass/69 laminates.[5]

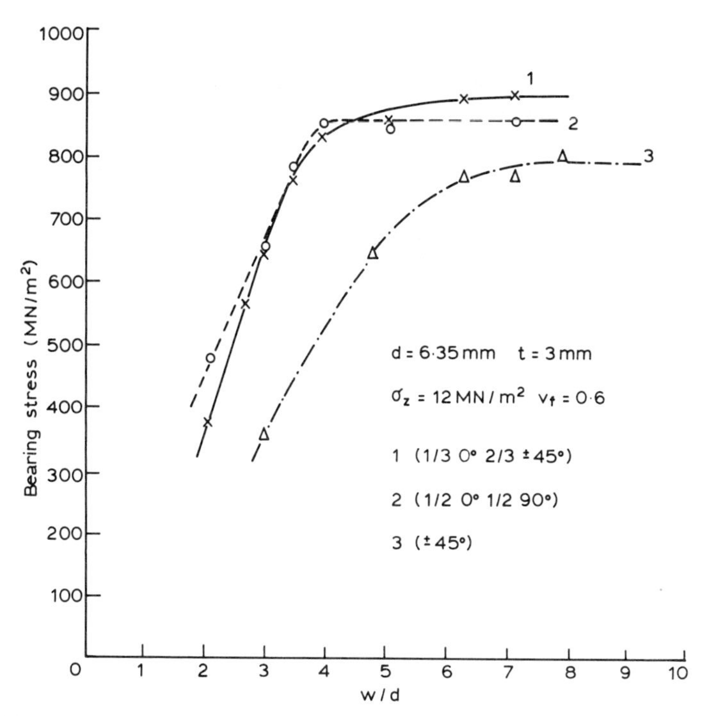

Fig. 16. Variation of bearing stress with *w/d* for various E Glass/913 laminates.[5]

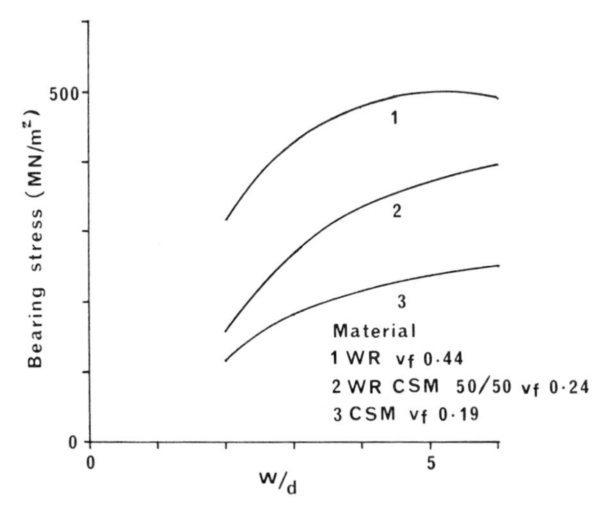

Fig. 17. Variation of bearing stress with *w/d* for GRP.[17]

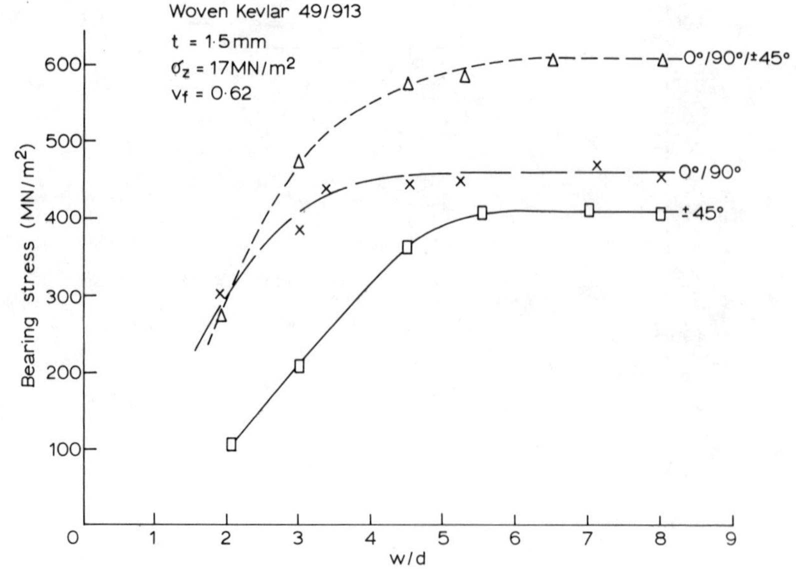

Fig. 18. Variation of bearing stress with w/d for KFRP.[11]

failure. For full joint efficiency, end distance will have to be chosen carefully to ensure that the maximum bearing strength capability is realised. The influence of end distance on bearing strength at failure for different CFRP and GRP laminates is given in Figs 19–24.

2.8.3. Hole size

The effect of hole size on joint strengths has been evaluated for all of the three main failure modes. For hole sizes of 6·35–12·7 mm it has been demonstrated by Collings[1] for $0/\pm 45°$ CFRP laminates, and by Matthews and Kretsis[5] for $0/\pm 45°$ GFRP laminates, that hole size has little influence on the net tensile strength, but, as might be expected, gross tensile strength is reduced as hole size increases.

Shear strengths of both CFRP and GFRP do not appear to be sensitive to hole diameter for the same range of hole sizes and lay-ups. Similarly the bearing strength of CFRP is unaffected, for the range of holes of 6·35–12·7 mm diameter, provided sufficient lateral constraint is available from the clamping action of the bolt. This is not the case for GFRP,[5] where it seems the low modulus of glass can favour instability effects that are independent of lateral constraint for values of $d/t > 3$. These effects may be occurring on the microscopic scale, but for values of $d/t > 3$

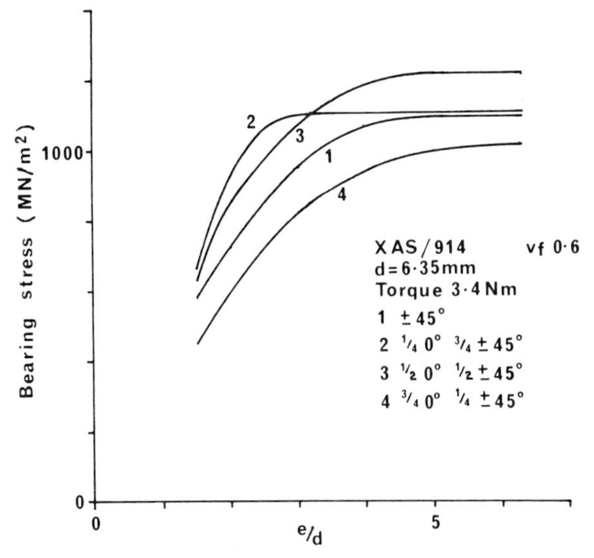

Fig. 19. Variation of bearing stress with e/d for $0/\pm 45°$ CFRP.[2]

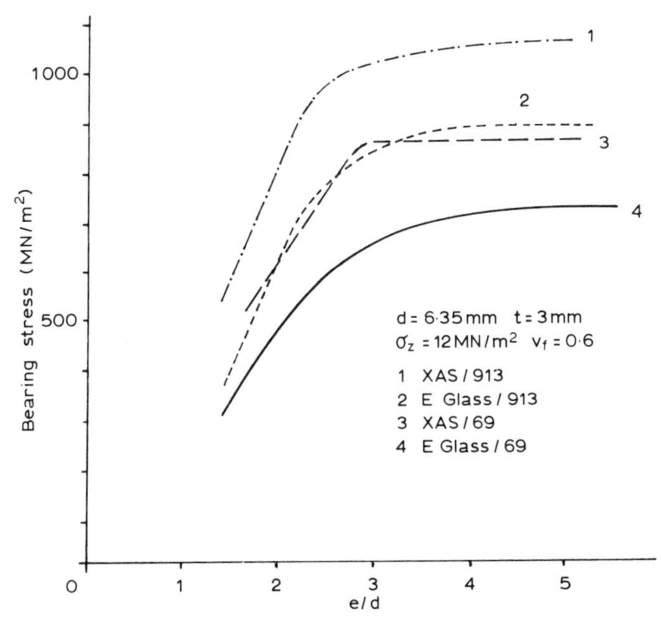

Fig. 20. Variation of bearing stress with e/d for $\frac{1}{3}$ 0°, $\frac{2}{3}$ $\pm 45°$ laminates.[5]

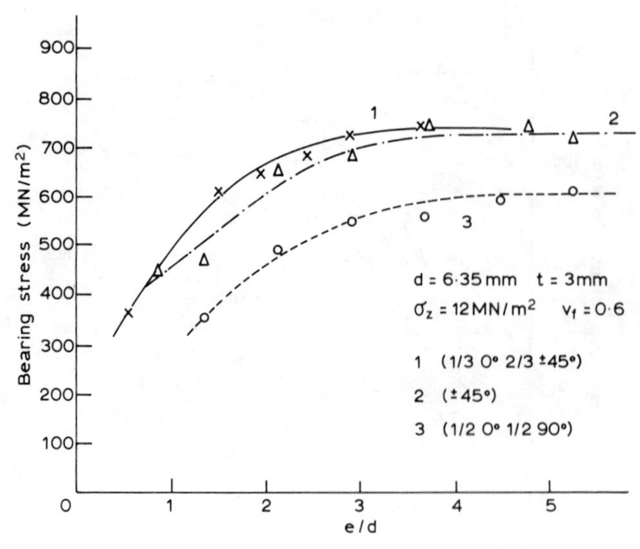

Fig. 21. Variation of bearing stress with *e/d* for various E Glass/69 laminates.[5]

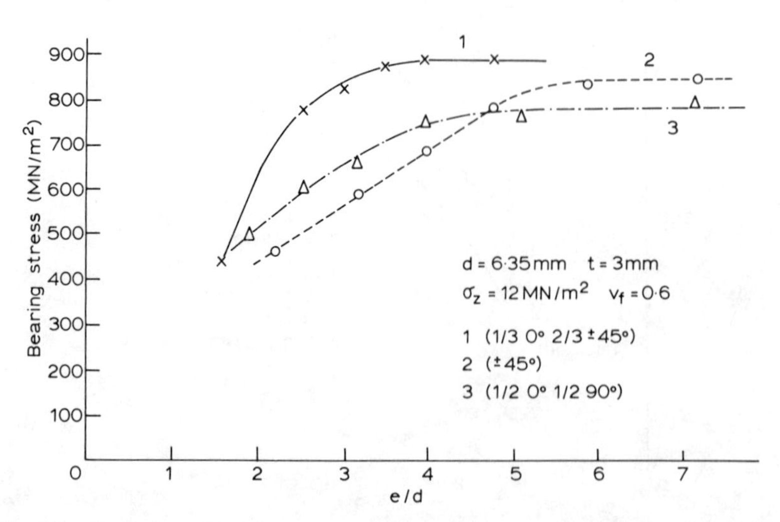

Fig. 22. Variation of bearing stress with *e/d* for various E Glass/913 laminates.[5]

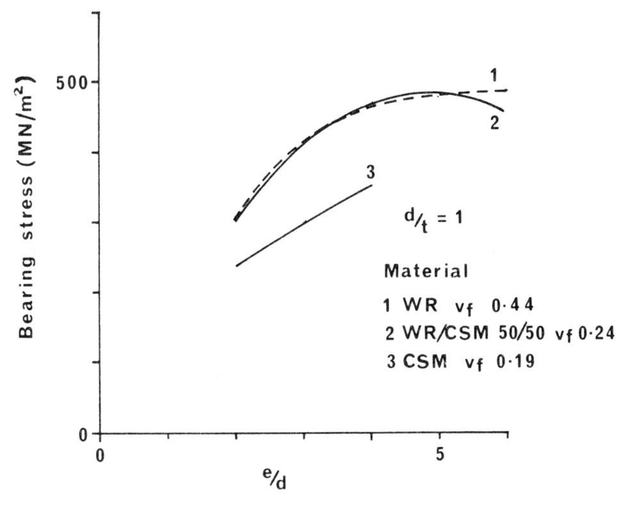

Fig. 23. Variation of bearing stress with e/d for GRP.[17]

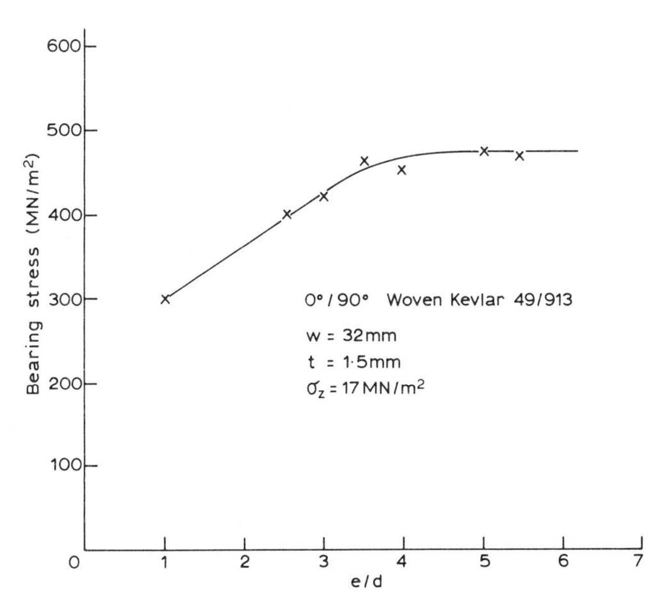

Fig. 24. Variation of bearing stress with e/d for KFRP.[11]

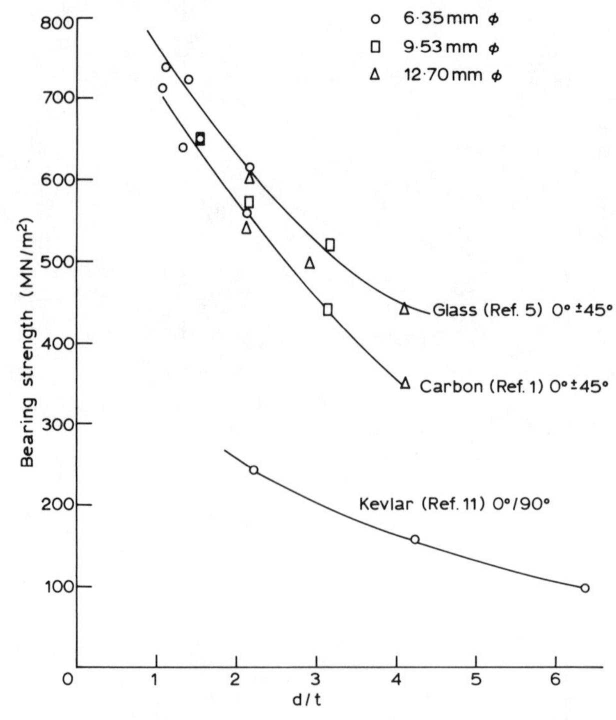

Fig. 25. Variation of bearing stress with d/t for a laminate without lateral constraint.

considerable out-of-plane buckling occurs for $0/\pm 45°$ lay-ups. It is suggested therefore that designers use low values of d/t for GFRP joints. Naturally there is a limit below which certain fasteners would fail in shear and this limit will depend upon the bolt material. Nevertheless the recommendation is that values of $d/t < 1{\cdot}5$ should not be used. For rivets where clamping is light and for pinned joints where no lateral constraint exists, the effect of hole diameter will need to be considered. The effect of hole diameter on bearing strength for free pins is given in Fig. 25.

2.9. STATIC STRENGTH OF BOLTED JOINTS

The various factors affecting the strength of bolted connections and their failure modes have been described and discussed in the previous sections.

This section deals with strength data drawn from a number of experimental investigations carried out on a variety of fibres, material forms and laminate configurations. No attempt has been made to cover all permutations of materials and lay-ups; instead, sufficient data have been presented to enable designers to interpolate to meet their own requirements.

2.9.1. Tensile strength

The results of single hole tests enable a clear picture to emerge of the likely performance of bolted joints in CFRP and GFRP made from warp material and KFRP made from woven material. It is evident from Figs 26–30 that the net tensile strength is dependent on specimen width and ply orientation, and that $0/\pm45°$ laminates exhibit the best tensile performance.

The net tensile stress concentration factor at failure around a hole in a $\pm45°$ CFRP laminate is generally low, in the order of 1·5, whereas in a unidirectional (0°) CFRP laminate the corresponding value is 7. A major contributory factor for this difference is the large difference in anisotropy between the two laminates, so it is to be expected that $0/\pm45°$

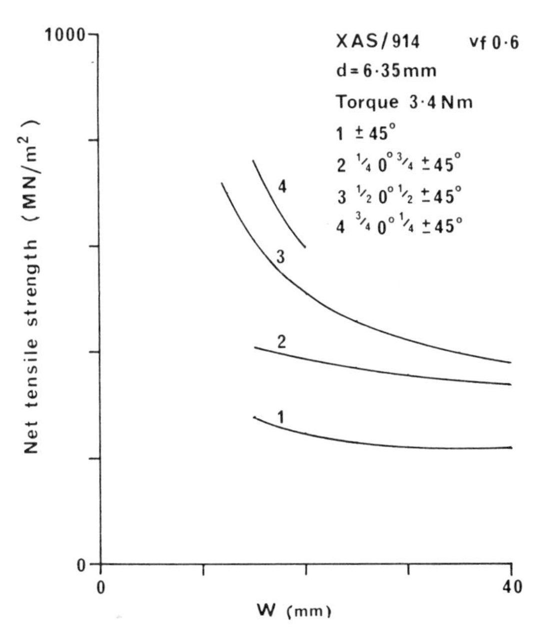

FIG. 26. Variation of net tensile strength with width for $0/\pm45°$ CFRP.[2]

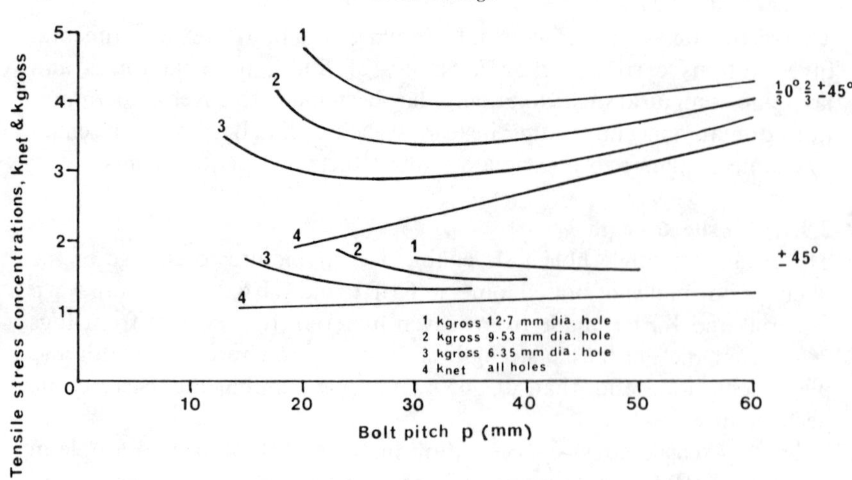

FIG. 27. Variation of stress concentration factors with bolt pitch.[23]

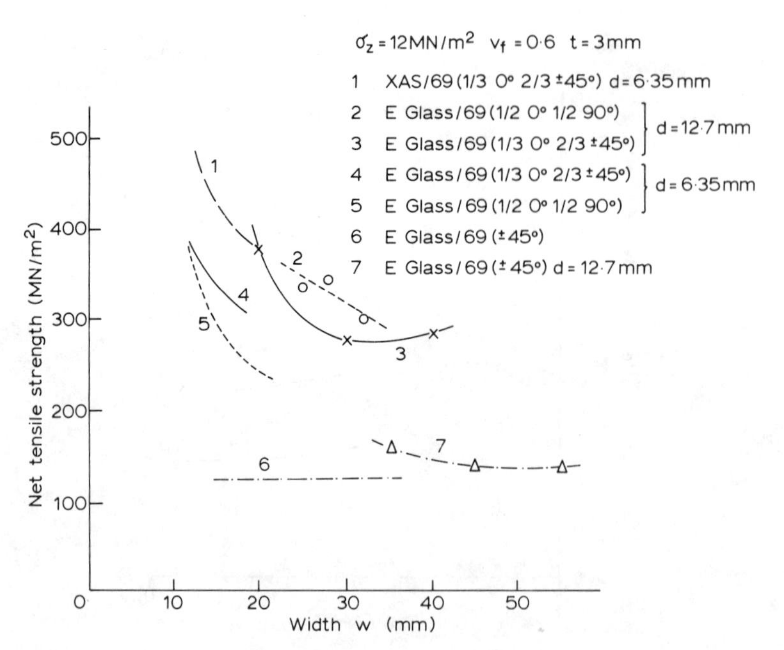

FIG. 28. Variation of net tensile strength with width.[5]

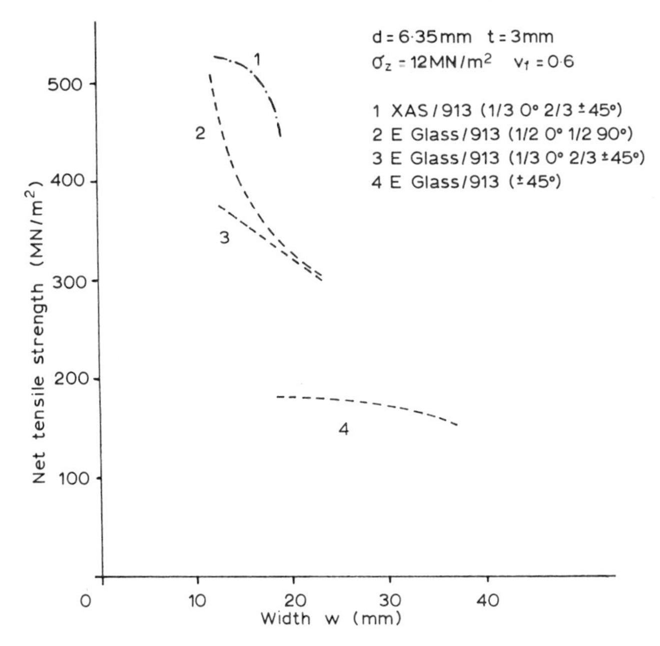

Fig. 29. Variation of net tensile strength with width.[5]

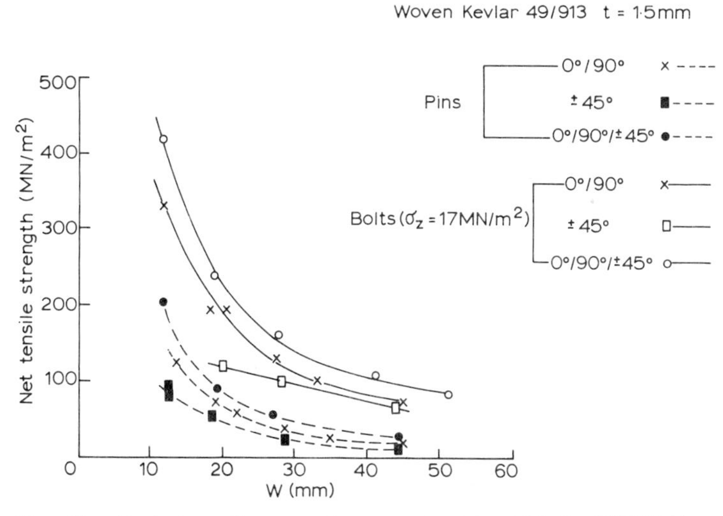

Fig. 30. Variation of net tensile strength with width for KFRP.[11]

laminates exhibit stress concentrations that lie somewhere between the two; experimental evidence confirms this.[1] The indications are that ±45° material imparts a significant non-linear, or pseudo-plastic behaviour, and this clearly enhances joint performance in what is normally considered to be a rather brittle material.

Although ±45° laminates have the advantage of low stress concentrations, their tensile strength is poor compared with that of 0° material. Thus, quite high proportions of 0° plies must be included if a material of high strength and therefore high overall performance is to be achieved. Clearly a balance has to be struck between, on the one hand, the advantages of high strength stemming from the 0° plies and, on the other, the advantages of low stress concentrations and pseudo-plasticity contributions of the ±45° plies. In fact, from experimental work,[23] an optimum appears to exist for CFRP when the ratio of 0° to ±45° plies is about 2:1 (Fig. 31), although even quite small amounts of 0° plies can cause a significant improvement in the overall joint tensile strength. It is worth noting that the gross failing stress is relatively constant over a wide range of proportions of 0° to ±45° plies and this indicates that the tensile stress concentrations vary proportionally with the mean tensile strength of the plain laminate. Such a result has implications in joint design since it appears that a range of stress concentrations can be obtained with little change in the load-carrying capacity of the joint.

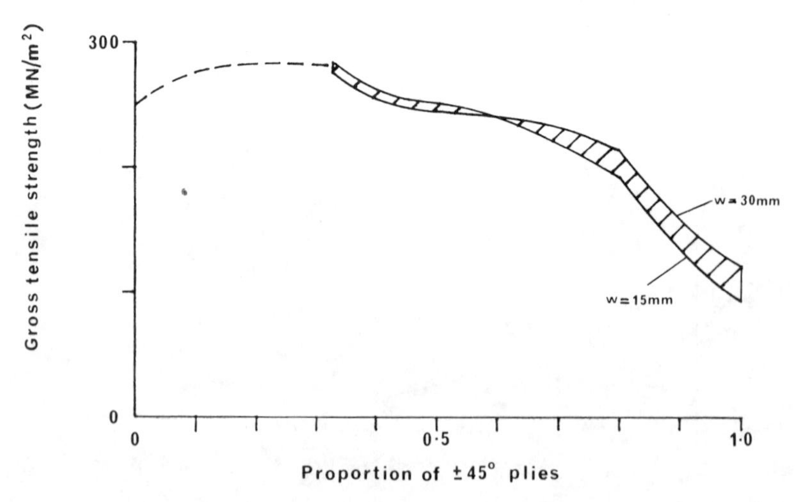

Fig. 31. Variation of gross tensile strength with proportion of ±45° plies.[23]

Two other laminates that have been studied are $0/\pm 60°$ and $0/90°$. Both show evidence of joint softening but not to the same extent as $0/\pm 45°$. This is to be expected since $\pm 45°$ plies exhibit a beneficial combination of elastic properties, together with high failure strains, not shown by other ply orientations.

Although FRP suffers a large loss of efficiency due to the presence of large stress concentrations, there is still a distinct advantage to be gained over conventional materials. Expressed in terms of the ratio of specific strengths of CFRP to metals, the potential varies from 1·62 to 1·87 when compared with L71 aluminium alloy and from 2·15 to 2·53 when compared with S96 steel. A comparison of the specific tensile strength properties of CFRP with those of metals is made in Table 3; it is made on the basis that in general metals exhibit some yielding, and therefore the stress concentrations in the metals are negligible. However, such situations are not realistic since metal structures are normally designed to work within the elastic region and it is arguably more correct to make comparisons that include stress concentrations in the metal joints. On such a basis CFRP joints would compare even more favourably.

2.9.2. Shear strength

As might be predicted from a knowledge of the in-plane shear strength of multidirectional FRP laminates, the shear strength is dependent upon ply orientation. As shown in Fig. 32, the shear strength of a $\pm 45°$ CFRP laminate is about 100 MN/m^2 and that of a 0° CFRP laminate is about 23 MN/m^2. Combining 0° plies with $\pm 45°$ plies increases the shear performance to give strengths greater than either the 0° or the $\pm 45°$ laminates alone. A reduction in the degree of anisotropy of a laminate together with a strengthening of potentially weak planes of failure leads to a decrease in the magnitude of the stress concentration and an increase

TABLE 3
Specific Tensile Strengths[1]

Material	*Tensile strength (MN/m²)*	*Specific gravity (g/cm³)*	*Specific tensile strength*
Steel S96	927	7·85	118
Aluminium alloy L71	432	2·7	160
CFRP XAS/914, $0/\pm 45°$; ($\frac{1}{2}0°$, $\frac{1}{2}\pm 45°$; $w = 25$ mm)	460	1·54	299

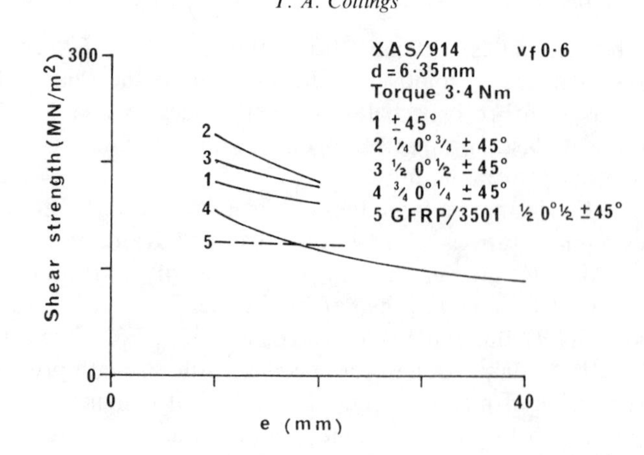

Fig. 32. Variation of shear strength with end distance e.[2]

in shear strength. The results of shear tests carried out on various lay-ups show, for $0/\pm 45^\circ$ CFRP laminates (Fig. 32), that laminate configurations containing 25–50% $\pm 45^\circ$ plies give optimum shear strength performance.

Comparison of specific shear strengths with other materials is made in Table 4, where it can be seen that CFRP and steel have similar specific shear strengths but that CFRP is about 40% better than aluminium alloy. It should be noted that for isotropic materials shear strength is not easily determined. In aluminium alloy tensile failure occurs in a shear mode on the maximum shear plane, which is at 45° to the load axis, and the quoted shear strength is usually taken to be half the tensile strength. In the case of steel this type of failure does not occur and for several reasons the shear strength is taken to be two-thirds of the tensile strength. It is with these values of shear strength that CFRP has been compared.

TABLE 4
Specific Shear Strengths[1]

Material	*Shear strength (MN/m^2)*	*Specific gravity (g/cm^3)*	*Specific shear strength*
Steel S96	618	7·85	88
Aluminium alloy L71	216	2·7	80
CFRP XAS/914, $0/\pm 45^\circ$ ($\frac{1}{2}0^\circ$, $\frac{1}{2}\pm 45^\circ$; $e = 20$ mm)	175	1·54	114

The joint shear strength of GFRP and CFRP has been investigated by Matthews and Kretsis[5] and is presented in Figs 33 and 34. Here it has been shown that $0/\pm 45°$ lay-ups are also to be preferred for GFRP to give a high shear strength performance and that the change of strength with end distance is similar to that of CFRP. Some limited work on the shear strength of KFRP (woven) has been reported by Matthews and Kalkanis[11] and this is presented in Fig. 35.

2.9.3. Bearing strength

The effect of ply orientation on the bearing strength of $0/\pm 45°$ CFRP laminates that are adequately constrained normal to the plane of the laminate is not as pronounced as might be expected from the knowledge that the compressive strength of $\pm 45°$ laminates is low. Unlike compression, where the major strength contribution is from the 0° plies, $\pm 45°$ plies contribute as large a part to the overall bearing strength as do 0° plies. This is made evident by comparing the ultimate bearing strength of a 0° laminate with that of a $\pm 45°$ laminate; here the values are 910 MN/m² and 1050 MN/m² respectively for an XAS/914 fibre/resin

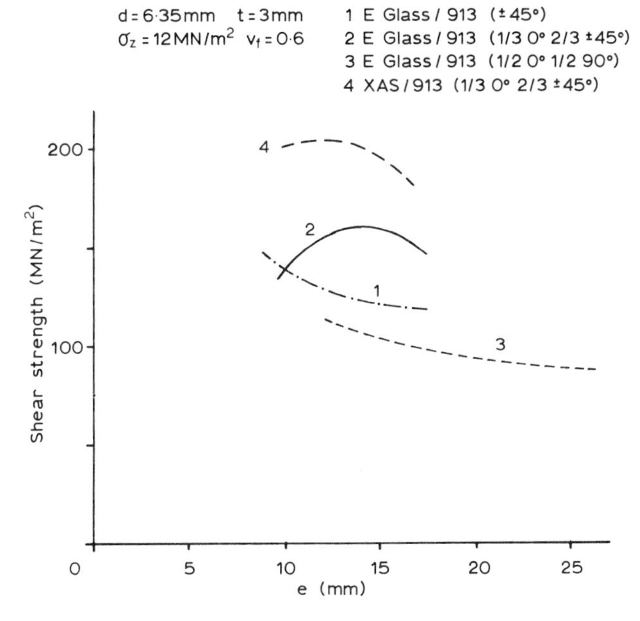

Fig. 33. Shear strength of GFRP and CFRP laminates.[5]

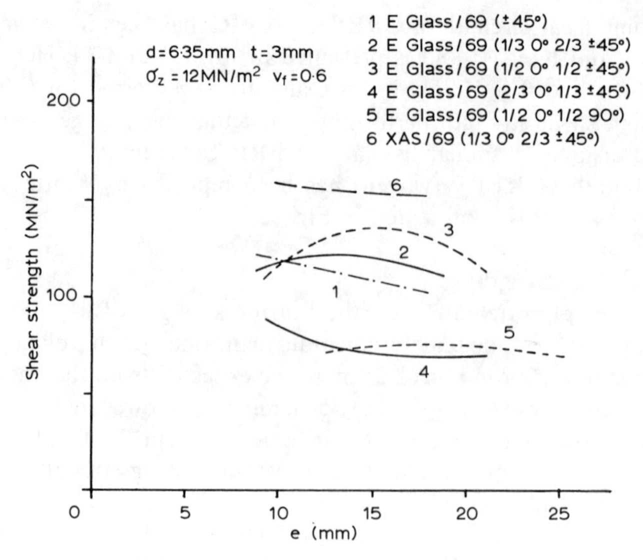

Fig. 34. Shear strength of GFRP and CFRP laminates.[5]

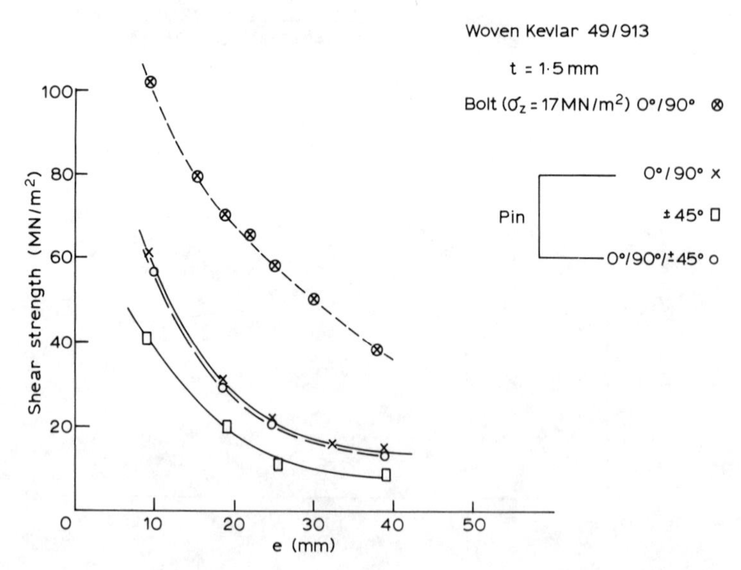

Fig. 35. Variation of shear strength with end distance *e* for KFRP.[11]

system with a 0·6 fibre volume fraction. Clearly the mode of failure of a ±45° laminate in bearing is different from that in compression. It has been shown by Collings[9] that for a loaded pin in a 0/±45° laminate, laterally constrained to at least 22 MN/m^2, failure occurs in a mode in which both fibres and matrix are broken, which is typical of both longitudinal compression and constrained transverse compression. Since it is well known[24] that fibres that are fully constrained in one transverse direction and loaded in compression in the other transverse direction fail at a stress level equal to or greater than that of longitudinal compression, it is not surprising that high strengths are achieved with ±45° laminates in bolt bearing. However, for optimum bearing performance a significant proportion of 0° material is still required and, in fact, an optimum appears to exist when the 0° plies account for about 25–55% of the total laminate thickness (Fig. 36), depending upon the fibre/resin system. This is also true for 90/±45° laminates (Fig. 37), but in this case there is a significantly higher percentage increase in bearing strength for small proportions of ±45° plies. The magnitude of these improvements is somewhat greater than could reasonably be expected using a simple law of mixtures and the individual laminate strengths. It has been shown by Collings[9] that

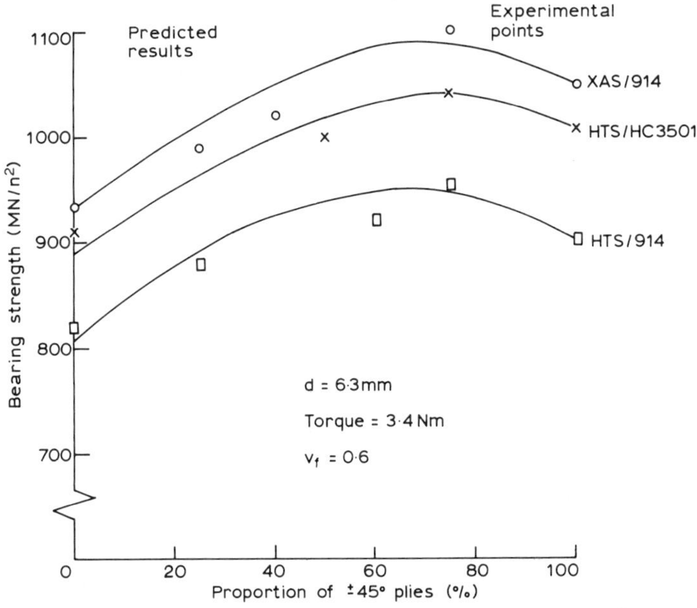

Fig. 36. Bearing strength of 0/±45° CFRP laminates.[9]

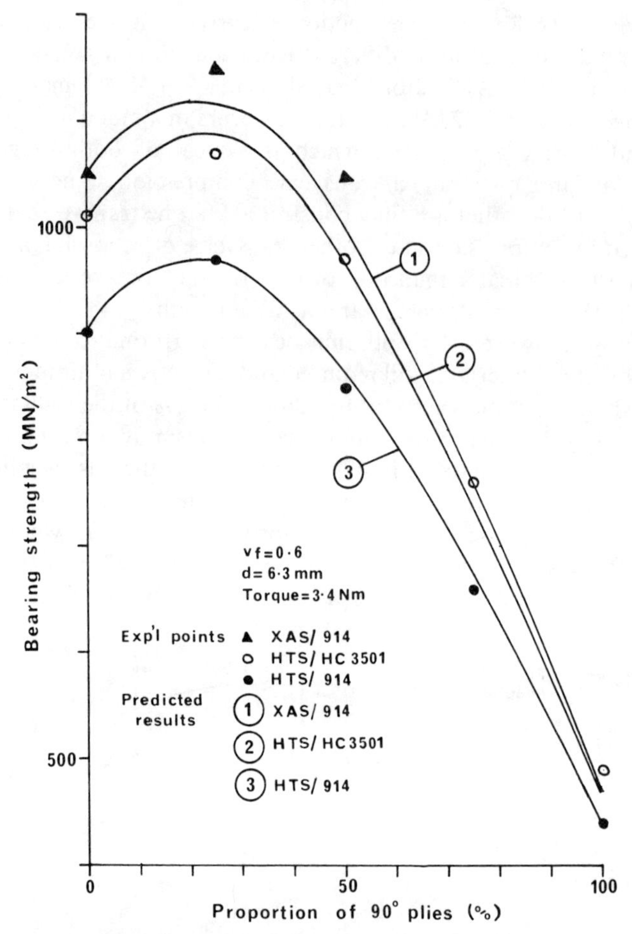

Fig. 37. Bearing strength of 90/±45° laminates.[9]

for both 0° and 90° laminates the inclusion of even small amounts of ±45° plies will change the mode of failure. The reason for this change is the increase in transverse integrity provided by the ±45° and the 90° plies to an extent which prevents failure occurring in longitudinal splitting of the laminate parallel to the direction of the 0° plies. Thus a minimum reinforcement of the 0° plies with about 25% ±45° or 90° plies is necessary if sufficient transverse strength is to be imparted to allow adequate bearing strengths to be achieved. It can be seen from Fig. 38 that the variation in bearing strength of 0/90° laminates with the

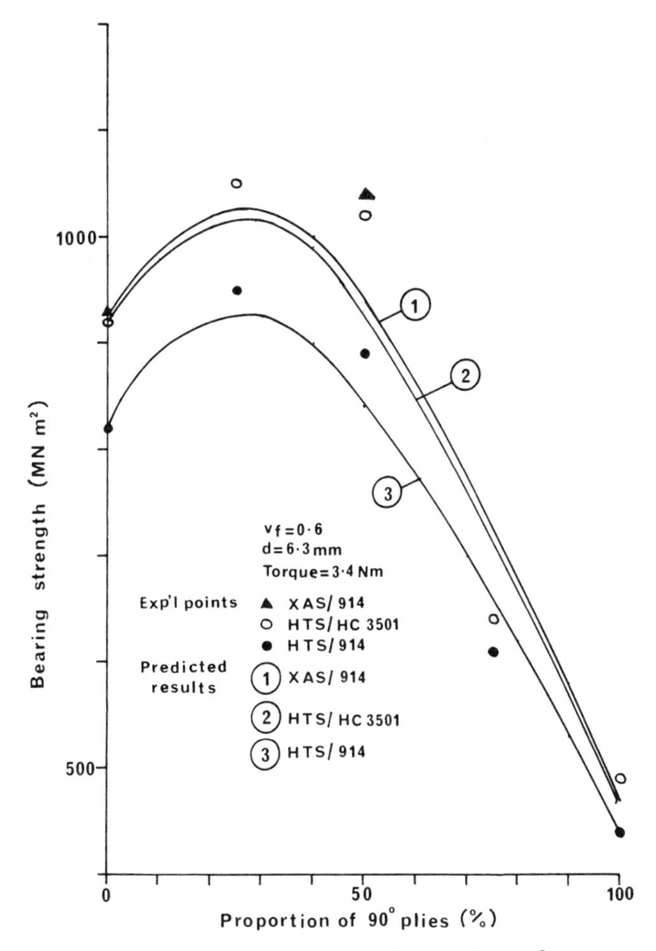

FIG. 38. Bearing strength of 0/90° laminates.[9]

proportion of 0° plies is similar to that of 90/±45° laminates with the proportion of ±45° plies. Thus it appears that the addition of either 0° or ±45° plies to 90° plies produces a very similar change in the magnitude of the bearing strength.

Ply stacking sequence, i.e. the order in which plies are distributed through the laminate thickness, can have a significant effect on the bearing strength. Some measure of the magnitude of this effect has been determined in refs 1, 19 and 20. However, for other structural reasons it is unlikely that highly 'blocked' lay-ups, mentioned in Section 2.6, will find many applications in practice so this factor is of no real consequence except in determining the way joints might be locally built-up or reinforced. The effect of thickness on the bearing strength of $0/\pm45°$ laminates is small provided adequate lateral constraint is applied ($22\,MN/m^2$). If lateral constraint is not applied, then large variations in bearing strength can exist and when bearing strength is plotted against the ratio of hole diameter to laminate thickness, as in Fig. 25, it can be seen that there is a steady fall-off in strength as the ratio of diameter/thickness increases. Such a result is similar to that observed in metals where the efficiency of joints has been found to depend on the diameter/thickness ratio. In Fig. 25 the results were obtained using three hole sizes and three laminate thicknesses. The potential advantage offered by CFRP over both aluminium alloy and steel is clearly indicated in Table 5, where a comparison of specific strengths is made.

The bearing strengths of GFRP and KFRP have been reported in refs

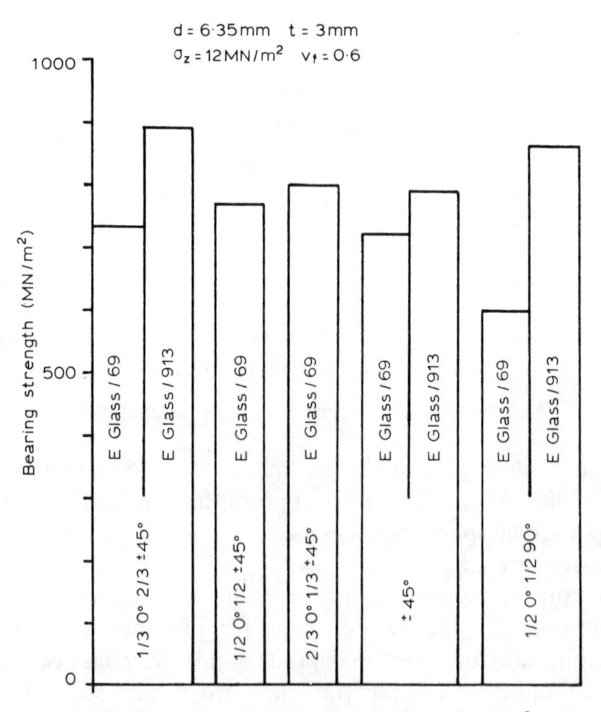

FIG. 39. Bearing strength of GFRP laminates.[5]

TABLE 5
Specific Bearing Strengths[1,25]

Material	*Bearing strength (MN/m^2)*	*Specific gravity (g/cm^3)*	*Specific bearing strength*
Steel S96	973	7·85	124
Aluminium alloy L71	425	2·7	157
CFRP XAS/914, 0/+45° ($\frac{1}{2}$0°, $\frac{1}{2}\pm45°$; $d = 6·35$ mm)	1 070	1·54	695
KFRP ($\frac{1}{2}$0°, $\frac{1}{2}\pm45°$)	460	1·35	340
KFRP (0·7 0°, 0·3 $\pm$45°)	370	1·36	272

5 and 11 and the results are presented in Figs 39 and 40. Here again, the emphasis is made that sufficient lateral constraint must be applied in order to prevent delamination at the hole edge if an optimum bearing strength is to be achieved. The advantage offered by KFRP over steel and aluminium alloy metals is shown in Table 5.

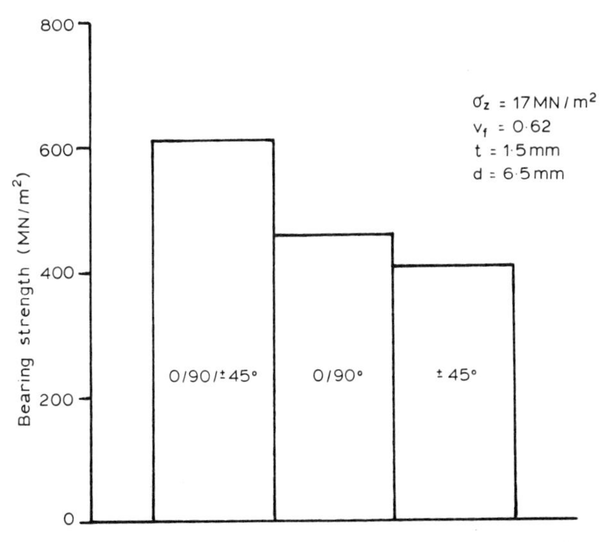

Fig. 40. Bearing strength of woven 'Kevlar' 49/913.[11]

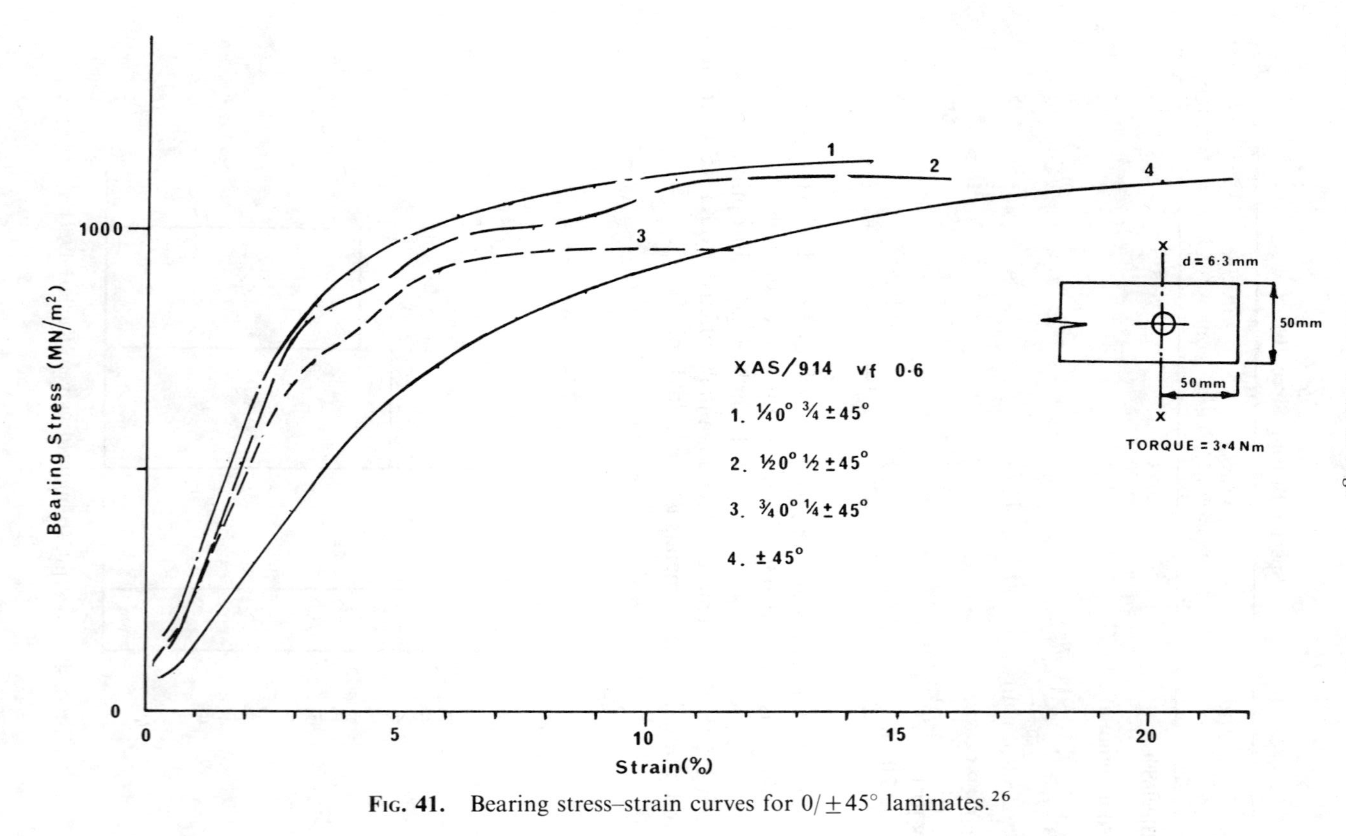

FIG. 41. Bearing stress–strain curves for 0/±45° laminates.[26]

2.10. BEARING DEFLECTION

FRP are normally assumed to have relatively linear elastic properties to failure when loaded in the tensile mode although evidence exists to suggest that this is not the case in compression. Bearing failure is essentially one of compression and from the evidence supplied by failed specimens it is demonstrated that although gross damage is apparent well before ultimate failure, the load-carrying capability of the laminate is not impaired. Since significant hole elongation is evident before failure, special consideration in the choice of the design stress level will be needed if joint hammering in fatigue environments and lost motion in control activation linkages are to be avoided. In addition, since the curves of hole deflection (strain) and stress become non-linear at low stress levels (Fig. 41), deflection behaviour will have a significant influence in deciding the stress level suitable for joints that are to be used in a fatigue environment. The deflection Δ of a hole is taken as the movement of the pin centre in the load direction and measured relative to a fixed point on the laminate (marked XX in Fig. 41). This deflection is conveniently expressed as an effective bearing strain ε_b by relating it to the hole diameter as follows

$$\varepsilon_b = \frac{\Delta}{d} \tag{2.7}$$

It is suggested from the evidence provided by ref. 26 that, for $0/\pm 45°$ laminates made from an XAS/914 fibre/resin system with a 0·6 fibre volume fraction, the ultimate permissible stress will lie between 600 MN/m^2 and 750 MN/m^2. Even at these stress levels CFRP can still show, on a specific strength basis, an advantage over most other structural materials.

2.11. OFF-AXIS LOADING

Ideally, joints are designed so that the major load resultant axis passes through the centroid of the fastener group, thereby avoiding complexity of moments. In reality many situations occur in which the eccentric loading of fastener groups cannot be avoided. Whilst for isotropic materials this is tolerable from a strength point of view, it is not necessarily so in the case of FRP material in which the strength and stiffness properties vary, in the plane of the laminate, with the angle the load axis makes to the principal fibre axis. Some investigations have been carried out to study the effect of rotating the load axis and measuring the resultant

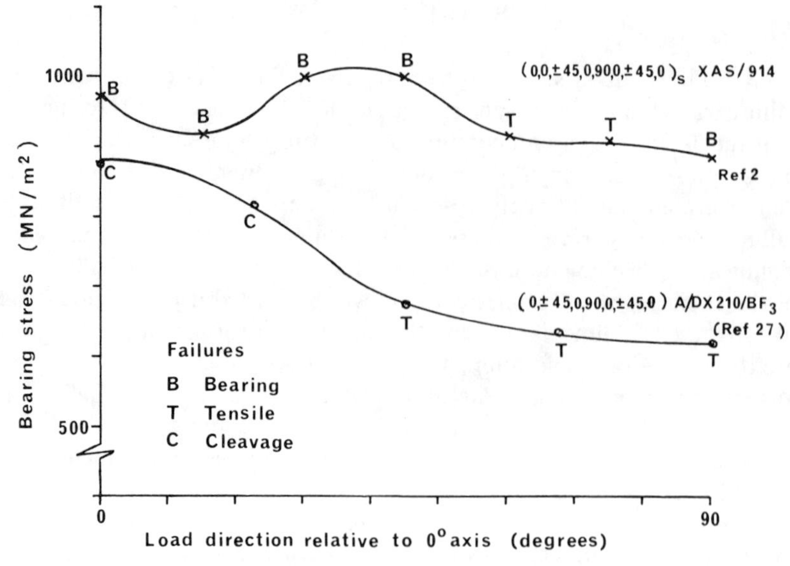

FIG. 42. Off-axis loading of CFRP.

bearing strength capability of a single bolt in a particular laminate configuration. The results of these investigations are given in Fig. 42. Here the maximum bearing strength achieved at failure is plotted as a function of the loading angle. The conclusion drawn from Fig. 42 is that, for the lay-ups investigated, bearing strength is not as dependent as tensile strength upon fibre direction.

2.12. EFFECT OF TEMPERATURE AND MOISTURE

It is well known[28] that the exposure of FRP to a hot wet environment results in the absorption of moisture (water) into the resin matrix. The result of this moisture absorption is to plasticise the resin and to lower the glass transition temperature and thus influence directly the material strength properties that are matrix-dependent. Of these properties, the ones most likely to be affected are shear and bearing strengths. At the present time evidence exists only for the bearing strength property from tests carried out by Kim and Whitney[29] in which they investigated the effects of temperature and humidity and the combination of the two for several laminate configurations. Their conclusions were that temperature

TABLE 6
Bearing Strengths of T300/5208[29]

Laminate[a]	*Test condition*[b]	*Ultimate bearing strength* (MN/m^2)
$(0_2, \pm 45)_{2S}$	RT and D	587
	RT and W	524
	127°C and D	441
	127°C and W	363
$(90_2, \pm 45)_{2S}$	RT and D	516
	RT and W	462
	127°C and D	350
	127°C and W	316
$(0, 90, \pm 45)_{2S}$	RT and D	619
	RT and W	535
	127°C and D	416
	127°C and W	371

[a] $d = 3{\cdot}18$ mm; $w/d = 4$; $e/d = 6{\cdot}5$.
[b] RT, room temperature; D, dry; W, wet.

appeared to have a more significant effect than moisture on bearing strength and that the presence of both moisture and temperature produced a further loss in strength of the order of 10% of that due to temperature alone. In all the laminates tested the greatest strength reduction due to both temperature and humidity was 40%. Table 6 lists the results obtained from ref. 29.

2.13. COMPARISON OF SINGLE-HOLE AND MULTI-HOLE JOINTS

The preceding sections have been concerned with the behaviour of single holes, and failure has been defined without the complication of interaction effects due to the proximity of holes. In most applications the joint geometry of concern is one of a flat plate containing one or two rows of fasteners along an edge. In such situations an important consideration is the possible interaction due to the general loading and grouping of the fasteners.

Joint geometry is generally chosen such that potential tensile or shear failures occur simultaneously and at a mean stress as close as possible to the bearing failure stress. On this basis it is generally agreed between investigators that, when these requirements plus those of fastener spacing to permit fastener installation are satisfied, the effect of interaction between fasteners is low. Pyner and Matthews,[30] who were investigating multihole joints in GFRP laminates using a pitch/diameter ratio of 4 ($d = 6{\cdot}35$ mm), showed that the strength of both single rows and double rows (with holes in tandem) could be predicted from a knowledge of single-hole data. For CFRP laminates Collings[1] demonstrated, with very few exceptions, that there is no adverse interaction between holes and, therefore, no loss in efficiency as the number of holes is increased. Figure 43 shows, for various bolt pitches, that the total load carried by a multi-hole joint in CFRP can be predicted from single-hole data.

In the interest of joint efficiency single-row bolts are to be preferred to multi-row bolts since multi-row bolted joints offer no significant strength increase for the same joint geometry. Indeed, as also discussed in Chapter 6, use of a single row of fasteners at a closer pitch is to be preferred since it provides the most efficient joint geometry (the tensile strength is known to increase with reduced pitch). This poses the possibility of a catastrophic tension failure associated with high loads. If bearing is the likely mode of failure, then multi-row fasteners would of course be used. The joint geometry that has been demonstrated to be most successful for $0/\pm 45°$

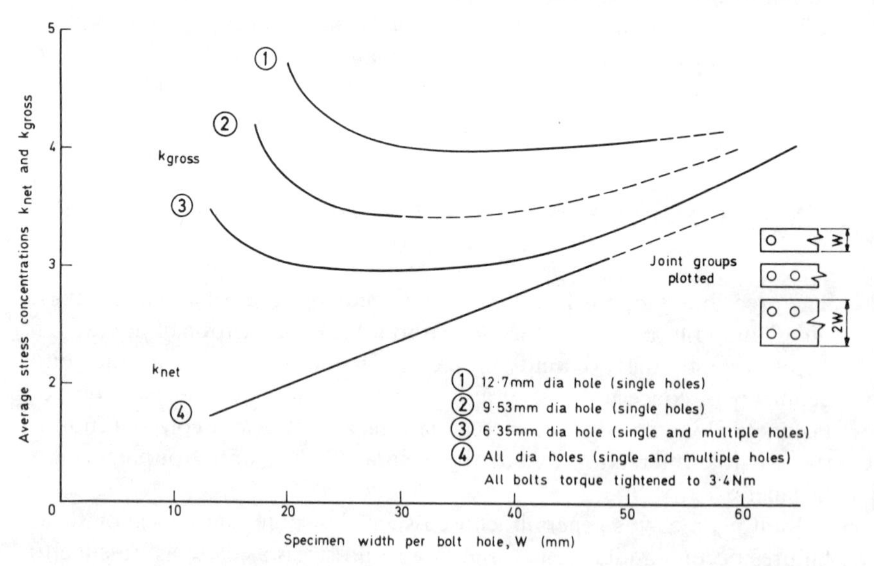

FIG. 43. Average tensile stress concentration factors ($0/\pm 45°$ laminate).[1]

CFRP laminates is the 'square array'. This consists of one or two rows of fasteners with the fasteners in tandem in the same direction as the major load.

2.14. FATIGUE STRENGTH

The prediction of the fatigue behaviour based on mechanical properties is not possible at the present time due to the numerous factors which need to be known that can influence fatigue life. These include geometry, fibre orientation, sequence of lay-up, laminate cure, temperature and humidity. As a consequence *S–N* curves are usually derived from joints that represent both the manufactured quality and construction of a structure.

Using open-hole behaviour to give an approximate measure of joint performance, a comparison of an aluminium alloy and a composite has been made by Heath-Smith[31] in Fig. 44. Here full load transfer joints (full load through the bolt) are presented by points 3–5 for 33% 0° composites and points 6–8 for a 50% 0° composite. Filled holes with

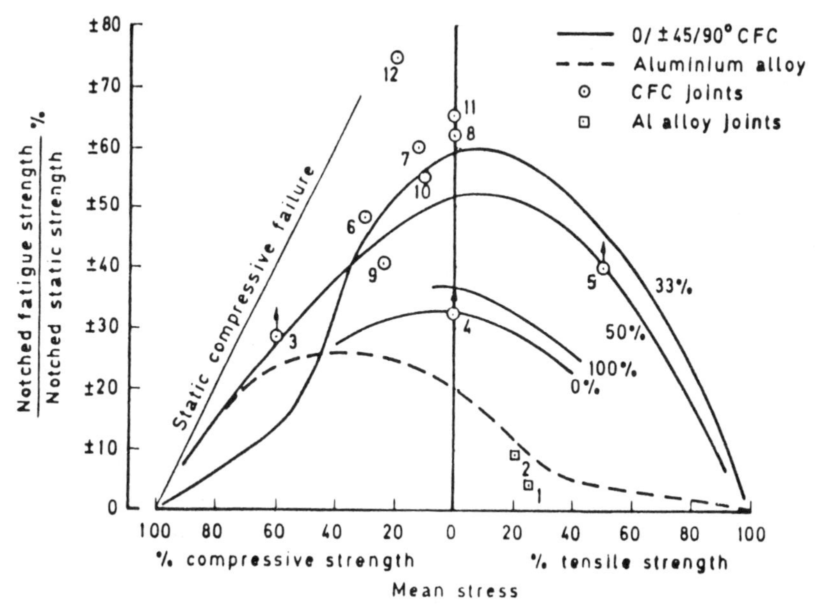

FIG. 44. Comparison of notched aluminium alloy and CFRP.[31]

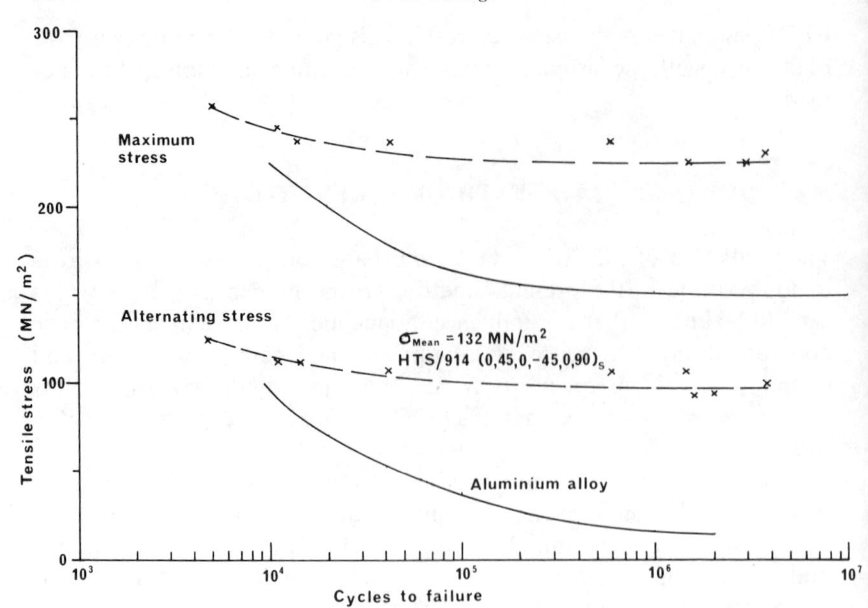

FIG. 45. Fatigue of 6 mm 100° countersunk bolted joints (single shear).[32]

clamped interference bolts are given by points 9–11. This figure demonstrates that the fatigue/static performance of bolted joints in CFRP can be better than that of open holes; thus mechanical joints in composites are unlikely to be fatigue critical, in contrast with the same situation in aluminium alloy. Further proof of this is given by Clayton and Jones[32] (Fig. 45), who were investigating the fatigue behaviour of countersunk bolts in CFRP. Here again the indications are that, in general, fatigue will not be a controlling design parameter for bolted joints in CFRP provided the static requisites are satisfied.

Experimental results by Garbo and Ogonowski[19] show that residual strengths are generally equal to or greater than non-fatigued specimens, but in most cases the specimens had acquired hole elongations of 5% of the hole diameter. This elongation exceeds that normally allowed for in metallic joints. Therefore the design constraint that will feature in the fatigue of bolted composite joints will be the amount of hole elongation that can be suffered (see also Section 2.10, on bearing deflection). Results of some fatigue testing reported in ref. 33 are given in Fig. 46 in the form of S–N curves for flush head and also for protruding head bolts for various CFRP configurations.

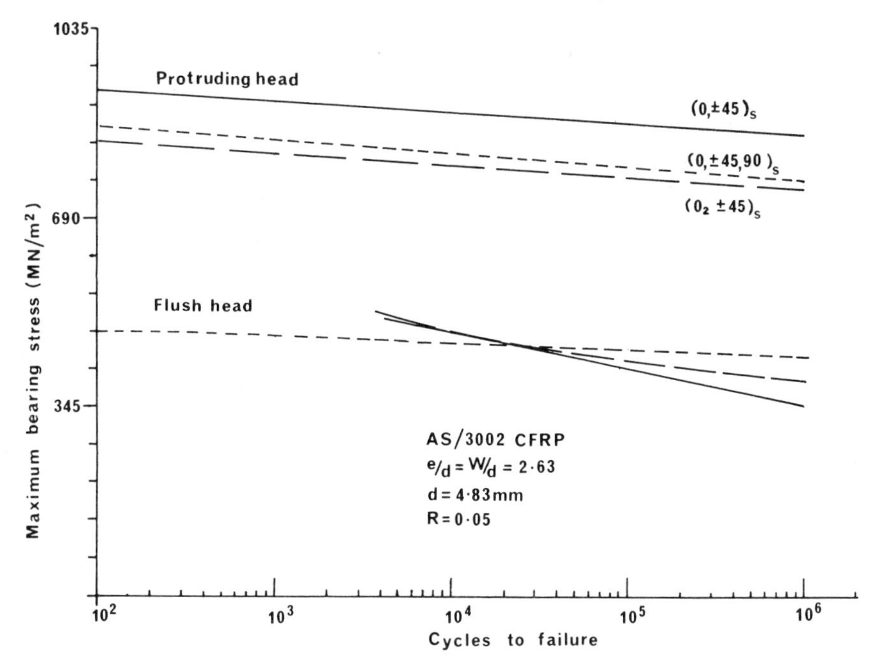

FIG. 46. *S–N* curves for CFRP-to-steel single lap joint.[33]

2.15. HOLE PREPARATION

CFRP and GFRP can be drilled using standard machinery and employing some of the methodology already developed for the machining of more conventional materials. However, because of the complicated property differences of FRP, mainly associated with poor shear strength and the abrasive nature of the material, the use of special cutting tools and cutting techniques are a prerequisite to high-quality hole production. Because of the abrasive nature of FRP, high-speed steel tools are not recommended since cutting edges are readily dulled, resulting in increased heating and tearing of the laminate surface and delamination at the hole edge. Preliminary studies show that tungsten carbide or diamond coated core drills can be successfully employed for increasing tool life, reducing heating and improving the quality of hole production. With all laminated materials one of the main consequences of poor machining is drill break-through. This causes delamination of the rear surface plies due to the poor in-plane shear strength and peel strength of the material. To avoid this problem it is necessary to use some backing to the laminate, such as wood to give support to the rear surface of the laminate. Such backing can, however, be avoided by the use of 'controlled-feed' drills.

Generally the design of twist drills for conventional materials can be successfully applied to composite materials provided debris removal is adequate. Ideally, pilot holes can reduce the punching force required to push the point of the drill into the material. Some drill manufacturers recommend (for GRP) an included angle of 55–60° for thin laminates and 90–100° for thick sections, as compared with the more usual angle of 120° used for most metallic materials.

For the drilling of KFRP laminates a different approach has been found necessary. Because of the weak bonding between the planes of molecules of the fibre there is a tendency for fibres to split into fibrils, rather than for debonding to occur at the fibre/resin interface. This failure mechanism, together with the low compressive stiffness of the fibre, tends to cause the fibres to recede into the resin matrix rather than to be cut off during the machining process. For these reasons it is difficult to achieve a clean cut edge when machining KFRP. It has been found,[34] however, that multidirectional laminates are more readily cut than are unidirectional ones, and that plain woven materials are easier to cut than are satin weaves. It has also been shown[35] that the addition of one layer of 120 style glass fabric to the laminate surface improves the quality of the cut edge. Other attempts to improve the machining of KFRP have looked at improved designs in the cutting edges of drills, and a successful design has been reported in ref. 16.

Since a great deal of dust will result from drilling, vacuum extraction should be used in order to protect the health of the machine operator and to reduce the problem of messy waste material collection. During the drilling operation it is good practice to back off the drill frequently; this helps in removing debris from the cutting face and helps to prevent overheating at the cutting face. This is essential for FRP that are made of non-heat-conducting fibres.

Speed and feed of operation will alter with drill size and the fibre reinforcement used; nevertheless, a rule-of-thumb measure of cutting speed is 20 m/min and of feed about 10–20 mm/min. It should be emphasised that rubbing the tool against the material should be avoided during the drilling process, otherwise the cutting edge of the tool will be rapidly dulled and excessive overheating will result. For this reason a continuous positive feed should be maintained during drilling. Coolants can be used to improve drilling, especially for thick sections, although the removal of the residue of dust from the coolant can be a problem.

2.16. HOLE QUALITY

As mentioned in Section 2.15, the preparation of holes in composite materials is usually carried out by conventional drilling techniques. Such techniques do however produce some hole defects that may have to be suffered and accepted, provided structural integrity is not severely impaired, on the basis of reduced cost of the end product. Generally the defects most commonly found due to drilling are oversize holes, delamination of surface plies and tearing or chipping of the hole bearing surface.

Such imperfections in hole quality have been investigated by Pengra and Wood[36] for a Narmco 5208/T300 CFRP using both a $(\pm 45/0000/\pm 45/000/\pm 45/\pm 45/00^\circ)_s$ and a $(45/0/-45/90/90/-45/0/45^\circ)_3$ lay-up. It was demonstrated, for a pin loaded hole with finger-tight lateral constraint, that delamination of surface plies had no influence upon the static pin bearing strength or pin bearing endurance under cyclic loading. Oversize holes (e.g. a 5·0 mm-diameter pin in a 5·03 mm-diameter hole) lowered the pin-bearing endurance (stress) level under cyclic loading but did not change its endurance behaviour (i.e. the shape of the *S–N* curve). Hole tearing or chipping reduced both the static and cyclic endurance characteristics.

Since finger-tight lateral constraint is not typical of the level of constraint that would normally be applied, some improvement in structural integrity can be envisaged for defects that include tearing or chipping of the hole-bearing surface, for joints that have been fully torque-tightened.

2.17. CONCLUSIONS

On a strength-to-weight basis, mechanically fastened joints offer an effective method for joining FRP components. As a general design philosophy the best overall joint performance, for all of the three main failure modes, is obtained with $0/\pm 45^\circ$ laminates. Maximum joint strength is obtained by providing the necessary geometry to suppress the tensile and shear modes of failure so that failure will occur in bearing. To develop a maximum bearing strength, sufficient lateral constraint across the thickness of the laminate must be provided. For this reason bolts are to be preferred to rivets, and rivets to other forms of mechanical fastenings.

Although mechanically fastened joints in FRP share the same basic failure modes as metals, the mechanisms by which damage initiates and propagates can be fundamentally different and so classical metals failure criteria are not always applicable. An understanding of these mechanisms is therefore essential before any reliable predictive capability can be devised. To this end the work in refs 9, 37 and 38 has attempted to predict failure for each of the three main failure modes.

REFERENCES

1. Collings, T. A., *Composites*, January 1977, 43.
2. Collings, T. A., Royal Aircraft Establishment, Farnborough, UK, unpublished work.
3. Potter, R. T., *Proc. R. Soc. London*, 1978, **A361**, 325.
4. Collings, T. A. and Mead, D. L., Royal Aircraft Establishment, Farnborough, UK, unpublished work.
5. Kretsis, G. and Matthews, F. L., *Composites*, April 1985, 92.
6. Godwin, E. W., Matthews, F. L. and Kilty, P. F., *Composites*, July 1982, 268.
7. Rothman, E. A. and Molter, G. E., STP460, February 1969, ASTM, Philadelphia, Pa, USA.
8. Whitney, J. M., Stansbarger, D. L. and Howell, H. B., *J. Comp. Materials*, January 1971, **5**, 24.
9. Collings, T. A., *Composites*, July 1982, 241.
10. Stockdale, J. H. and Matthews, F. L., *Composites*, January 1976, 34.
11. Matthews, F. L. and Kalkanis, P., *Proc. 5th Int. Conf. on Composite Materials*, San Diego, USA, July/August 1985.
12. Matthews, F. L., Nixon, A. and Want, G. R., *Proc. Reinforced Plastics Congress*, Brighton, UK, November 1976, The British Plastics Federation, London.
13. Matthews, F. L. and Leong, W. K., *Proc. 3rd Int. Conf. on Composite Materials*, Paris, August 1980, Pergamon Press, Oxford, UK.
14. Lubin, G., *Handbook of Fiberglass and Advanced Plastics Composites*, 1969, Van Nostrand Reinhold, New York, USA.
15. *Design Data Fibreglass Composites*, Fibreglass Ltd, St Helens, UK.
16. Pinzelli, R., *Int. Conf. Fibre-Reinforced Plastics*, Liverpool, UK, April 1984.
17. Matthews, F. L., Godwin, E. W. and Kilty, P. F., Technical Note TN 82–105, 1982, Aeronautics Department, Imperial College, London, UK.
18. Wardle, M. W., presented at *Int. Symp. Composites: materials and Engineering*, University of Delaware, Newark, Delaware, USA, September 1984, in press.
19. Garbo, S. P. and Ogonowski, J. M., AFWAL-TR-81-3041, Vol. 1.

20. Quinn, W. J. and Matthews, F. L., *J. Composite Materials*, April 1977, **11**, 139–45.
21. Shivakumar, K. N. and Crews, J. H., Technical Memorandum 83268, January 1982, NASA, Langley, Virginia, USA.
22. Waddoups, M. E., Eisemann, J. R., Kaminski, B. E., *J. Composite Materials*, October 1971, **5**, 446.
23. Collings, T. A., *Proc. Designing with Fibre-Reinforced Materials*, London, UK, 27–28 September 1977, Institute of Mechanical Engineers, London.
24. Collings, T. A., *Composites*, May 1974, 108.
25. De Koning, C. A. M. and Van Dreumel, W. H. M., *Kevlar in Aircraft*, *Tech. Symp. IV*, Geneva, Switzerland, October 1982.
26. Collings, T. A. and Beauchamp, M. J., *Composites*, January 1984, 33–8.
27. Matthews, F. L. and Hirst, I. R., *Proc. Symp. Jointing in Fibre-Reinforced Plastics*, Imperial College, London, 1978, IPC Science and Technology Press, Guildford, Surrey, UK.
28. Shen, C. and Springer, G. S., *J. Composite Materials*, 1977, **11**, 2.
29. Kim, R. Y. and Whitney, J. M., *J. Composite Materials*, April 1976, **10**, 149.
30. Pyner, G. R. and Matthews, F. L., *J. Composite Materials*, July 1979, **13**, 232.
31. Heath-Smith, J. R., Technical Report TR 79085, 1979, Royal Aircraft Establishment, Farnborough, UK.
32. Clayton, G. and Jones, D. P., Report SON (P) 149, 1976, British Aircraft Corporation, Warton, Preston, UK.
33. *Engineering Design Handbook Darcom-P706-316*, March 1979, US Army, Alexandria, USA.
34. Koning, W. and Schmitz-Justen, C., *Kevlar in Aircraft*, *Tech. Symp. IV*, Geneva, Switzerland, October 1982.
35. Bosco, A., Favini, A. and Galanti, O., *2nd Int. Conf. SAMPE European Chapter*, Stresa, Italy, June 1982.
36. Pengra, J. J. and Wood, R. E., Paper No. 80-0777, 1980, AIAA, New York, USA.
37. Wilson, D. W. and Pipes, R. B., *Composite Structures*, 1981, Elsevier Applied Science, London, p. 34.
38. Hart-Smith, L. J., *Fourth Conference on Fibrous Composites in Structural Design*, San Diego, California, USA, November 1978.

Chapter 3

Theoretical Stress Analysis of Mechanically Fastened Joints

F. L. MATTHEWS

Centre for Composite Materials,
Imperial College of Science and Technology, London, UK

3.1. INTRODUCTION

The ultimate objective of analytical techniques, whether 'classical', continuum, or numerical methods, is to provide a rational basis for design. Because of the complex nature of the analysis of mechanically fastened joints in composites, current design methods for joint strength are frequently semi-empirical.

The complexity arises from a number of sources. Firstly there is the nature of the composite material itself—should the fibre/matrix combination be considered on a micro or macro level as far as strength and stiffness are concerned? Next there is the influence of stacking sequence and lay-up. The usual geometric factors such as width, end distance, pitch, diameter, thickness must also be considered. The fit of the fastener in the hole and the magnitude of the friction force between hole and fastener should not be ignored. The contact arc between fastener and hole is a non-linear consideration and fastener flexibility and clamp-up force introduces three-dimensional effects. Also, non-linear material behaviour should be accounted for. In a multi-fastener joint the influence of neighbouring fasteners and the manner is which the various holes take up load all affect the behaviour of the joint. Finally the choice of an appropriate failure criterion must be made.

Although parametric studies can perhaps be best achieved using classical methods it is clear that such an approach would soon become intractable as more and more of the factors listed above are included. On the other hand, whilst it is technically feasible that the problem, including all the factors, could be solved using numerical techniques such as finite element methods, the cost would be prohibitive.

In order to predict strength accurately, a sufficiently detailed stress distribution which includes fibres, matrix and interface, must be available. In deriving a suitable analysis a compromise must be reached between acceptable cost and ease of manipulation and the inclusion of all significant effects. Because of these limitations it is normal to represent both the laminate and its constituent laminae as homogeneous single-phase materials with the appropriate elastic properties. When deciding what to include in such analyses the results of experimental investigations should always be borne in mind. In this context the work described in Chapter 2 is particularly important.

The current chapter describes many of the approaches used at present, the results obtained and, where possible, their success in matching experimental data. Although important in their own right, and used by several workers to confirm their theoretical results, experimental methods of stress analysis (strain gauges, photoelasticity, Moiré grids, coherent light and speckle methods) will not be discussed here.

3.2. CLASSICAL METHODS OF STRESS ANALYSIS

3.2.1. Background

A closed-form solution for the elastic stress distribution in a composite plate, containing either a loaded or unloaded hole, can be obtained from two-dimensional anisotropic theory of elasticity, as given, for example, by Lekhnitskii.[1] The solution is strictly valid only for homogeneous media and to be applied to layered composites must be combined with classical lamination theory,[2,3]

In this approach the solution satisfies compatibility and equilibrium by deriving a stress function which satisfies the generalised biharmonic equation for anisotropic materials as, for example, shown by Savin,[4] i.e.

$$S_{22}\frac{\partial^4 F}{\partial x^4} - 2S_{26}\frac{\partial^4 F}{\partial x^3\,\partial y} + (2S_{12} + S_{66})\frac{\partial^4 F}{\partial x^2\,\partial y^2} - 2S_{16}\frac{\partial^4 F}{\partial x\,\partial y^3} + S_{11}\frac{\partial^4 F}{\partial y^4} = 0 \tag{3.1}$$

where S_{11}, S_{12}, etc., are the normal laminate compliances in the principal material directions as given, for example, in ref. 2, F is the stress function and x and y are Cartesian co-ordinates.

The stress function can be written as

$$F = 2\,\mathrm{Re}\,\{\mathrm{F}_1(z_1) + \mathrm{F}_2(z_2)\} \tag{3.2}$$

where $\mathrm{F}_1(z_1)$ and $\mathrm{F}_2(z_2)$ are analytic functions of the complex variables $z_1 = x + \mathrm{R}_{1_\phi} y$ and $z_2 = x + \mathrm{R}_{2_\phi} y$.

The complex constants R_{1_ϕ} and R_{2_ϕ} are defined by

$$\mathrm{R}_{k_\phi} = \frac{\mathrm{R}_k \cos\phi - \sin\phi}{\mathrm{R}_k \sin\phi + \cos\phi}; \qquad k = 1, 2 \tag{3.3}$$

where ϕ denotes the angle between the co-ordinate axes and material symmetry axes, as shown in Fig. 1, and R_k may be solved from

$$\left.\begin{aligned} \mathrm{R}_1^2\mathrm{R}_2^2 &= \frac{S_{22}}{S_{11}} \\ \mathrm{R}_1^2 + \mathrm{R}_2^2 &= -\frac{(2S_{12} + S_{66})}{S_{11}} \end{aligned}\right\} \tag{3.4}$$

Introducing the functions

$$\phi_1(z_1) = \frac{\partial F(z_1)}{\partial z_1} \qquad \text{and} \qquad \phi_2(z_2) = \frac{\partial F(z_2)}{\partial z_2} \tag{3.5}$$

will give the stresses once $\phi_1(z_1)$ and $\phi_2(z_2)$ are found, i.e.

$$\left.\begin{aligned} \sigma_{xx} &= 2\,\mathrm{Re}\sum_{k=1}^{2} \mathrm{R}_{k\phi}^2\,\phi_k'(z_k) \\ \sigma_{yy} &= 2\,\mathrm{Re}\sum_{k=1}^{2} \phi_k'(z_k) \\ \sigma_{xy} &= -2\,\mathrm{Re}\sum_{k=1}^{2} \mathrm{R}_{k\phi}\phi_k'(z_k) \end{aligned}\right\} \tag{3.6}$$

where the 'prime' superscript denotes differentiation with respect to the complex arguments.

Likewise the displacements u and v in the x and y directions respectively may be obtained from

$$\left.\begin{aligned} u &= 2\,\mathrm{Re}\sum_{k=1}^{2} u_{k\phi}\phi_k(z_k) + \mathrm{K}_1 y + \mathrm{K}_2 \\ v &= 2\,\mathrm{Re}\sum_{k=1}^{2} v_{k\phi}\phi_k(z_k) + \mathrm{K}_3 x + \mathrm{K}_4 \end{aligned}\right\} \tag{3.7}$$

where K_1–K_4 are constants of integration, K_2 and K_4 representing rigid body translation of the plate and K_1 and K_3 being zero if rigid body rotation is prevented.

In eqn (3.7),

$$\left.\begin{aligned} u_{k\phi} &= S_{11\phi}\mathrm{R}_{k\phi}^2 + S_{12\phi} - S_{16\phi}\mathrm{R}_{k\phi} \\ v_{k\phi} &= S_{12\phi}\mathrm{R}_{k\phi} + S_{22\phi}/\mathrm{R}_{k\phi} - S_{26\phi} \end{aligned}\right\} \quad k = 1, 2 \tag{3.8}$$

where the $S_{ij\phi}$ are material compliances in the co-ordinate directions.

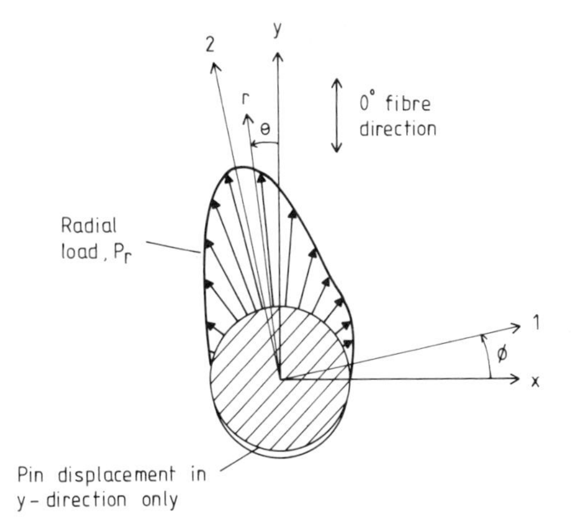

Fig. 1. Definition of global (x, y) axes, material symmetry $(1, 2)$ axes, co-ordinates r and θ and fibre direction.

3.2.2. Frictionless pin-loaded hole

The stress distribution around a hole loaded by a frictionless rigid pin is treated in a series of papers by De Jong, from whose work the above equations are summarised. In ref. 5 the case of an infinite plate is treated, i.e. the pin-load is equilibrated at infinity by a uniform distribution of infinitely small forces. In ref. 6 this work is extended to cover the effect of arbitrary load direction and in ref. 7 plates of finite width are considered.

For the case of a pin-loaded hole in an infinite plate, the complex functions of eqn (3.5) are given by:[6]

$$\phi_k(z_k) = \mathrm{A}_k \ln \zeta_k + \phi_k^{\circ}(z_k); \qquad k = 1, 2 \tag{3.9}$$

where

$$\zeta_k = \frac{z_k + \sqrt{z_k^2 - \mathrm{R}_{k_\phi}^2 - 1}}{1 - i\mathrm{R}_{k_\phi}} \tag{3.10}$$

and the function $\phi_k^{\circ}(z_k)$ is holomorphic outside the hole.

The radial load distribution on the edge of the hole must fulfil the requirements

$$\left.\begin{array}{lll} P_r \neq 0 & \text{for} & -\dfrac{\pi}{2} \leq \theta \leq \dfrac{\pi}{2} \\ P_r = 0 & \text{for} & \dfrac{\pi}{2} < \theta < \dfrac{3\pi}{2} \end{array}\right\} \tag{3.11}$$

where P_r and θ are defined in Fig. 1.

A suitable expression for P_r is found by multiplying the sine series $p_0\sum_{n=1,2}^{\infty} a_n \cos n\theta$, continuous around the hole boundary, by a step function

$$\frac{1}{2}+\frac{2}{\pi}\sum_{m=1,3}^{\infty}\frac{\cos m\theta}{m}=\begin{matrix}1 \text{ for } -\dfrac{\pi}{2}<\theta<\dfrac{\pi}{2}\\ 0 \text{ for } \dfrac{\pi}{2}<\theta<\dfrac{3\pi}{2}\end{matrix} \tag{3.12}$$

The final expression for P_r is then

$$P_r = p_0\left[\frac{1}{2}\sum_{n=1,2}^{\infty} a_n \cos n\theta + \frac{1}{\pi}\left\{\sum_{m=1,3}^{\infty}\frac{a_n}{n}+\sum_{m,n}^{*} a_n\left(\frac{1}{n-m}+\frac{1}{n+m}\right)\sin m\theta\right\}\right] \tag{3.13}$$

where

$$\sum_{m,n}^{*}=\sum_{n=1,3}^{\infty}\times\sum_{m=2,4}^{\infty}+\sum_{n=2,4}^{\infty}\times\sum_{m=1,3}^{\infty}$$

Limiting the contact region between pin and hole to the loaded, upper, half of the hole $(-(\pi/2)\leq\theta\leq(\pi/2))$ is an unnecessary restriction, as acknowledged by the author.[6] In fact the contact angle is an unknown and can be found, using an iterative technique, from the condition that tractions between pin and plate are physically impossible. In order to reduce computational effort, De Jong maintains a constant contact angle within the range $-79°\leq\theta\leq 79°$, these values being based on preliminary calculations.

The two components of $\phi_k(z_k)$ can be expressed in terms of the unknown coefficients a_n by application of stress boundary conditions. Finally the a_n values are found from the displacement condition at the hole boundary. At any point around the contact region the displacement can be considered as composed of two parts: a part equal to the (prescribed) movement of the pin as a rigid body and a part relative to the pin. In the contact area the radial displacement of the plate relative to the pin must be zero.

The displacement condition is evaluated at a number of points around the hole boundary resulting in an equation of the form

$$[a_{n_0}]\{a_n\}=\{a_0\} \tag{3.14}$$

where the coefficients a_θ and a_{n_0} are known.

The required a_n values are then found following inversion of the matrix $[a_{n\theta}]$. In De Jong's work, e.g. ref. 6, 22 angular positions are used, thus yielding 22 values of a_n. De Jong makes no mention of any special techniques used in the solution of eqn (3.14).

A similar approach using complex variables is also adopted by Oplinger and Gandhi,[8] Oplinger,[9] Gandhi[10] and Oplinger.[11] These authors use a least-squares collocation method to ensure that the boundary conditions are satisfied with sufficient accuracy. This method, which requires that the error function, implicit in any approximate solution, vanishes at a number of specified points on the boundary, is described by Sokolnikoff.[12]

Oplinger and Gandhi[8] consider directly a plate of finite width, imposing the boundary condition of zero resultant force on the unloaded edges, as opposed to De Jong[7] who superimposes his infinite plate results[5] with those for a stretched plate containing an unloaded hole. The latter approach is strictly invalid because of the non-linearity between the load and contact arc. Gandhi[10] obtains poor results for narrow plates or those in which the hole is close to the end. This is because of the convergence characteristics of the Laurent series used to represent the orthotropic stress functions.

Analyses are often undertaken in which the distribution of radial pressure between the pin and hole boundary is assumed to vary cosinusoidally. Whilst this approach may be convenient it can lead to serious errors in the stress distribution for certain geometries, materials and lay-ups. This is well illustrated by Oplinger.[11] As his results in Fig. 2 show, in

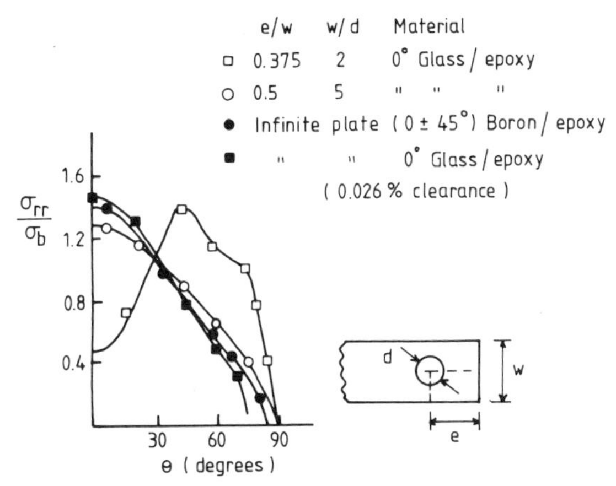

Fig. 2. Variation of radial stress (σ_{rr}) around hole boundary, as a function of bearing stress (σ_b), with material, lay-up and geometry.[11]

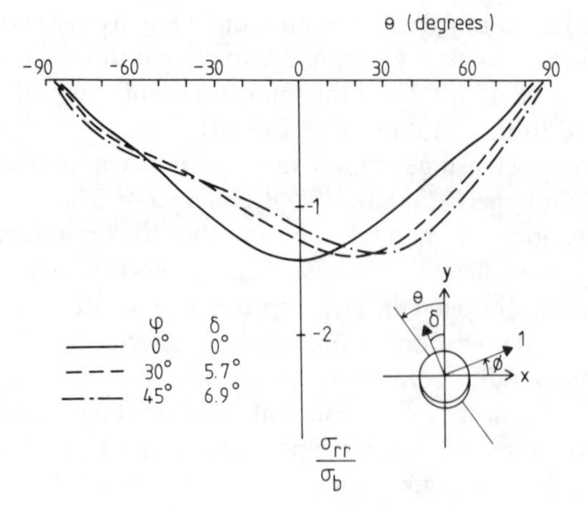

FIG. 3. Influence of pin-load direction (δ) on radial stress around hole boundary in a $(0 \pm 45°)_s$ CFRP laminate.[6]

some instances the radial stress can have a distribution which is far from a cosine shape. De Jong confirms this with his investigations into the effect of pin load direction in infinite plates.[6] Whilst the distribution is close to cosinusoidal for quasi-isotropic CFRP, as the degree of orthotropy changes the shape of the radial stress distribution can alter significantly. This is illustrated in Fig. 3, which shows the influence of load direction in a $(0/\pm 45°)_s$ CFRP laminate. De Jong also shows that the radial stress distribution in a $(\pm 45°)_s$ CFRP laminate does not depend on load direction and can be closely approximated by a cosine distribution.

The way in which the peak radial stress shifts with load direction is also shown in Fig. 3. The position of the peak stress has implications about the strength of the configuration. In general De Jong found that peak radial stresses occurred near points on the edge of the hole where the direction of the highest Young's modulus of the plate material is perpendicular to the hole boundary. The maximum tangential stresses do not necessarily occur in the plate net area perpendicular to the load direction. For a $(\pm 45°)_s$ laminate, compressive tangential stresses were found to occur over some regions of the hole boundary.

The influence of lug geometry, principally width (w) and end distance (e) is investigated by both De Jong[7] and Oplinger.[9,11] The stress distributions are found to become independent of geometry for ratios of w/d and e/d greater than about 4 for the cases considered. This is consistent with experimental observations (see Chapter 2).

3.2.3. Pin-loaded hole with friction

Although the importance of friction between the pin surface and hole boundary is generally acknowledged, the topic has not been widely treated. The inclusion of this effect introduces considerable complication to the analysis since, in addition to determining the arc of contact, it is now necessary to establish regions of slip and non-slip within the contact region. When obtaining numerical results there is the subsidiary difficulty that the coefficient of friction, which will vary around the circumference, is not accurately known. The length of the contact arc and the size of the slip (and hence non-slip) region must be determined iteratively. In cases where the material principal axes are not coincident with loading direction the contact area and non-slip area are not symmetric with respect to the co-ordinate directions. This introduces the additional complication that the first and last point of both areas must be found separately.

Because of difficulties with convergence, De Jong[13] introduces a 'release-area' which has no physical significance but makes the problem mathematically more tractable by ensuring that the functions representing radial and tangential stress vary smoothly between the slip and non-slip regions. To minimise computational effort all his calculations are based on a fixed contact area and, except for one case, the non-slip region is omitted. Oplinger and Gandhi[8] and Oplinger[9] also ignore the non-slip region although they do solve iteratively for the contact angle. All these workers produce results for a range of values of friction coefficient (μ).

Hyer and Klang[14] use an iterative method for finding the contact arc and non-slip region, based on the 'correctness' of the radial and shear stress distributions, respectively, on the hole boundary. The results were found to be sensitive to the number of collocation points used to satisfy the boundary conditions, but acceptable comparisons with data from other authors were possible within the limits of the computer system used.

A somewhat different approach is taken by Zhang and Ueng.[15,16] They use the superposition of a pin-loaded semi-infinite sheet with an infinite plate containing an unloaded hole and subjected to an arbitrary bidirectional stress field. The correctness of this approach is not at all clear and although friction is included no allowance is made for a contact arc $<180°$, nor for slip and non-slip regions. Their results show only general agreement with those of De Jong.[13]

The inclusion of friction has a significant effect on the stress distribution. The greatest changes occur at the top of the hole and the general trends are for both radial and tangential stresses to reduce as μ is increased. Very high values of $\mu(>1)$ also produce large changes in stress at the net

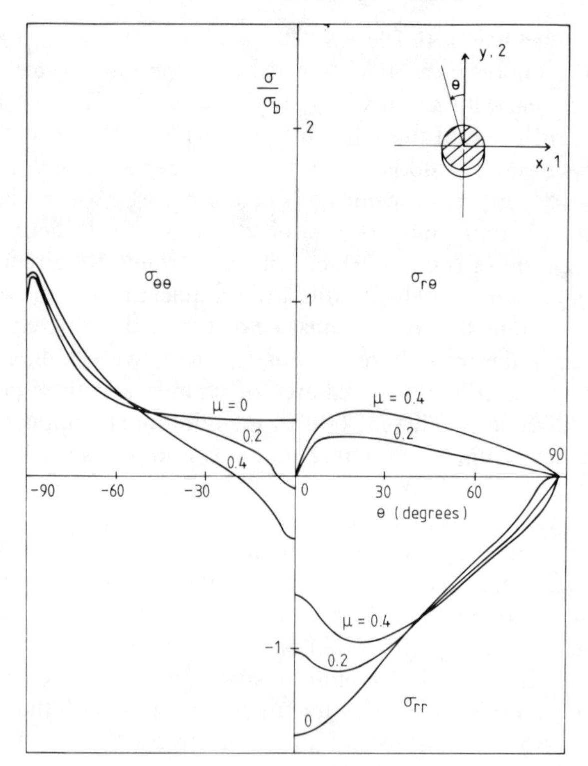

Fig. 4. The effect of friction on the stress distribution around a pin-loaded hole in a $(0_2/\pm45°)_s$ CFRP laminate.[13]

section. Typical results from De Jong[13] are shown in Fig. 4. It is also seen that for all cases in which friction is considered the representation of radial pressure by a cosine distribution would be totally invalid.

Hyer and Klang[14] also consider the influence of pin flexibility, an effect ignored by most authors. They conclude that for aluminium alloy or steel pins in CFRP the stress distributions are very close to those obtained for a rigid pin. A far more important parameter is the initial clearance between the pin and the hole. As this is increased the position of maximum tangential stress moves towards $\theta = 0$ and more of the hole boundary becomes highly loaded. At the same time the maximum radial stress increases and is confined to a smaller region.

3.2.4. Multi-pin joints

A row of pins is considered by Oplinger and Gandhi[8] and Oplinger[9] using the same approach as for a single pin. A strip containing a single hole is isolated from the array and appropriate boundary conditions applied to the edge of the strip. It is found that in the single-pin configuration net section tension is higher (by up to 10%) and shear-out and bearing stresses are lower (by as much as 20%) than for the row of pins. This has implications for the results obtained from material evaluation tests which frequently use single-pin coupons.

Oplinger[9,11] also gives results for a line of fasteners. In such a configuration part of the load on the joint will be carried by the material surrounding the pin, the so-called 'by-pass' load. The level of by-pass load is found to have a significant effect on the radial pressure distribution around the pin, as shown in Fig. 5. Clearly for such situations the assumption of a cosine distribution would lead to serious errors. As with a row of holes the net section tension stress is found to be less when compared with the single pin case.

3.2.5. Three-dimensional analyses

No analyses of mechanically fastened joints which include through-thickness effects appear to have been undertaken. Harris *et al.*[17] outline a method which has been applied to bolted joints in metal. In this approach the bolt is modelled as a short beam on elastic foundations, the foundation stiffness coming from a planar analysis. Such a model is capable of accounting for fastener flexibility, head rotation and countersink effects, as well as stress variation and non-uniform foundation modulus through the thickness. The latter effect is particularly pertinent for composites in which the in-plane stiffness can vary from layer to layer.

It seems, however, that the approach has not so far been applied to composites. Clearly to apply the classical analytical methods described in Section 3.2.2 on a layer-by-layer basis, involving iteration for a different contact angle in each layer, would be a formidable task.

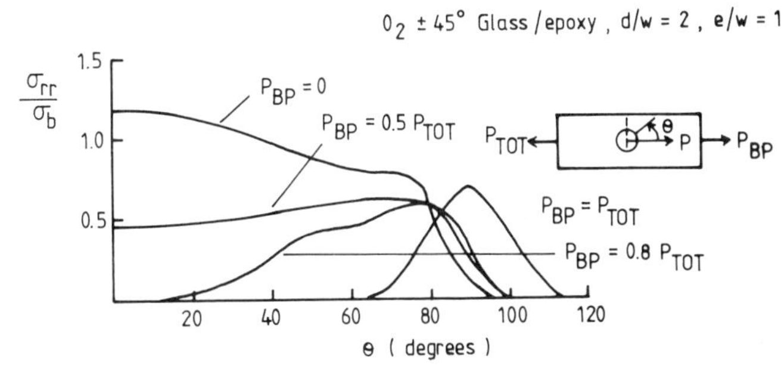

FIG. 5. The effect of by-pass load (P_{BP}) on hole boundary radial stress distribution. Total load, P_{TOT}, and pin-load, P, are related by $P = P_{TOT} - P_{BP}$.[11]

3.3. NUMERICAL METHODS OF STRESS ANALYSIS

3.3.1. Background

Although in a few cases boundary element methods have been used (see below), the bulk of a very considerable body of literature uses the finite element method.

The 'displacement method', which is used in the majority of cases, requires the formation of the stiffness matrix **K** to represent the stiffness of the structure. This is achieved by forming a model, or idealisation, of the structure or component, composed of a finite number of discrete elements. This 'mesh' of elements is imagined to be connected at nodal points. The forces that act at the element nodes are related to the corresponding nodal displacements by the element stiffness matrix **k**.

It may be shown, for example,[18] that **k** has the form

$$\mathbf{k} = \int_V [\alpha]^t \bar{\mathbf{K}} [\alpha] \, \mathrm{d}V \tag{3.15}$$

The integration over the volume V is usually performed numerically using a technique such as Gaussian quadrature.

In eqn (3.15), $\bar{\mathbf{K}}$ is a matrix of elastic constants and $[\alpha]$ is determined from the co-ordinates of the element's nodes and the interpolation function describing the assumed variation of displacement across the element. For so-called 'isoparametric' elements the displacement function is the same as that describing the shape of the element.

A well-established assembly process produces **K** from the element stiffnesses, the resulting equation being

$$\mathbf{R} = \mathbf{K}\mathbf{r} \tag{3.16}$$

The formal solution of eqn (3.16) gives the unknown nodal displacements **r** of the assembled elements, in terms of the corresponding applied nodal forces **R** as

$$\mathbf{r} = \mathbf{K}^{-1}\mathbf{R} \tag{3.17}$$

In practice it is too costly to find the inverse of **K** and methods such as Gaussian elimination or Cholesky decomposition[19] are used to solve the simultaneous set of equations (3.16).

The dependence of **k** on co-ordinates means that bodies can only be approximately represented.

The finite element method is potentially more powerful than classical methods in that it is easier to undertake three-dimensional analyses.

However, this advantage may not be realised in practice because of the enormous cost of such investigations. Hence most analyses of joints will be of two-dimensional models. For these, triangular or quadrilateral 'membrane' elements would be used, i.e. it is assumed that the element is in a state of plane stress and no bending occurs. The assumed variation of in-plane direct and shear stresses across the element could be constant, linear or parabolic corresponding respectively to linear, parabolic or cubic displacement variations. In such idealisations through-thickness stresses are, of course, neglected.

The stresses obtained will be averaged across the thickness and it will therefore be necessary to use these results in combination with lamination theory if the layer stresses are required, as is also necessary when using classical methods of stress analysis. Likewise the elastic constants for the model, needed as input data for the finite element program, will be found by applying lamination theory to the chosen material and lay-up.

Although it is relatively easy to get results using finite element methods it should be emphasised that the accuracy of the stresses depends critically on the choice of mesh layout, particularly in regions of rapid stress variation, and treatment of boundary conditions.

3.3.2. Frictionless pin-loaded hole

Whereas in the classical methods described above the laminate was loaded by imposing a displacement of the rigid pin, a number of finite element investigations impose a cosinusoidal distribution of radial pressure around the hole boundary. As seen previously this can lead to errors for some laminates and geometries. The work of Wong and Matthews[20] and Fu-Kuo Chang *et al.*[21] uses this type of applied loading. In the case of Wong and Matthews preliminary investigations showed only minimal differences with results arising from imposed pin displacements. Chang *et al.* use this approach since it avoids the difficulties inherent in determining the exact distribution and because the stresses away from the hole boundary, in which they are principally interested, are assumed not to be significantly affected by details of the boundary loads.

Other workers, for example Oplinger,[11] Crews *et al.*[22] and Soni,[23] adopt the correct approach by moving the pin through a specified distance, the far end of the plate being fixed, and ensuring that there is no relative radial displacement between plate and pin. In contrast Wilson and Pipes[24] consider the pin to be held and impose a uniform uniaxial stress across the plate remote from the hole. Around the hole boundary, on the loaded side of the pin, the displacements parallel to the applied stress are taken as zero[24] in what would seem to be an invalid boundary condition for frictionless contact. Because the joint is taken as symmetric about the load direction only one-half of the plate is considered by the workers mentioned above, who all use meshes in which the element size is reduced in the vicinity of the hole boundary. The exact number of elements employed clearly depends on the length of plate and the values of w/d and e/d being modelled. Wong and Matthews[20] typically take 96 constant-strain (four-noded) quadrilateral elements with an in-house general purpose package, FINEL. With this mesh the side length of the smallest element is about $d/6$. This mesh is shown in Fig. 6a. Fu-Kuo Chang *et al.*[21] also use isoparametric four-noded elements although their mesh, which contains 306 elements, is much finer at the hole edge, as seen in Fig. 6b. They use a specially written computer program.

Oplinger,[11] using specially developed finite element codes, also employs quadrilateral elements with a mesh density similar to that of Wong and Matthews. Soni,[23] with the widely-used NASTRAN system, uses a mesh containing 372 quadrilateral and triangular constant-strain elements. The mesh used by Wilson and Pipes[24] contains even more, largely rectangular, elements and the analysis is performed with the SAPV system.

The finest and most sophisticated mesh is that used by Crews *et al.*,[22] whose analysis is performed with the BEND finite element system. Their mesh combines linear-strain (six-noded) triangles in a $2{\cdot}5d$ by $5d$ region around the hole, with constant-strain (three-noded) triangles elsewhere. The side length of the smallest element is about $0{\cdot}02d$. No mention is made of the effect on the results that is presumably caused by lack of inter-element compatibility at the junction of these two regions in the mesh. Unusually, Crews *et al.* also use finite elements to model the pin, which is connected to the laminate by short ($0{\cdot}002d$ in length) spring elements having high axial stiffness and zero transverse stiffness, thus transferring only radial load. These springs allow the pin and laminate to separate along the contact arc. Having performed a computer run, any springs sustaining a tensile load have their stiffness put to zero and the run is repeated. This procedure is repeated until convergence is obtained. Oplinger[11] is another worker to allow for the length of contact arc being less than 180°, although no details of his method are given.

In all the analyses cited above the validity of the chosen meshes is confirmed by comparing results with those of a known standard situation, usually an isotropic plate in tension containing an unloaded central hole. In all cases satisfactory comparisons are reported. In some instances the results for the pin-loaded cases are compared also with classical analyses

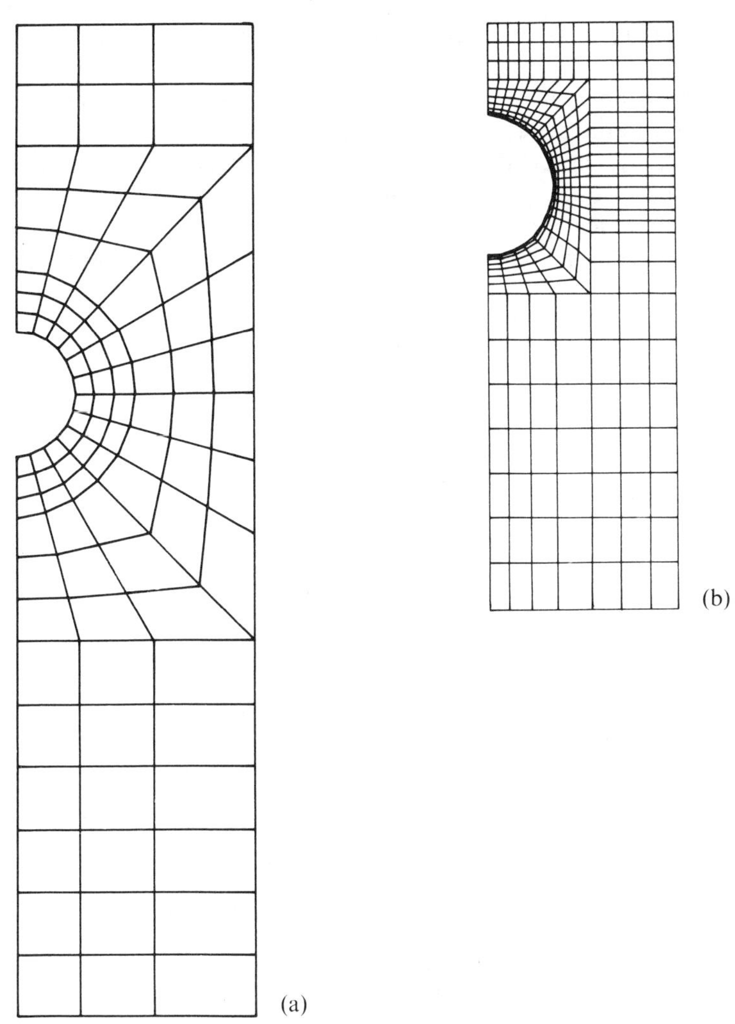

FIG. 6. Typical finite element meshes: (a) from ref. 20; (b) from ref. 21.

as in Wong and Matthews[20] and Crews *et al.*[22] Both these groups compare some of their results with data given by De Jong[7] and close agreement is found.

3.3.3. Pin-loaded hole with friction

Although a large number of contact problems have been solved using finite elements, the use of the method for mechanically fastened joints is rare. The only work in this field appears to be the related efforts of Wilkinson *et al.*,[25] Rahman[26] and Rowlands *et al.*[27]

As already noted in Section 3.2.3, this problem is extremely difficult to analyse due to the need to account for regions of slip and non-slip within the contact arc, the length of which varies non-linearly with load. Unlike the approaches using classical methods by De Jong[13] and Oplinger and Gandhi[8] in which the region of non-slip is omitted, the approach of Wilkinson *et al.*[25] retains all these regions. Using finite elements introduces

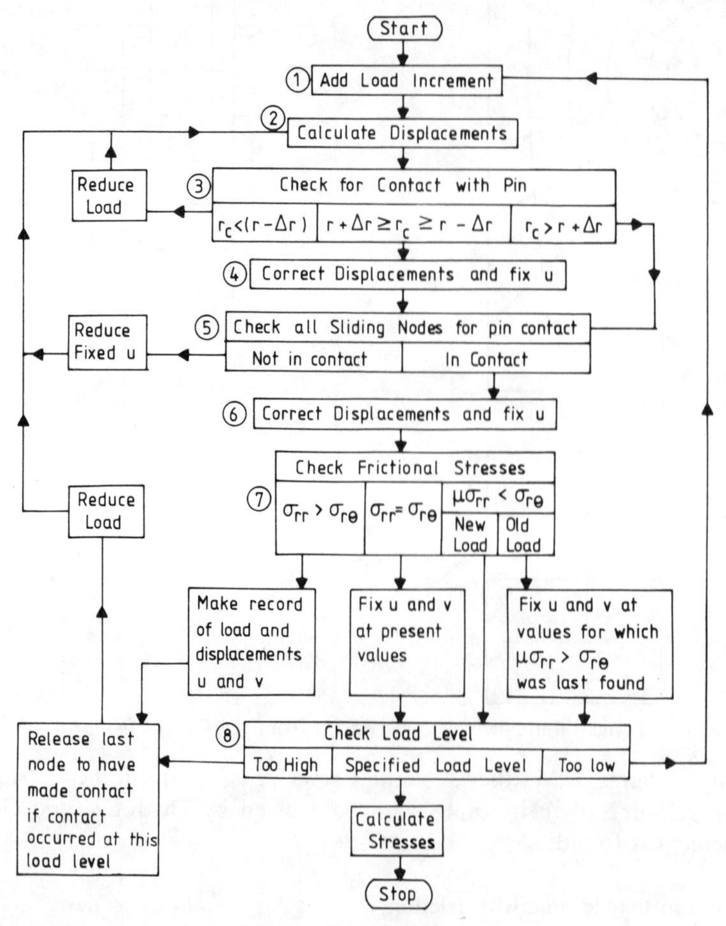

FIG. 7. Flow chart of iteration procedure to determine contact arc and slip and non-slip regions.[25]

particular difficulties since, after slipping, the nodes of contacting elements, originally coincident, will have moved relative to each other and their new positions have to be found. This particular aspect is avoided by Wilkinson *et al.*,[25] who consider the pin to be rigid and thus do not need to model it. They model the plate with isoparametric eight-noded quadrilaterals (linear-strain elements) which conform to the curved boundary of the hole. A typical mesh, for one-half of the plate, consists of 110 elements, the smallest of which has a side length of $0{\cdot}017d$. The pin centre is considered fixed and the load applied as a constant stress on one end of the plate. The coefficient of friction is assumed constant around the contact arc.

Some aspects of the incremental iterative solution procedure, which is outlined in the flow chart of Fig. 7, should be noted. Step 3 of the procedure checks the position (r_c) of each node, in the contact region, with respect to the pin surface (radius r) within a specified tolerance (Δr). For example, if $r_c < (r - \Delta r)$ the node is within the pin and after reducing the load the procedure returns to step 2, and so on. To obtain smooth stress distributions step 4 (and 6) corrects displacements so that the node is exactly on the pin surface. Step 5 checks the positions of sliding nodes.

Typically the load was applied in 25 increments with the tolerance limit Δr set at $8 \times 10^{-5}r$. Clearly both these factors affect the computing time, which was anyway kept to a minimum by the use of efficient equation-solving techniques and judicious nodal numbering. Even so, a normal run involved the solution of 600–730 simultaneous equations from 80 to 130 times.

The results of the above analysis confirm the importance of including friction if accurate stresses are required. Rowlands *et al.* show that for a low-stiffness material such as timber the differences between radial pressure with and without friction included are much larger than for a stiffer material such as GFRP.[27] Their results with zero friction also confirm the sensitivity of radial stress distribution to laminate stiffness; thus a stiffer material which cannot wrap itself so easily around the pin will have a steeper (less cosine-like) distribution.

3.3.4. Multi-pin joints

As with classical methods, very few workers have analysed multi-pin joints using finite element methods. Oplinger[11] and Rowlands *et al.*[27] consider a line of two pins, whereas Wong and Matthews[20] investigate a two-pin row. Chang *et al.*[28] consider both the two-pin line and row. As in their earlier work,[21] the latter authors assume a cosine distribution of radial pressure on the loaded half-circumference of a single pin or two pins in a row, whereas for the two-pin line the length of contact arc, different on each pin, is found in the analysis. There seems no good reason why these authors did not adopt the latter approach throughout. It should also be noted that the boundary conditions used by Chang *et al.*[28] are equivalent to the composite laminate being pinned to a rigid plate; this may or may not be representative of actual joints.

The results of these investigations confirm the classical analyses showing that peak stresses depend on fastener spacing and that the distribution of radial pressure around pins in a line is particularly sensitive to the level of by-pass load.

3.3.5. Three-dimensional analyses

As mentioned earlier, although technically feasible three-dimensional analyses are exceedingly expensive, even when techniques such as sub-structuring are used, and this has restricted the number of such investigations. The majority of those that have been performed, as with two-dimensional studies, have been for open holes rather than bolted holes.

Although not directly related, results from the large body of literature available on the through-thickness stress distribution at the free edge of a laminated plate are of relevance to the analysis of bolted joints. The accuracy of finite element solutions in the presence of the stress singularity that can exist at an interface between plies is of particular importance.

Raju *et al.*[29] consider this general plane strain problem and conclude that the finite element displacement method yields accurate stresses everywhere except in the two elements closest to a stress discontinuity or singularity. These elements can, of course, be made arbitrarily small by suitably defining the mesh.

Raju and Crews[30] examine the stress distribution around open holes in $(0/90°)_s$ and $(90/0°)_s$ laminates. Due to the symmetry they consider only one-eighth of the laminate. The mesh used has a two-dimensional region away from the hole, a transition region and a three-dimensional region near the hole. The mesh in the latter region contains almost 1300 elements and is of increasing fineness as the hole edge/ply interface junction is approached, the dimensions of the smallest 20-noded 'brick' element being an order of magnitude smaller than typical current fibre diameters. At this level any assumptions about a homogeneous material would be completely invalid. The results of the three-dimensional analysis are compared with those of a pseudo three-dimensional analysis applied to radial planes using data from a two-dimensional investigation to provide the effective applied loads. The two sets of stresses are extremely close and in view of the consequent reduction in computing time the application of this approximate method to bolted joints could be very worthwhile. Clearly it is impractical to employ such fine meshes as Raju and Crews[30] to problems containing a large number of layers.

It appears that the only workers who have so far looked at the stresses around a bolted hole are Matthews *et al.*[31] In order to reduce the number of elements involved, a special element was developed to represent several plies. The new element is derived from a standard 20-noded (linear-strain) brick element by modifying the integration process for the element stiffness to account for the variation from ply to ply of the in-plane elastic properties. The ply stresses are determined from the nodal displacements and clearly the accuracy of the element will decrease as the number of layers increases, the assumed quadratic displacement variation becoming increasingly unrepresentative of the actual through-thickness distribution.

Matthews *et al.*[31] examine a $(0/\pm 45/0°)_s$ laminate, the periphery of the hole being loaded by an arrangement of rigid bars pin-jointed between the hole centre and the loaded side of the hole ($90 \geq \theta \geq -90°$). Load is applied by imposing a displacement to the centre node. The length of the contact arc is found iteratively, bars which sustained a tensile force on the first run having their stiffness set to zero for the next run. The so-determined arc compares closely with the values found by De Jong[7] and Crews *et al.*[22] The inclusion of through-thickness stresses in the analysis allows the investigation of the hole loaded by a plain pin (no through-thickness restraint), a bolt with 'finger-tight' washers (through-thickness expansion prevented), or a tightened bolt (compressive force applied through-thickness). The results show that the through-thickness direct and shear stresses differ widely and confirm the experimental observations that the failure mode changes between these three situations.

A particular difficulty with the standard form of finite element is the inability to satisfy equilibrium at a traction-free edge, although this condition can be approximated if a sufficiently fine mesh is used. This is a deficiency of the element used by Matthews *et al.*,[31] but the stresses within the laminate do not appear to be significantly affected. The situation can be improved by using hybrid elements in which stress as well as displacement variations are specified. Suitably chosen functions allow the stress-free boundary condition to be satisfied exactly. Whilst elements of this type have been used to model a laminate in tension,[32] there is no evidence of their use for the loaded hole problem.

Shivakumar and Crews[33,34] also consider the through-thickness clamping force imposed by tightening the bolt. Their purpose is to determine the variation of this force with time in an unloaded laminate. They are not therefore specifically interested in the stress distribution within the laminate. The reason for this investigation is the suspicion that the viscoelastic properties of typical matrices may lead to a relaxation of the clamping force with a consequent reduction in the joint strength (see Chapter 2).

They modelled a double-lap joint as axisymmetric about the bolt axis using four-noded quadrilateral elements for both bolt and laminate. The bolt was represented by elastic elements and the laminate by viscoelastic elements. The viscoelastic properties were obtained from micromechanics procedures based on the behaviour of fibres and matrix, the latter being assumed linearly viscoelastic.

The results from the finite element analysis were used with a least-squares regression technique to obtain a power law describing the variation of clamping force with time. Experimental data from a quasi-isotropic carbon/epoxy laminate showed good correlation with the predictions.

3.3.6. Boundary element methods

In recent years there has been a rapid growth in applications of boundary element methods, particularly to problems in soil and fluid mechanics.

As we have seen (Section 3.3.1), with finite element methods the interior of a body is idealised as an assembly of discrete elements over which the unknown displacements are represented approximately by linear, quadratic, etc., variations. Thus the governing equation of internal equilibrium is satisfied approximately.

With the boundary element method, the surface of the body is discretised as an assembly of elements over which the variation of the unknowns are approximated. Having determined the values of the unknowns on the boundary the governing equation, in the interior, is satisfied exactly. As with finite elements, boundary elements can have assumed linear, quadratic, etc., variations of the unknowns.

The formulation of the method, as described by Brebbia,[35] leads to a matrix equation similar in form to that (eqn (3.16)) of the finite element method, i.e.

$$\mathbf{A}X = \mathbf{F} \tag{3.18}$$

where X contains the unknowns on the boundary. An essential difference with the finite element method is that whereas the stiffness matrix $\mathbf{K}$ is banded, the matrix **A** is fully populated, which means that the method may be less efficient computationally.

The boundary element method is very well suited to solving 'infinite body' problems. Many such problems arise in soil and fluid mechanics; in solid mechanics a point load applied to a semi-infinite plate can be cited as an example. When the two methods are applied to the same problem the boundary element method is found to need fewer elements, and far less data preparation, and to result in a smaller system of equations. In the vicinity of singularities neither method gives good results.

The boundary element method becomes relatively more expensive than the finite element method when applied to components with a high surface area/volume ratio.[36] Likewise, their use is not advised for components having a complex geometry or one with an extreme shape, e.g. very long and thin.

Apart from one instance that will be mentioned later, the method does not appear to have been applied to bolted composite joints. This is in spite of the fact that the technique is able to cope readily with frictional contact between elastic bodies.[37]

3.4. LAMINATE STRESS DISTRIBUTIONS

3.4.1. Introduction

The ultimate purpose of stress analyses, be they classical or numerical methods, is to allow prediction of joint strength based on the application of an appropriate failure criterion. Before discussing such predictions in Section 3.5, the stress distributions on which they are based will be presented. Only stresses appropriate to the basic failure modes, bearing, tension and shear-out, will be examined, particular attention being paid to the effects of lay-up and geometry.

3.4.2. Stresses around the hole boundary

The radial pressure distribution between pin and hole has already been discussed at some length. However, since the maximum value of this stress is generally considered to trigger bearing failure, further consideration is worthwhile. Also the variations of tangential stress and shear stress in the laminate at the hole boundary are important because of their relationship to tensile and shear-out failure respectively.

Crews *et al.*[22] present extensive results on stress distributions and the influence of changing width and end distance. They consider six CFRP lay-ups; quasi-isotropic, 0°, 90°, $(0/90°)_s$ $(\pm 45°)_s$ and $(0/\pm 45°)_s$. It is shown that the tangential stress ($\sigma_{\theta\theta}$) and shear stress ($\sigma_{r\theta}$) are more sensitive than the radial stress (σ_{rr}) to changes in both w/d (for fixed e/d) and e/d (for fixed w/d). The magnitudes and positions of the maximum stresses are dependent on lay-up, as are the lengths of the contact arc. For all lay-ups the maximum tangential and shear stresses decrease markedly as w/d and e/d increase, whereas maximum radial (i.e. bearing) stress changes only slightly, a result also given by Wong and Matthews.[20] This indicates that bearing is likely to be the critical failure mode if width and end distance are sufficiently large—a conclusion supported by experimental observation. Contact stresses are also affected by the clearance between pin and hole. The maximum radial stress increases as the clearance increases,[14,27] indicating that a stronger joint will be obtained with a close fitting pin, an observation confirmed experimentally.

Crews *et al.*[22] show that the maximum $\sigma_{\theta\theta}$ occurs in the 0° laminate followed by, in decreasing order, 90°, $(0/90°)_s$ and $(\pm 45°)_s$ roughly equal

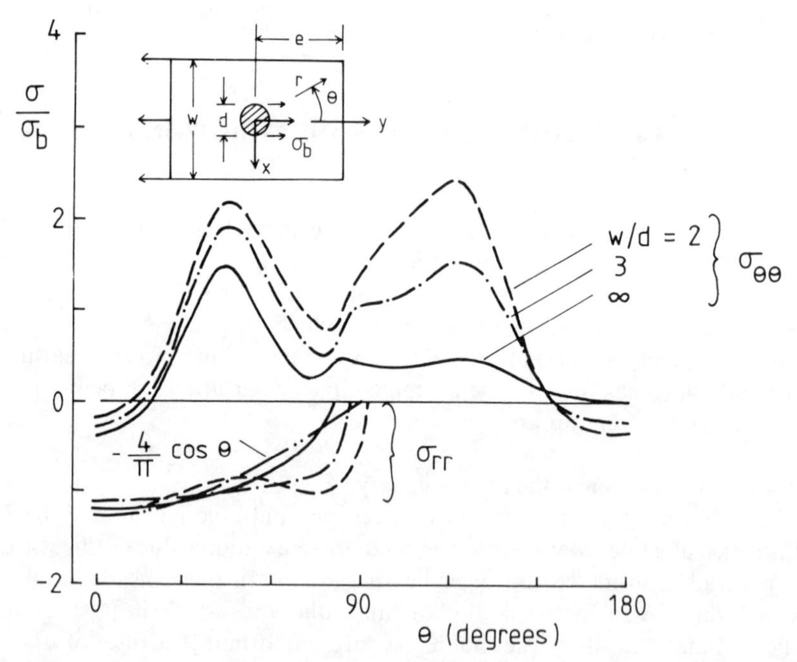

Fig. 8. Effect of width on stresses around hole boundary for a pin-loaded $\pm(45°)_s$ CFRP laminate.[22]

and finally $(0/\pm 45°)_s$ and quasi-isotropic roughly equal. Maximum σ_{rr} also occurs in the 0° lay-up followed again by the 90° laminate; all other lay-ups have similar values. Typical distributions are shown in Fig. 8 for the effect of w/d on a $(\pm 45°)_s$ lay-up and in Fig. 9 for the effect of e/d on a $(0/\pm 45°)_s$ lay-up.

The results of Crews *et al.*[22] are confirmed by De Jong[7] Fu-Kuo Chang *et al.*,[21] and Soni[23] although the latter authors present their results in terms of stresses in the co-ordinate (x, y) directions (see Fig. 1).

De Jong[7] and Soni[23] also show the distribution of shear stress (σ_{xy}) around the hole boundary. For quasi-isotropic[7] and $(0/90/\pm 30°)_s$[23] laminates the maximum shear stress occurs at $\theta = 45°$. For unidirectional (0°) lay-ups the peak is at $\theta \simeq 20°$ and for $(0_4/\pm 45°)_s$ at $\theta \simeq 30°$.[7]

3.4.3. Stresses along the co-ordinate (x, y) axes

Several workers give variations in direct and shear stresses along co-ordinate axes with origin at the hole centre (Fig. 1). De Jong,[7] for example,

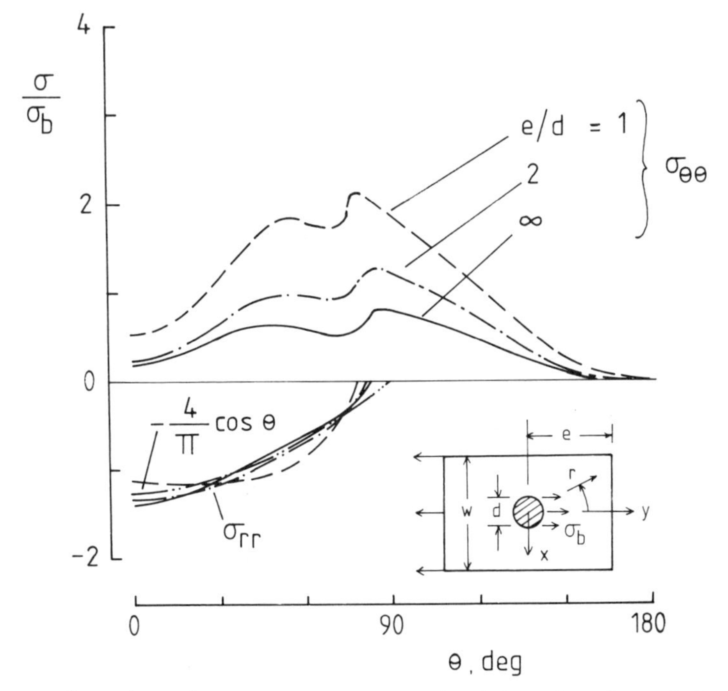

Fig. 9. Effect of end distance on stresses around hole boundary for a pin-loaded $(0/\pm 45°)_s$ CFRP laminate.[22]

shows that the tensile stress (σ_{yy}) along the x-axis decreases rapidly, reaching quite low values within one radius of the hole edge. This distribution, which has implications for the application of failure criteria to the tensile mode, is confirmed by York *et al.*,[38] whose analytical method is the same as that of Wilson and Pipes,[24] already mentioned. Typical results from York *et al.* are shown in Fig. 10 for a CFRP laminate with $(45/0/-45/0_2/-45/0/45/0_2/90°)_s$ lay-up.

In contrast to the rapid variation along the x-axis, σ_{yy} varies quite slowly along the y-axis on the loaded side of the hole. The rate of change depends on the lay-up but could be 50%, or more, of the maximum at one radius from the hole edge.[7]

3.4.4. Distribution along the shear-out planes

The distribution, on the loaded side of the hole, of shear stress along the shear-out planes ($y = \pm d/2$) is of interest when predicting shear-out failure. Distributions are given by Crews *et al.*[22] for the six laminates described earlier and by Wilson and Pipes[24] for the same laminate as York *et al.*[38] Since neither of these groups considers the same lay-up, comparison is difficult. Although they show the same general trends,

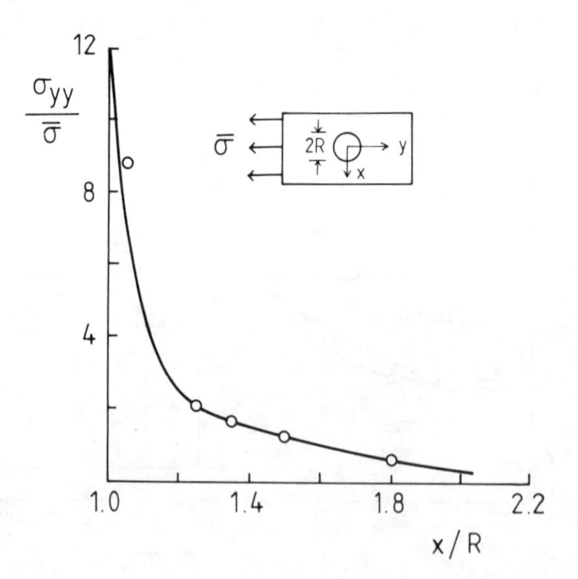

Fig. 10. Variation of σ_{yy} along the x-axis for a pin-loaded $(45/0/-45/0_2/-45/0/45/0_2/90°)_s$ CFRP laminate.[38]

Crews *et al.*[22] indicate that these shear stresses are very sensitive to e/d changes and relatively insensitive to w/d change, whereas Wilson and Pipes[24] show the opposite. However, as noted above, the boundary conditions imposed by Wilson and Pipes are different from those of Crews *et al.* What is clear is that the peak in the distribution occurs within one radius of the axis. Distributions for a $(0/\pm 45°)_s$ laminate from Crews *et al.*[22] are shown in Fig. 11.

It is interesting to relate these distributions to the variation of shear stress around the hole boundary described earlier (Section 3.4.2). For a similar laminate to that represented in Fig. 11, De Jong[7] shows the peak shear stress to occur at $\theta \simeq 30°$. Thus one can imagine that the locus of maximum shear stress in this case would be along the 30° radius, joining thc shear-out plane near the point where the shear on this plane is a maximum. This point will be close to the edge of the washer in a bolted joint.

A theoretical result for the maximum shear stress given by Oplinger[9,11] shows that the locus is close to, but not coincident with, the shear-out plane. The need to include non-linear shear behaviour is emphasised. Failure paths following such a locus are observed experimentally.[39]

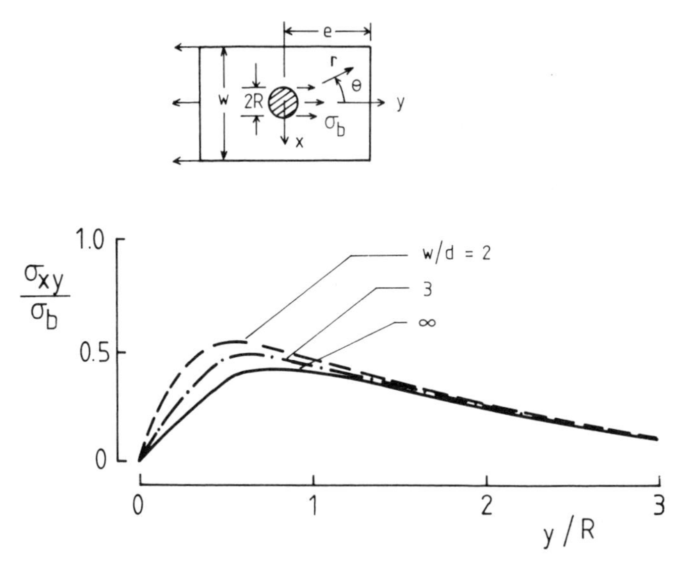

Fig. 11. Variation of shear stress, σ_{xy}, along the shear-out plane ($x = R$) for a pin-loaded $(0/\pm 45°)_s$ CFRP laminate.[22]

3.4.5. Three-dimensional effects

The results described so far in this section arise from two-dimensional analyses and thus refer effectively to pin-loaded joints. As described in Section 3.3.5, the analysis of Matthews *et al.*[31] includes through-thickness tensile and shear stresses. Their results for a $(0/\pm 45/0°)_s$ CFRP laminate agree in general terms with other workers when through-thickness averages of the in-plane stresses are considered.

Their analysis shows high through-thickness tensile stresses (σ_{zz}) on the loaded side of the pin-loaded hole. When combined with the local high in-plane compressive stresses one can expect the typical 'brush-like' failure mode observed experimentally.

When a bolt with finger-tight washers is considered (through-thickness expansion prevented) the tensile magnitude of σ_{zz} are reduced. This is consistent with the observed increase in failure load compared with a pin joint.

Simulating a tightened bolted joint (by imposing through-thickness compressive strains) gives compressive σ_{zz} in the vicinity of the hole. However, the in-plane compressive stress (σ_{yy}) in the outer ply becomes significantly higher at the washer edge, as does the interlaminar shear stress σ_{zx}, the latter reaching 22% of bearing stress in the +45° layer. The high compressive σ_{yy} is consistent with outer ply delamination at the washer edge and the high σ_{zx} is consistent with through-thickness shear cracks. Both these phenomena are observed on failed specimens.[31]

3.5. STRENGTH PREDICTION

3.5.1. Introduction

Methods available for the prediction of joint strength vary from the so-called 'phenomenological', in which the underlying failure mechanisms are not considered, through techniques involving the comparison of stresses with basic material strengths, to the methods of fracture mechanics in which the micromechanisms of failure are considered.

The need for predictive capabilities is clear when one considers the large range of fibre and matrix combinations available. To test every likely combination, even if practicable, would be prohibitively expensive. A method which is simple to use and needs only the measurement of a few basic parameters has obvious attraction, provided it is reasonably universally applicable, i.e. the basic parameters do not change with lay-up for a given fibre/matrix system. At present we are some way from achieving this ideal goal.

The phenomenological technique of, for example, Hart-Smith[40,41] has instinctive appeal for designers because of its simple and direct approach and will be discussed in detail in Chapter 6. The method is semi-empirical and is based on calculated stress concentration factors for elastic isotropic materials, which are assumed to be linearly related to the factors for fibre composites. Such linear correlation is based on experimental evidence for a limited range of materials.

The following sections will discuss the application to mechanically fastened joints of several methods, i.e. failure theories, two-parameter models, stress gradient and fracture mechanics methods, and combinations of these.

3.5.2. Use of failure theories

Of the multiplicity of failure theories that have been proposed for fibre-reinforced composites very few have been applied to joints.

The simplest are the limit theories in which separate maximum stresses, or strains, are compared with allowable values.[2] Failure may be said to have occurred in the corresponding mode when any one of the stresses, or strains, exceeds its appropriate limit. It is usual to consider shear and tension or compression along and normal to the fibres. There is assumed to be no interaction between the various stress, or strain, components.

The more sophisticated, interactive, theories do allow for stress component interaction although they do not directly predict the mode of failure. The latter is achieved indirectly by examining all the stress components, that closest to its corresponding strength indicating the likely failure mode. One of the earliest of such criteria was the Tsai–Hill distortional energy criterion.[2] At failure this criterion has the form:

$$\left(\frac{\sigma_1}{X}\right)^2 + \left(\frac{\sigma_2}{Y}\right)^2 + \left(\frac{\sigma_{12}}{S}\right)^2 - \frac{\sigma_1\sigma_2}{X^2} = 1 \tag{3.19}$$

Where X, Y and S are the principal strengths corresponding respectively to the direct stresses σ_1 and σ_2 and the shear stress σ_{12} in the orthotropic lamina. The direct strengths can be tensile or compressive, as appropriate.

The Tsai–Hill criterion is segmentally quadratic but a modification by Hoffman[42] allows for changes from tension to compression in a continuous manner.

An improvement on the above interactive methods is the Tsai–Wu

TABLE 1
Comparison of Strength Predictions[44]

Failure mode[a]	*Laminate*	*Material*[b]	*Percentage composition*			$\frac{e}{d}:\frac{s}{d}$	*Predicted failure load/ Experimental failure loads*		
			±45°	*0°*	*90°*		*Tsai–Hill*	*Maximum stress*	*Maximum strain*
T, T, C	$(\pm 45^\circ)$	G/E	100	0	0	2·4	0·47	0·47	0·62
T	$[(\pm 45/\bar{0}^\circ)_s/\overline{90}^\circ]_s$	B/E	72·7	18·2	9·1	2·4	0·76	0·58	0·49
T	$(\pm 45_5/90^\circ_6)$	B/E	62·5	0	37·5	3·4	0·93	0·94	1·00
S	$[\pm 45/0_6/\overline{90}^\circ)_s]_s$	B/E	13·3	80·0	6·7	6·6	0·58	0·61	1·89
S	$(0_6/\pm 45^\circ_5)$	B/E	62·5	37·5	0	2·5:7·53	0·98	1·10	0·95
B	$(0_6/\pm 45^\circ_5)$	B/E	62·5	37·5	0	4:7·53	~0·65	0·58	0·49

[a] T, tension; C, combination; S, shear-out; B, bearing.
[b] G/E, graphite/epoxy; B/E, boron/epoxy.

tensor polynomial criterion.[43] Failure by this criterion may be expressed as:

$$F_{11}\sigma_1^2 + F_{22}\sigma_2^2 + F_{66}\sigma_{12}^2 + 2F_{12}\sigma_1\sigma_2 + F_1\sigma_1 + F_2\sigma_2 + F_6\sigma_{12} = 1 \quad (3.20)$$

where the F_{ij} are various strength parameters.

When using any of the above criteria for laminated composites it is normal to apply them on a ply-by-ply basis, on the assumption that the behaviour of a single unidirectional ply within the laminate is the same as that of an isolated ply. Clearly this assumption is invalid as adjacent plies will provide constraints, thus modifying the behaviour. Having decided that a ply has failed it is still necessary to define how such failures affect the stiffness of the laminate.

Waszczak and Cruse[44] consider the maximum stress, maximum strain and the Tsai–Hill criteria in conjunction with stresses determined by the finite element method. Ultimate load was calculated in an iterative fashion. If any layer within an element was deemed to have failed, according to the appropriate criterion, it was removed from the analysis and the stresses redistributed amongst the remaining laminae. If no further laminae fail then the load is increased. The procedure is repeated until total laminate failure occurs.

It was found that the Tsai–Hill distortional energy criterion always gave conservative predictions for the laminates considered, being between 53% and 2% too low. This was not the case with the maximum stress and maximum strain criteria as shown in Table 1. Also, the Tsai–Hill criterion always predicted the failure mode correctly, whereas the other two criteria were unsuccessful in this, presumably because they ignore stress interactions in a complex stress state.

Waszczak and Cruse[45] report similar success in a programme aimed at optimising multi-bolted joints. In this case, however, they use only the maximum stress criterion and the stresses are obtained from anisotropic elasticity theory. This approach is much more efficient than using finite elements because of the large number of times the stress analysis is used in the optimisation procedure.

Oplinger and Gandhi[8] use the Hoffman criterion with anisotropic elasticity theory to predict the strength of a pin-loaded $(0_2/\pm 45^\circ)_s$ CFRP laminate. In contrast to Waszczak and Cruse,[44] who use distortional energy contours to predict failure mode, Oplinger and Gandhi use predicted pin failure load for each lamina plotted around the hole periphery. Comparison is then made with corresponding ply strengths. Figure 12 shows how consideration of friction indicates that failure will

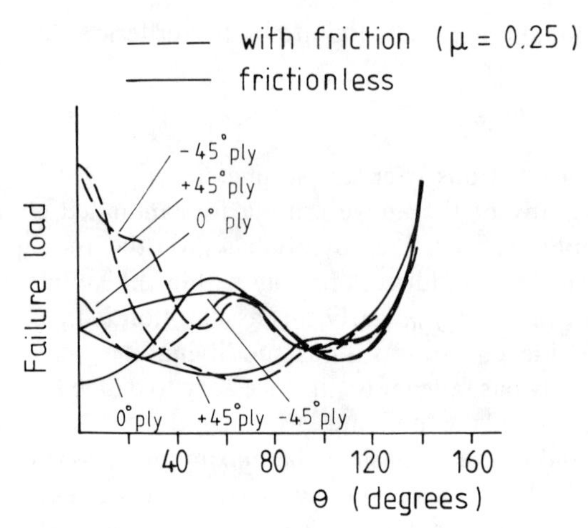

Fig. 12. Effect of including friction on the prediction of failure in a pin-loaded $(0_2/\pm45°)_s$ CFRP laminate.[8]

change from cleavage, lowest load in 0° ply at $\theta = 0°$ due to hoop stress, to shear-out, lowest load in +45° layer at $\theta = 65°$. The large divergence between predicted and measured values for shear modes indicates that non-linear stress–strain behaviour must be included for realistic predictions (see also ref. 9).

The sensitivity of some laminates to cleavage failure and the importance of including friction in the analysis[8] is confirmed by De Jong.[13] De Jong who uses the Tsai–Hill criterion with anisotropic elasticity, shows that maximum bearing strength occurs with a moderate value (0·2) of the coefficient of friction (μ). Very high values of μ cause the mode to change from bearing to net tension, thus indicating the need to use smooth pins in actual joints.

A pseudo-isotropic $(0/\pm60°)$ lay-up of 1:1 glass/carbon hybrid is investigated by Humphris[46] using stresses determined by a finite element analysis. The Tsai–Hill criterion is applied on a ply-by-ply basis in conjunction with subsidiary criteria to check splitting normal to the fibres due either to direct or to shear stress.

After failure of a layer has occurred the [A] matrix of the laminate, from classical lamination theory[2] (stiffness matrix relating in-plane forces and strains), is recalculated for the remaining laminae. This is repeated as necessary at a given load level, complete failure for an element being defined as [A] becoming singular. Application of this procedure allowed progression of the damage front, across the net section, to be delineated.

It is clearly shown that the point of maximum stress moves away from the hole edge as load increases, giving a predicted failure load some 60% higher than indicated using stress concentration factors. The predicted ultimate load was 15% lower than the measured value. Again the need to include non-linear elasticity and through-thickness stresses to improve predictions is emphasised.

The Tsai–Wu failure criterion is favoured by Soni[47] in a programme to predict the strength of a large range of CFRP lay-ups of varying geometry. No modification to laminate compliance was made following ply failures, on the basis that plies which fail before the last ply do not disintegrate, but share their load with intact plies. The accuracy of this assumption must depend upon element size and the number of points of assessment within an element—in this case only the element centroid is used. Soni defines ultimate strength as the failure strength of the strongest ply at the weakest point of the laminate. In addition to loaded holes, unloaded and partially loaded holes (i.e. by-pass load present) are examined. The failure mode is determined by consideration of the stress components at the weakest point. Predictions are always conservative with differences from less than 10% to about 50% below measured strengths, depending on lay-up and geometry. The worst results were for $(0_2/\pm45^\circ)_s$ laminates failing in bearing and $(90_2/\pm45^\circ)_s$ laminates failing in tension. Similar accuracy was obtained both for fully loaded and partially loaded holes.

Chiang and Rowlands[48] also employ the Tsai–Wu criterion for pin-loaded holes, accounting for friction, pin clearance, finite specimen size and, most importantly, non-linear stress–strain behaviour of the laminate. Their analysis, which is based on earlier work,[25–27] uses a finite element method in which the structural stiffness is up-dated at each increment of load. Elastic and shear moduli and Poisson's ratio are taken as functions of strain and are represented by cubic splines fitted to experimental data.

Stress distributions for spruce show close agreement with experimental data using Moiré techniques. Peak stresses from a linear elastic analysis are some 30% higher.[48] A series of calculations for GFRP emphasises the importance of using non-linear material characteristics, especially for $(0/90^\circ)$ and $(\pm45^\circ)$ lay-ups. It is shown that stresses, strength, failure load and mode all depend on the assumed stress–strain variation.

3.5.3. Use of two-parameter models

Classical concepts of stress concentration factors cannot explain the fact that the tensile strength of a laminate containing a central circular hole depends on hole size, strength being lower for a larger hole. Whitney and Nuismer[49] proposed two approaches, the point stress and the average stress methods, to explain this phenomenon. These are often referred to as two-parameter models since the strength is expressed in terms of two experimentally determined parameters.

In the point stress criterion, failure is said to occur when the stress (σ_y) in the direction of the load equals the unnotched laminate strength (σ_0) at some distance d_0 from the hole edge (measured perpendicular to the load): σ_0 and d_0 are the parameters to be determined.

In the average stress criterion, failure is said to occur when the average stress over some distance a_{0_t} equals the unnotched laminate tensile strength, i.e.

$$\sigma_0 = \frac{1}{a_{0_t}} \int_R^{R+a_{0_t}} \sigma_y(x, 0)\, \mathrm{d}x \tag{3.21}$$

R being the hole radius. In this case σ_0 and a_{0_t} have to be determined.

To be universally applicable one would expect d_0 and a_{0_t} to be constant for all hole sizes and lay-ups in a given fibre/resin system. The evidence indicates that this is not the case. Certainly laminate quality and combined stresses affect the value of a_0,[50] and d_0 appears to vary with notch size.[24]

Nuismer and Labor[51] extended the average stress approach to compressive loading, replacing a_{0_t} by a_{0_c} but still using eqn (3.21). Typically a_{0_c} is larger than a_{0_t} (6·2 mm compared with 2·3 mm for the laminates tested). In addition to compression testing of unloaded holes, Nuismer and Labor also test several specimens containing partially loaded holes within an overall compressive field. The failure criterion is used in conjunction with a finite element stress analysis, a_{0_c} being taken as tangential to the bottom of the hole and normal to the load, although this is acknowledged as being an approximation since the maximum compressive stress occurs along the hole boundary around an arc of some 30°. Around this arc the local strength will vary, a fact not allowed for in the analysis. Close agreement between test and prediction is shown for a very small number of specimens.

The average stress criterion is used by Agarwal[52] to predict the strength of bolted joints in a number of CFRP laminates. The NASTRAN finite element system is used to determine the stress distribution around the hole when loaded by a frictionless pin. The unnotched laminate strength σ_0 is determined from the maximum strain theory using stresses in each lamina found from lamination theory in terms of strain allowables, the minimum value of stress then giving the corresponding laminate allowable strength. The latter will of course depend on the values used for limiting strains, there being considerable uncertainty about the shear strain.

In applying the average stress criterion, Agarwal[52] assesses the failure mode by integrating along radial lines, as shown in Fig. 13. His treatment of a_{0_c} for bearing failure, as essentially parallel to the applied tensile load, is in contrast to that of Nuismer and Labor,[51] who, as explained above, take this length normal to the applied load, which in their case is compressive. Values used for a_{0_t} and a_{0_c} are taken as those of Nuismer *et al.*[49-51] for a different CFRP system, whilst a_{0_s} is found from experiment. Numerical tests show that predicted strengths are very sensitive to the values of a_0. In general, predicted values were higher than test data, differences being greatest for laminates involving considerable shear deformation ((0/90°) and (±45°) lay-ups). Failure modes are usually predicted successfully.

Ramkumar,[53] who again uses the average stress criterion, also comments on the importance of the values of a_0 used for prediction. He also notes the inability of a two-dimensional analysis, such as he uses, to allow for delaminations between plies. He shows close agreement between predictions and tests, which is particularly surprising since the latter are, unusually, on single shear specimens.

Curtis and Grant[54] apply the average stress criterion, successfully demonstrating the change of failure mode that occurs with changing geometry. They, like Ramkumar,[53] acknowledge that the method does not permit the effect of delaminations to be studied.

Wilson and Pipes[24] concentrate on the shear-out failure mode which they treat using the point stress criterion. The stress distribution along

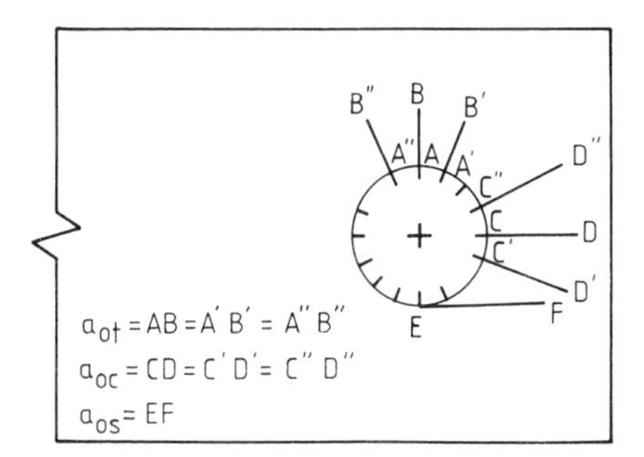

Fig. 13. Definition of characteristic lengths a_{0_t} (tension), a_{0_c} (compression) and a_{0_s} (shear) for use with the average stress criterion.[52]

the shear-out plane, determined by a finite element analysis, is fitted by a polynomial. The dependence of the constants on w/d and e/d is included in the function, thus allowing parametric studies to be easily undertaken. The results from such studies show that the shear stress concentration decreases as e/d increases and increases as w/d increases, confirming the work of Oplinger.[9] The general applicability of the approach is not confirmed since only one laminate was tested.

In a repetition of the approach of Wilson and Pipes,[24] York *et al.*[38] consider the tensile failure mode in bolted joints. The predictions are compared with rather more experimental data than were used in the shear-out study and show that, provided the constants defining the variation of d_0 with hole size are well chosen, correlations are within a few per cent.

3.5.4. Use of stress gradient method

A model based on the sequential fracture of 0° fibres by Potter[55] provides a convincing failure mechanism for notched laminates. It is shown that failure will occur if the gradient of tensile stress, along a line normal to the applied load, is less than a critical value that depends on the properties of the fibre, matrix and interface. In effect this criterion says that if the stress gradient is too high, the second fibre will only be lightly loaded when the first fibre, adjacent to the notch, fails so that the limited load that can be transferred into it from the newly broken fibre will be insufficient to cause its failure.

Curtis and Grant[54] consider this method in their study of the failure of notched laminates, although they apply it only to unloaded holes. When average laminate stresses are used the critical gradient is found to vary with lay-up. The authors suggest that a better approach may be to apply the method on a ply-by-ply basis.

No other workers appear to have used this technique.

3.5.5. Use of fracture mechanics

The application of fracture mechanics methods to predicting crack growth in metals is well documented. In essence the behaviour is expressed by an equation of the form

$$K = \sigma\sqrt{\pi a} \tag{3.22}$$

where K is the stress intensity factor, σ the applied stress and a the half crack length. If the critical value of K is known (the fracture toughness) the critical crack length that will propagate under a given stress can be calculated. It is presumed that the crack always grows in a self-similar fashion.

Many workers have applied fracture mechanics principles to composites, with mixed success. One major problem is that mode of propagation can change as the crack grows, i.e. self-similarity is not always assured.

Eisenmann[56] is one of the few workers so far to have used fracture mechanics to determine the strength of bolted joints. He considers one fastener from an array subjected to a general set of two-dimensional stresses with an arbitrary fastener load. The stress intensity factor is evaluated for a $(0_2/\pm 45°)$ lay-up, at eight points equispaced around the hole boundary, as the linear superposition of factors corresponding to unit shear and direct stresses and unit bearing stresses in the co-ordinate directions. This approach is justified by the experimental observation that, at these points, through-thickness cracks propagate radially in a Mode I (opening) fashion for a short distance before changing to another mode.

Fracture toughness is measured on tensile and bending specimens specially fabricated so that they represent laminate properties tangential to the hole at the chosen points. Stress intensities are calculated for several values of crack length using the boundary integral method and are applicable to isotropic materials. These are corrected for orthotropy using expressions for elastic stress concentrations in combination with appropriate elastic constants. Even then the expressions obtained are inexact because the hole boundary conditions are not correctly represented. Excellent correlation is shown between test and prediction for various combinations of bearing and by-pass load. However, the approximations in the stress intensity factors necessitates the inclusion of empirical correction factors. The need for such factors clearly limits the generality of the approach.

3.5.6. Use of combined methods

The idea of combining a failure criterion, such as Tsai–Hill, with a characteristic length has been applied to unloaded holes by several authors. Only Chang *et al.*[21] seem to have used this approach for joints. Their approach combines the failure theory of Yamada and Sun[57] with a 'characteristic curve' based on the dimension a_0 that appears in the formulation of the average stress criterion.[49–51]

The failure theory of Yamada and Sun is a quadratic theory, involving only the shear stress and the direct stress along the fibres, and taking the form

$$\left(\frac{\sigma_1}{X}\right)^2 + \left(\frac{\sigma_{12}}{S_c}\right)^2 = 1 \tag{3.23}$$

This equation is based on the fact that at failure plies will have already failed in the transverse direction and thus σ_2 can be neglected. The shear strength S_c is different (higher) from the shear strength of a single ply and is an attempt to allow for the support given by adjacent laminae; it is determined by tests on cross-ply laminates.

Chang *et al.*[21] apply the above theory on a ply-by-ply basis using stresses determined from a finite element analysis. Friction is ignored. If any ply is deemed to have failed, at any point on the characteristic curve, the joint is taken as failed. The location of the failure is taken as an indicator of failure mode. Results of the analysis are compared with experimental data of other workers and generally agree within 10% although this clearly depends on values chosen for the characteristic distance. Also, for laminates containing a large proportion of 0° fibres the predictions are sensitive to the value of S_c. Best accuracy is achieved for quasi-isotropic lay-ups.

Further work by the same authors applies the same approach to multi-pin joints.[28] This paper, which seems to use a different characteristic curve for each hole, contradicts the earlier work[21] in which this curve is taken as material-dependent only. Strength predictions agree closely with experimental data except for (0/90°) and (±45°) lay-ups, i.e. those in which significant non-linearity would be expected.

Chang *et al.* address non-linear effects in a later paper[58] by including a non-linear shear stress–strain relation with their previous analysis.[21,28] They show improved prediction of failure mode and smaller differences between predicted and measured strengths (25% as opposed to 40% lower) for (0/90°) and (±45°) lay-ups. For other lay-ups the differences between results for linear and non-linear analyses were very small.

3.6. CONCLUSIONS

It is apparent from the work described above that an accurate stress distribution will only be obtained if pin/hole friction, material non-linearity and three-dimensional effects are all accounted for. Such a comprehensive analysis would have to be undertaken using numerical techniques and, to date, has not been attempted.

No universal method of strength prediction is yet available. Although good results are demonstrated for particular failure modes using two-dimensional analyses, this success depends on experimentally determined 'constants' and the ability of the approach to 'read across' to other materials or lay-ups is not guaranteed.

REFERENCES

1. Lekhnitskii, S. G., *Theory of Elasticity of an Anisotropic Body*, 1981, Mir Publishers, Moscow, USSR.
2. Agarwal, B. D. and Broutman, L. J., *Analysis and Performance of Fiber Composites*, 1980, Wiley, New York, USA.
3. Jones, R. M., *Mechanics of Composite Materials*, 1975, Scripta Book Co., Washington DC, USA.
4. Savin, G. N., TT F-607, 1968, NASA, Washington DC, USA.
5. De Jong, T., Report LR-223, 1976, Aerospace Dept, Delft University of Technology, The Netherlands (in Dutch).
6. De Jong, T. and Vuil, H. A., Report LR-333, 1981, Aerospace Dept, Delft University of Technology, The Netherlands.
7. De Jong, T., *J. Composite Materials*, 1977, **11**, 313.
8. Oplinger, D. W. and Gandhi, K. R., AMMRC-MS-74-8, 1974, Army Materials and Mechanics Research Center, Watertown, Mass., USA.
9. Oplinger, D. W., *Proc. 4th Army Materials Technology Conference Advances in Joining Technology* (Eds J. J. Burke, A. E. Gorum and A. Tarpinian), 1975, Brook Hill Publishers, Chestnut Hill, Mass., USA.
10. Gandhi, K. R., AMMRC-TR-76-14, 1976, Army Materials and Mechanics Research Center, Watertown, Mass., USA.
11. Oplinger, D. W., *Proc. 4th Conf. on Fibrous Composites in Structural Design* (Eds E. M. Lenoe, D. W. Oplinger and J. J. Burke), 1980, Plenum Press, New York, USA.
12. Sokolnikoff, I. S., *Mathematical Theory of Elasticity*, 2nd edn, 1956, McGraw Hill, New York, USA.
13. De Jong, T., Report LR-350, 1982, Aerospace Dept, Delft University of Technology, The Netherlands.
14. Hyer, M. W. and Klang, E, C., VPI-E-84-17 (CCMS-84-02), 1984, Virginia Polytechnic Institute and State University, Blacksburg, Virginia, USA.
15. Zhang, K-D. and Ueng, C. E. S., *J. Composite Materials*, 1984, **18**, 432.
16. Zhang, K-D. and Ueng, C. E. S., *J. Composite Structures*, 1985, **3**, 119.
17. Harris, H. G., Ojalvo, I. U. and Hooson, R. E., AFFDL-TR-70-49, 1970, Wright–Patterson AFB, Ohio, USA.
18. Zienkiewicz, O. C., *The Finite Element Method in Engineering Science*, 2nd edn, 1971, McGraw Hill, New York, USA.
19. Fox, L., *An Introduction to Linear Numerical Algebra*, 1964, Clarendon Press, Oxford, UK.

20. Wong, C. M. and Matthews, F. L., *J. Composite Materials*, 1981, **15**, 389.
21. Chang, F-K., Scott, R. A. and Springer, G. S., *J. Composite Materials*, 1982, **16**, 470.
22. Crews, J. H., Hong, C. S. and Raju, I. S., TP 1862, 1981, NASA, Washington DC, USA.
23. Soni, S. R., *Proc. 1st Conf. on Composite Structures* (Ed. I. H. Marshall), 1981, Applied Science Publishers, London, UK.
24. Wilson, D. W. and Pipes, R. B., *Proc. 1st Conf. on Composite Structures* (Ed. I. H. Marshall), 1981, Applied Science Publishers, London, UK.
25. Wilkinson, T. L., Rowlands, R. E. and Cook, R. D., *Computers and Structures*, 1981, **14**, 123.
26. Rahman, M. U., Ph.D. Thesis, 1981, University of Wisconsin, Madison, Wisconsin, USA.
27. Rowlands, R. E., Rahman, M. U., Wilkinson, T. L. and Chiang, Y. I., *Composites*, 1982, **13**, 273.
28. Chang, F-K., Scott, R. A. and Springer, G. S., *J. Composite Materials*, 1984, **18**, 255.
29. Raju, I. S., Whitcomb, J. D. and Goree, J. G., TP 1751, 1980, NASA, Washington DC, USA.
30. Raju, I. S. and Crews, J. H., TM 83300, 1982, Washington DC, USA.
31. Matthews, F. L., Wong, C. M. and Chryssafitis, S., *Composites*, 1982, **13**, 316.
32. Spilker, R. L. and Chou, S. C., *J. Composite Materials*, 1980, **14**, 2.
33. Shivakumar, K. N. and Crews, J. H., TM 83268, 1982, NASA, Washington, DC, USA.
34. Shivakumar, K. N. and Crews, J. H., TM 84480, 1982, NASA, Washington, DC, USA.
35. Brebbia, C. A., *The Boundary Element Method for Engineers*, 1978, Pentech Press, Plymouth, Devon, UK.
36. Lachat, J. C., Ph.D. Thesis, 1975, University of Southampton, Southampton, UK.
37. Andersson, T., Report IKP-R-166, 1981, Inst. of Technology, Linköping, Sweden.
38. York, J. L., Wilson, D. W. and Pipes, R. B., *J. Reinforced Plastics and Composites*, 1982, **1**, 141.
39. Kretsis, G., M.Sc. Thesis, 1983, Aeronautics Dept, Imperial College, London, UK.
40. Hart-Smith, L. J., CR-144899, 1976, NASA, Washington DC, USA.
41. Hart-Smith, L. J., *Proc. 4th Conf. on Fibrous Composites in Structural Design* (Eds E. M. Lenoe, D. W. Oplinger and J. J. Burke), 1980, Plenum Press, New York, USA.
42. Hoffman, O. J., *J. Composite Materials*, 1967, **1**, 200.
43. Tsai, S. W. and Wu, E. M., *J. Composite Materials*, 1971, **5**, 58.
44. Waszczak, J. P. and Cruse, T. A., *J. Composite Materials*, 1971, **5**, 421.
45. Waszczak, J. P. and Cruse, T. A., AFML-TR-73-145, Vol. II, 1973, Wright-Patterson AFB, Ohio, USA.
46. Humphris, N. P., *Proc. Symp. Jointing in Fibre-Reinforced Plastics*, 1978, IPC Science and Technology Press, Guildford, Surrey, UK.

47. Soni, S. R., STP749, 1981, ASTM, Philadelphia, Pa, USA.
48. Chiang, Y. J. and Rowlands, R. E., *Composite Science and Technology*, 1986, in press.
49. Whitney, J. M. and Nuismer, R. J., *J. Composite Materials*, 1974, **8**, 253.
50. Nuismer, R. J. and Labor, J. D., *J. Composite Materials*, 1978, **12**, 238.
51. Nuismer, R. J. and Labor, J. D., *J. Composite Materials*, 1979, **13**, 49.
52. Agarwal, B. L., *AIAA Journal*, 1980, **18**, 1371.
53. Ramkumar, R. L., STP734, 1981, ASTM, Philadelphia PA, USA.
54. Curtis, A. R. and Grant, P., *J. Composite Structures*, 1984, **2**, 201.
55. Potter, R. T., TR 77023, 1977, Royal Aircraft Establishment, Farnborough, Hants, UK.
56. Eisenmann, J. R., TMX3377, 1976, NASA, Washington DC, USA.
57. Yamada, S. E. and Sun,. C. T., *J. Composite Materials*, 1978, **12**, 275.
58. Chang, F-K., Scott, R. A. and Springer, G. S., *J. Composite Materials*, 1984, **18**, 464.

Chapter 4

Experimentally Determined Strength of Adhesively Bonded Joints

K. LIECHTI

Department of Aerospace Engineering and Engineering Mechanics, University of Texas, Austin, Texas, USA

W. S. JOHNSON

Materials Division, NASA–Langley Research Center, Hampton, Virginia, USA

and

D. A. DILLARD

Department of Engineering Science and Mechanics, Virginia Polytechnic Institute and State University, Blacksburg, Virginia, USA

4.1. INTRODUCTION

The stress analyses described in Chapter 5 indicate that if the constitutive behaviour of the adhesive is assumed to be linear–elastic, then concentrations in the stresses will occur near the bond termination regions in lap-type joints. The severity of stress concentrations may be reduced with zero weight penalty by tapering the adherends in the single lap joint or considering modifications such as the scarf joint, double scarf joint or the double butt strap joint.[1] In reality, of course, the singular stresses will not occur because of the inelastic behaviour of the adhesive and so the stress field in the adhesive, under even relatively modest applied loads, consists of regions in which the stresses will have exceeded some (multiaxial) yield criterion and other regions, far removed from discontinuities, in which the stresses will be elastic and may even be uniform. The existence of higher stresses in the regions of non-uniformity near bond terminations means that failure will initiate in these regions. The initiation can be due to the onset of either the microdamage that occurs as the applied loads are increased under monotonic loading, or that which occurs for sufficiently high stress amplitudes under cyclic loading. The accumulation of such damage then gives rise to macroscopically observable cracks. The length of time between the initiation of failure, as indicated by the onset of microdamage, and the catastrophic crack growth that marks the final failure of the joint will generally be in the order of seconds for monotonically increasing loads at standard rates, or years under cyclic loading or static and cyclic loads in hostile environments. Since the concept of strength is associated with final failure of a component, it is clear that the definition of joint strength will depend on the particular conditions that led to its failure.

In the case where the time between initial and final failure is small, then a failure criterion or the joint strength may be defined in terms of limiting stresses and/or strains. Under these conditions the simplest failure criterion is often taken to be the load at failure divided by a nominal area such as the overlap of the adherends. The underlying assumption is that the shear stresses are uniform. If the non-uniform stresses are accounted for, then it has been suggested[2] that failure occurs when the maximum shear stress in the adhesive layer reaches the shear strength of the adhesive. However, other work[1,3,4] has indicated that a maximum stress criterion does not always apply, particularly for the more ductile adhesives which exhibit extensive plasticity in shear. For these cases a maximum shear strain or strain energy criterion has been proposed. In view of three-dimensional effects, it may be better to incorporate some multiaxial failure criterion.

For cases where a macroscopic flaw already exists, or where macroscopic crack growth rates for cracks initiating from bond terminations are small, fracture mechanics concepts may be used to characterise the strength of adhesively bonded joints in terms of a critical value of an appropriate fracture parameter. Crack face displacements are used to distinguish between three modes of fracture which are characterised by a tensile opening, an in-plane shear and an anti-plane shear, and designated as Mode I, Mode II and Mode III, respectively (Fig. 1). In situations where linear elastic fracture mechanics are applicable a number of fracture parameters in the form of stress intensity factors (K_I, K_{II} and K_{III}) and energy release rates (G_I, G_{II} and G_{III}) are most commonly used and are equivalent. When the extent of inelastic deformation near the crack front is relatively large, crack tip opening displacements, crack opening angles and the J-integral (in cases where deformation plasticity holds) are used

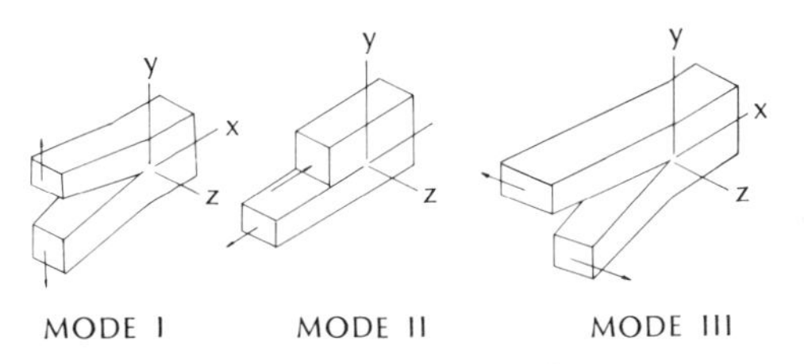

Fig. 1. Fracture modes.

as fracture parameters. Different crack paths may be accounted for by the cohesive fracture toughness of the adhesive (and adherends) and the adhesive fracture toughness associated with interfacial crack propagation. Cracks in adhesively bonded joints are often constrained to grow within the adhesive. Under such conditions, no single fracture mode is operative and crack growth is generally of a mixed-mode nature. Mixed-mode or effective fracture parameters which are combinations of the three modes are then used to characterise crack growth.

Early measurements of the fracture toughness of adhesives were based on analyses that essentially ignored the adhesive layer in a 'thickness-averaged' fracture mechanics approach, in which the bondline defined a weak material plane and energy release rates were obtained in terms of overall compliance changes in the specimens.[5,6] Fracture toughness then became a function of bond thickness.[7] More recently, finite element methods have permitted the adhesive layer and also cohesive and adhesive cracks to be included in the stress analyses that are used to determine fracture parameters.[8–10] When the adhesive layer is included in this way, the effects of variations in temperature and moisture on the stresses in the adhesive layer can be accounted for through viscoelastic stress analysis.[10] Because the fracture mechanics approach allows the details of the failure process to be analysed, and also because fatigue and environmental effects are likely to be the most critical design considerations, the emphasis in this chapter will be placed upon the determination of bond strength using the methods of fracture mechanics.

Composite adherends often have matrix materials much weaker than current structural adhesives. Therefore, rather than debonding in the joint, a failure may result within the composite in the form of delamination or interply failure. The exact nature of failure in an adhesively bonded composite joint will therefore depend on many variables such as the relative toughness of the adhesive and matrix materials, the orientation of the ply next to the bondline relative to the loading direction, the environment, the joint geometry, and the loading configuration. Some of these factors will be covered later in this chapter in Section 4.6.

4.2. SPECIMEN GEOMETRIES

The specimen geometries presented in this section are subdivided into those that may be used to determine modulus and strength properties (bearing in mind the points raised about strength in the introduction), fracture properties and durability.

4.2.1. Modulus and strength properties

Shear modulus data are often required for subsequent stress analyses of cracked joints. If it is assumed that the shear modulus of the adhesive is independent of the nature of the adherends surrounding it, joints involving metallic adherends may be used, which in turn allows the napkin ring[11] and thick adherend single lap shear specimens[12] to be employed. The napkin ring test produces very uniform stresses in the adhesive layer. However, it requires more effort to conduct and therefore a more popular specimen is the thick adherend single lap joint. The thicker adherends decrease (but do not eliminate entirely)[13] the peel stresses, particularly in regions far removed from bond terminations. If the relative displacements of the adherends are measured in the regions of uniformity, then reasonable shear modulus data may be obtained.

These specimens may also be used to obtain shear strength properties. The standard single lap joint specimen is also commonly used for such purposes. The strength values of the adhesive obtained from the latter specimen are much lower,[12] emphasising the effects of non-uniformity and questioning the meaning of strengths obtained in this way. At most, any such strength tests may only be used for comparisons of processing variables.

4.2.2. Fracture properties

Since the likelihood of voids and defects being present in the bondline is always high, test methods to assess the sensitivity to such defects of the geometry of a given adhesive joint geometry are desirable. The fracture toughness of a given adhesive can be used to predict the failure stress of a bonded joint with a known defect geometry. Various joint geometries and loading conditions can result in various combinations of debond tip-loading modes, ranging from pure Mode I to pure Mode II, to, most likely, combinations of both Mode I and Mode II (Mode III may also be present). Therefore several different specimens need to be tested in order to describe adequately this mixed-mode behaviour. Three popular specimens for fracture testing will be discussed. These specimens are also used to determine interlaminar fracture toughness of composites.[14]

4.2.2.1. Double cantilever beam specimen

The double cantilever beam (DCB) specimen was first suggested by Ripling *et al.*[5] to determine the adhesive fracture toughness. The DCB specimen results in a pure Mode I loading at the debond tip, thus G_{Ic} may be determined. Best results may be expected when unidirectional composite adherends are used in order to avoid the damage growing into the adherends.

4.2.2.2. Crack lap shear specimen
The crack lap shear (CLS) specimen was first used by Brussat *et al.*[15] It gives a mix of Mode I and Mode II loading. Depending on the relative stiffness of the adherends, the G_I/G_{II} ratio will vary from 0·20 to 0·40.

4.2.2.3. End notch flexure specimen
The end notch flexure (ENF) specimen was first suggested by Russell[16] for determination of pure Mode II interlaminar fracture toughness of composites. The specimen can also be used for testing adhesive joints. The ENF specimen is essentially a DCB specimen tested under three-point bending such that the debond tip is subject to pure shear.

All of the above specimens may be tested under a cyclic load and the debond growth measured. In this manner the debond propagation rate may be correlated with the strain energy release rate to provide data for designing cyclically loaded structures.

4.2.3. Durability
Single lap shear specimens and the Boeing wedge test have been widely utilised for durability studies with metal adherends. These specimen configurations have also been successfully applied for composite-to-metal and composite-to-composite durability tests.

The Boeing wedge test involves driving a wedge between two strap adherends and monitoring the crack tip position with time.[17] The wedge imposes a fixed displacement to the adherends and the energy stored in bending provides the driving force for crack growth. Because the available strain energy decreases as the fourth power of crack length, a very wide range of energy release rates is available with the specimen. This gives the specimen the ability to discriminate between the environmental integrity of a wide range of adhesive and surface preparation systems. Although it is a fracture type specimen in nature, quantitative analysis is not normally performed. Instead the specimen is used to provide a qualitative comparison of adhesives or surface preparation.

Because the wedge test is self-loading (i.e. requires no elaborate loading fixture), a large number of specimens can be easily tested in a variety of different environments. A 'good' adhesive permits the crack to grow slowly to a certain length and stop, presumably at some threshold strain energy release rate. A 'poor' adhesive would show faster crack growth and the crack might run out of the specimen, resulting in separation of the adherends. Because the available energy to drive a crack is so strongly affected by crack length, care should be exercised in interpreting crack growth data.

The single lap shear specimen is widely used in industry to determine joint durability because it is very simple to fabricate and may be tested quite easily in load chains.[18] Each specimen gives only one data point—total time to failure. As such it is less efficient than the wedge test specimen and there tends to be considerable data scatter. The complex stress state in the single lap shear specimen does provide a mixed-mode loading, whereas the wedge test is purely opening mode for identical adherends. However, if different adherend thicknesses and/or materials were used, a limited degree of mixed-mode loading could be obtained.

Because the environmental effect depends on diffusion of the environment into the adhesive bond, care must be used in comparing durability tests performed on different specimens or in extending the experimental results to production joints. Obviously the geometry and size of the specimen will play an important role in determining the life of a given joint.

4.2.4. Specimen fabrication

Adhesively bonded joint specimens must be fabricated under carefully controlled conditions to avoid excessive variability. One must ensure that production joints are fabricated under similar conditions to ensure that specimens and full-scale joints will have similar properties.

Bonding FRP materials may require special precautions to minimize surface contamination. When joints are co-cured with the composite laminate, contamination problems can usually be controlled by proper treatment of the splice plate and careful handling of the laminate and adhesive film. For joints which are not co-cured, contamination problems must be dealt with. Unlike metals which can be prepared in a variety of chemical baths, composite cleaning is normally done by abrasion and degreasing.

Relatively flat composite panels are usually fabricated from prepreg which is laid-up against plain weave glass cloth preimpregnated with a release agent. This release cloth does not drape well, so for surfaces with double curvature other techniques are required. Satin weave cloth has better draping characteristics than plain weave, but is still limited to relatively small curvatures. Metal moulds coated with a release agent containing polymer-fluorinated hydrocarbons or silicones are widely used for moulding complex geometries, as are silicone rubber moulds. In all of these fabrication techniques, some residual contamination remains. Using X-ray photoelectron spectroscopy (XPS) analysis, Parker and Waghorne[19] have shown that the contamination is heavy and uniform for steel and silicone rubber moulds, and is lighter and less uniform with glass release cloth. They have subjected laminates prepared against several surfaces to various abrasive procedures to determine the most effective technique for pretreating composite surfaces prior to bonding. As would be expected, their results indicate that the remaining contaminant decreases with the severity of abrasion. Interestingly, grit blasting with dry alumina grit (280 grade) did a very uniform and thorough job of removing contamination from release cloth or silicone rubber, and resulted in the strongest joints. Surfaces prepared by moulding against metal moulds coated with release agents were covered by a thick layer of contamination which was difficult to remove even with several passes of grit blasting. Surfaces with significant contamination resulted in joints with negligible strength.

In preparing bonded joints for experimental testing or production items, a number of subtle parameters must be carefully controlled. These include surface pretreatment, age of adhesive and adhesive open time, cure procedure, inclusions and voids within the bond,[20] bond thickness, and rate of testing. Bowditch and Stannard[21] have examined a number of these parameters and indicated significant variability. When a two-part acrylic adhesive was applied but bonding was delayed for 2 h, the resulting joint strength was only 20% of that of joints cured immediately after application. Glue line thickness has been shown to be an important parameter in joint strength. Anderson[22] summarises reasons why thin bondlines are advantageous in butt-type joints. For shear-type joints, however, there appears to be an optimal bond thickness.[23] For laboratory tests where a well controlled bond thickness is desired, precision glass microspheres can be used in the bondline.

4.3. MEASUREMENT OF BONDED JOINT DISPLACEMENT AND STRAIN FIELDS

The desire to conduct more detailed analyses of the failure processes in adhesively bonded joints leads naturally towards the use of more refined measurements than, for example, the load to failure. Such refinements are not only useful in the direct observation of fracture processes but also in the validation of numerical stress analyses which usually incorporate some level of assumption. The small dimensions of the adhesive thickness, three-dimensional effects and the discontinuities that arise at bond terminations and cracks offer quite a challenge to the experimentalist. Clearly, no one method will yield all the information that we require and so the picture of joint deformations and failure processes has to be developed using a number of techniques which may categorise broadly into pointwise and whole field methods.

Pointwise methods may be used in regions of uniform strain and have the advantage of being relatively easy to use, particularly where long-term measurements are to be made. The pointwise data may be obtained using displacement transducers mounted to the adherends through pickup points, or from properly placed strain gauges. One should bear in mind that although these devices are called pointwise, because they give only one measurement per transducer, they actually average the displacement over a finite and often relatively large gauge length.

The whole field methods allow displacements near bond terminations and cracks to be examined. The use of two such techniques is described in this chapter. The first is Moiré interferometry, which can be used to measure displacements on the surfaces and edges of composite-to-composite, composite-to-metal and metal-to-metal adhesively bonded joints. The method has the potential for application to cracked joints although three-dimensional effects, which may be important, cannot be determined from surface measurements. This aspect is addressed through the use of crack opening interferometry, which is the second technique described here. Transparent adherends are used in this technique so that three-dimensional effects in cracked joints may be examined by making measurements along the entire crack across the width of the specimen.

4.3.1. Pointwise methods

The Krieger gauge, developed at American Cyanamid, consists of twin LVDT transducers with attachment points for mounting to each adherend. This technique is an averaging approach for measuring the relative displacement across the bondline, and hence some indication of the adhesive properties.

Although the device has been used to a limited extent for metal thick adherend joints, the applicability to composite joints is not obvious. Because the attachment points are somewhat removed from the adhesive interface, a correction for shear within the adherend must be made. For metal joints, the correction is quite small and is relatively easy to predict. For laminated composites, the interlaminar shear strains are of the same order of magnitude as those in the adhesive and the correction becomes quite large and difficult to determine. Nonetheless, the device is useful for measuring adhesive properties and can be used in hostile environments to determine environmental effects.[24]

Brinson and Tuttle[25] have succeeded in embedding foil strain gauges within the adhesive layer of single lap joints. Gauges placed in the less highly loaded centre portion of the bond did not decrease the strength of the joints. These gauges measure a normal strain in the bond plane and, when placed in the centre of the bondline thickness, they appear to average the strains in the two adherends at that location. Although the practical applications of the technique are uncertain at the present time, the possibility of using embedded sensors for detecting changes in joints while in service is intriguing.

4.3.2. Moiré interferometry

Optical techniques offer the potential for full field displacement measurements. Post and his colleagues have pioneered the application of high sensitivity Moiré interferometry to practical displacement measurements for a variety of configurations. Of interest here is their work with bonded joints. The technique involves replicating a high-frequency reflective crossed-line grating on the specimen. A plane mirror perpendicular to the

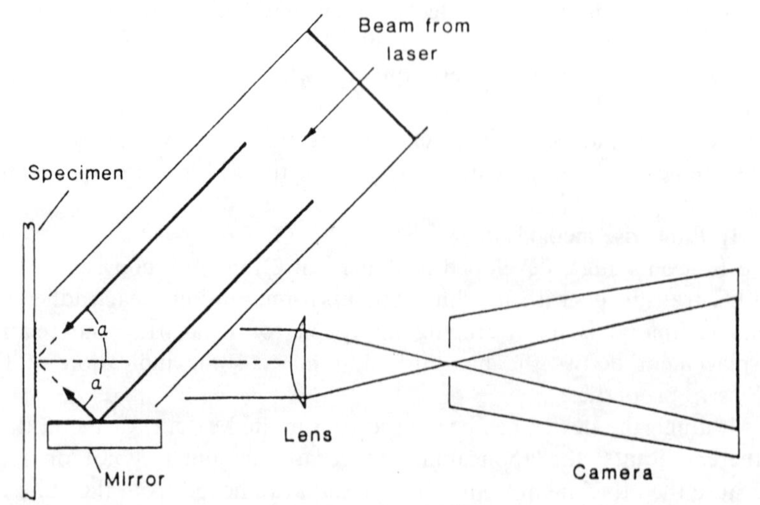

FIG. 2. Optical arrangement for Moiré interferometry.

specimen intercepts a portion of a collimated beam of laser light, resulting in two incident beams of light at the specimen surface. Because the light is coherent, planes of constructive and destructive interference are established which have the effect of a high-frequency 'virtual' reference grating. Interacting with the deformed specimen grating, this virtual grating forms a Moiré interference pattern which can be photographed. The optical set-up is illustrated in Fig. 2. By using a virtual grating to interrogate the specimen in two perpendicular directions, the u and v displacement fields are determined and the complete in-plane strain field can be computed.

Post *et al.*[26] have replicated a 1200 line/mm crossed-line grating on a thick adherend joint consisting of 9 mm-thick aluminium adherends bonded with FM-73M epoxy. After loading, the specimen was interrogated with a 2400 line/mm virtual reference grating and the resulting u-displacement field is shown in Fig. 3. The insets (b) and (c) of Fig. 3 show magnified

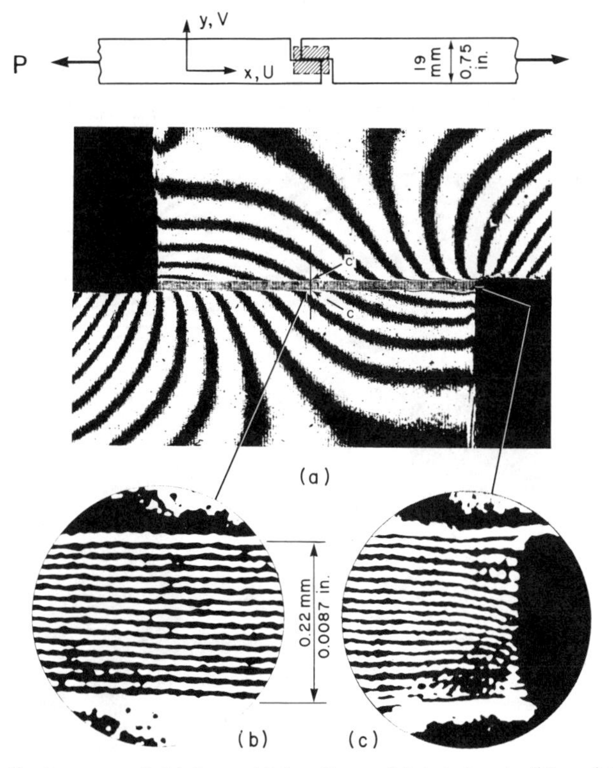

FIG. 3. U-displacement field for a thick adherend joint. Insets (b) and (c) show enlarged view of fringe pattern within the bondline.

(a)

(b)

Fig. 4. Displacement fields for thick adherend joint showing the soft epoxy bridges used for fringe continuity. (a) V-field. (b) U-field.

views of the displacements within the adhesive bondline. Fifteen fringes can be observed within the 0·22 mm-thick adhesive layer.

Total joint displacements are shown in Fig. 4. The Ω-shaped bridges are formed of a soft epoxy and provide a continuous path between adherends for counting fringes. This is needed when strains in the adhesive are so large that fringes in the adhesive cannot be resolved. The shape and soft epoxy material used for the bridges result in a very flexible connection which provides a fringe counting path without transferring enough load to affect the strain analysis. Figure 5 illustrates the displacement fields obtained for a $(\pm 45/0/90°)_{3s}$ AS4/3502 carbon/epoxy doubler bonded to an identical member with 9320 'Hysol' adhesive. The bond thickness was 0·76 mm. When loaded in bending, the resulting displacement field contours are less smooth than those obtained for homogeneous adherends as illustrated in Fig. 4. The explanation lies in the interlaminar shear deformations within the laminated adherend.[27] The interlaminar strains are very clearly seen in Fig. 6 for a 48-ply quasi-isotropic carbon/epoxy laminate loaded in three-point bending. Optical techniques such as these provide a unique opportunity to get full field displacements of adhesive joints and provide insights into the interlaminar deformations of composite adherends.

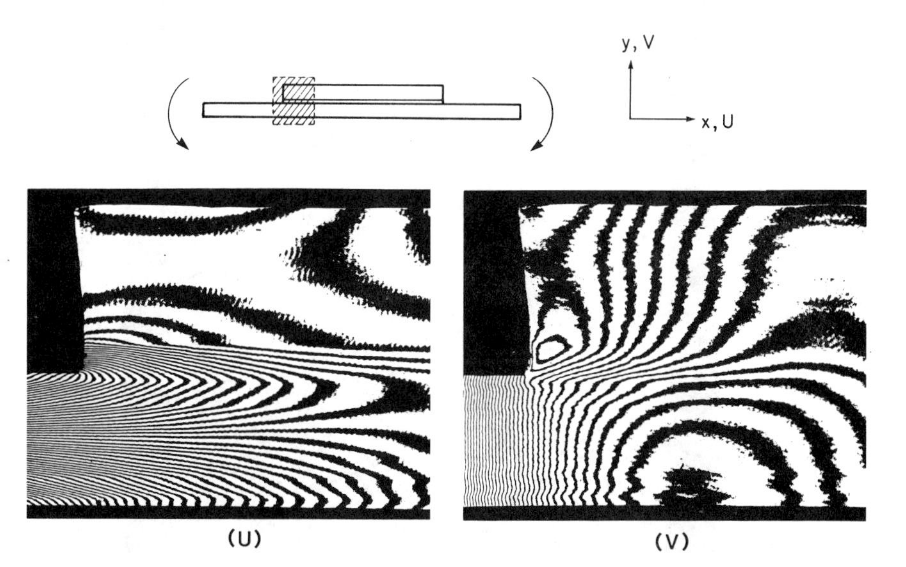

FIG. 5. *U*- and *V*-displacement fields for a doubler stiffened FRP specimen loaded in bending.

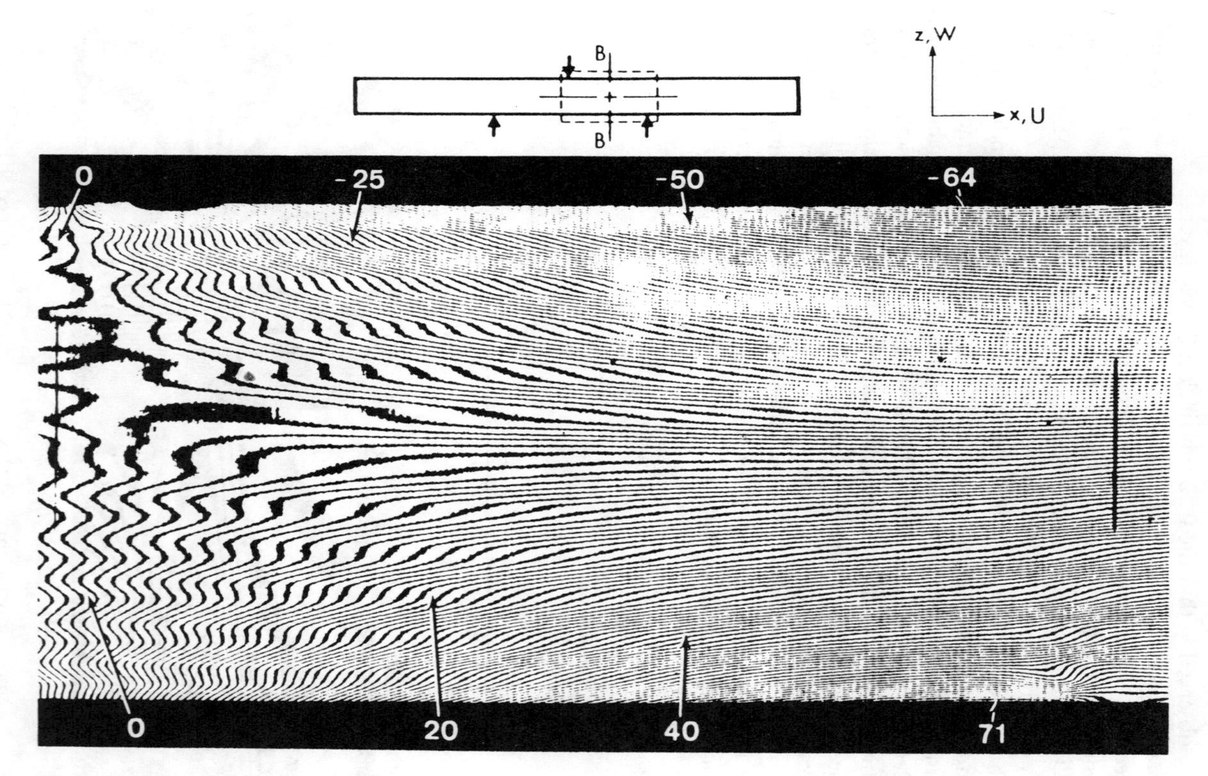

FIG. 6. *U*-displacement field for 48-ply quasi-isotropic composite beam loaded in three-point bending. Interlaminar deformations are clearly illustrated.

4.3.3. Crack opening interferometry

Crack opening interferometry has been used to measure directly the normal deformations in the crack front regions of cracked monolithic bodies subjected to Mode I loadings.[28–31] One problem that has limited the use of the method in these studies is that crack growth does not always result in crack surfaces of sufficient smoothness to produce interference effects. The method was adapted[32] to study interfacial crack growth in adhesively bonded joints where the surface of the adherends defining the interface was polished to sub-wavelength tolerances. The crack face separation can be resolved to within at least a half wavelength of the viewing light at any point in the field of view.

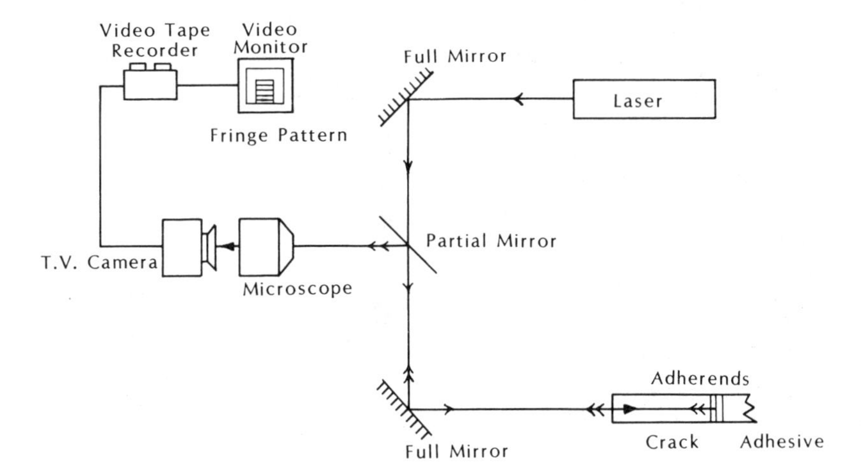

Fɪɢ. 7. Crack profile measurement system.

A schematic of the apparatus is shown in Fig. 7. Extensive mirror arrangements channel the incident laser beam to the specimen and beams reflected at the crack faces into a microscope. A closed-circuit television camera is mounted on the microscope and the changing fringe patterns are then recorded on video tape and displayed on a monitor. The microscope can be mounted on a traversing device so that crack motion can be tracked.

When the coherent and monochromatic beams reflected by each crack face are combined, an interference pattern is formed which consists of light and dark fringes corresponding to loci of constructive and destructive

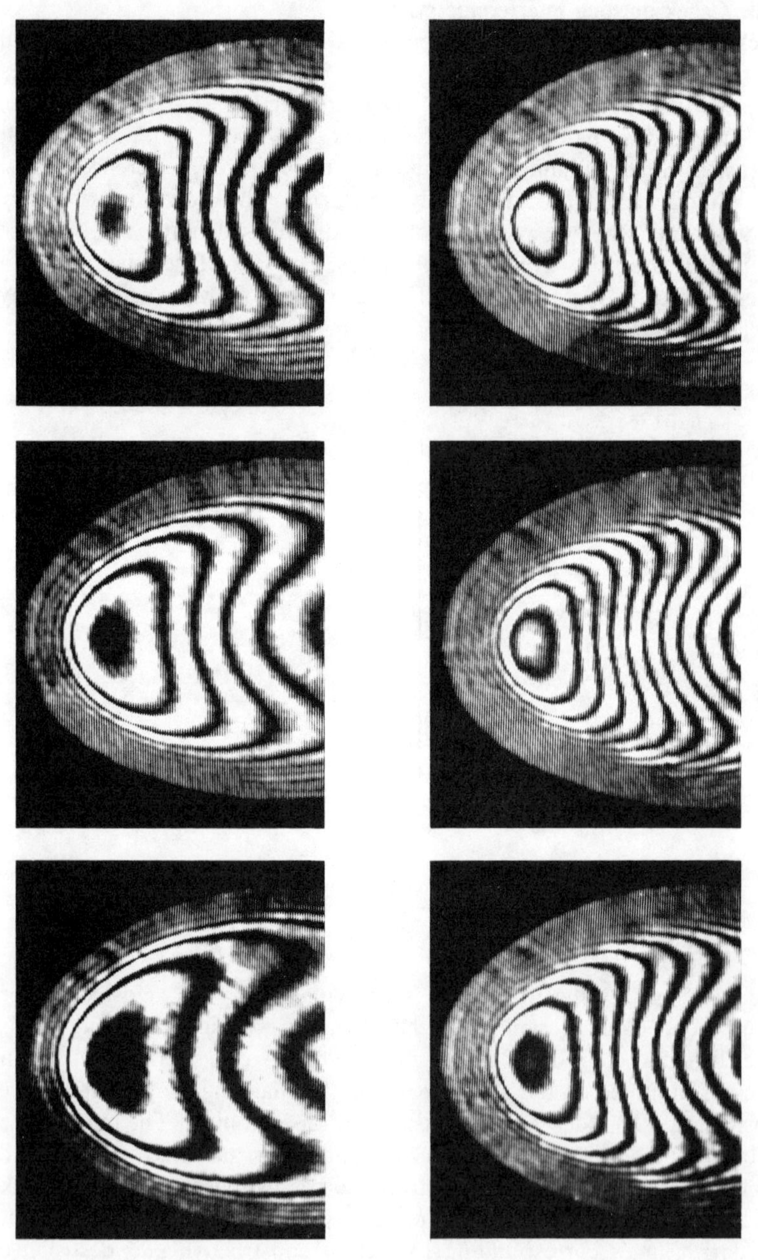

FIG. 8. Interference fringe patterns of an interface crack.

interference. At a given fringe of extinction, m say, counted from the crack front, the crack face separation distance, h, along that fringe is given by

$$h = \frac{m\lambda}{2n\cos\phi}; \quad m = 1, 2, 3 \ldots \tag{4.1}$$

where n is the refractive index of the medium separating the crack faces and ϕ is the angle of incidence of the incoming beam.

For sufficiently small angles between the crack faces, fringes can be observed with the naked eye. A microscope is necessary to resolve the higher fringe densities produced by larger angles. The fringe patterns are recorded on video equipment for subsequent data reduction. For normal incidence and an air-filled crack, each fringe corresponds to a crack face separation of $\lambda/2$. The spatial resolution in the plane of the crack was $6{\cdot}50\,\mu m$[37] and depends on the magnification that is available.

4.3.4. Some observations

At this time it is appropriate to note some characteristics of the fracture process that were observed so as to provide some physical insight for analytical modelling.

The system described above was used[33,34] to measure crack profiles in a butt joint having glass adherends and several adhesives including the model adhesive 'Solithane-13' (60% prepolymer, 40 wt% catalyst; Thiokol Chemical Corporation) and the structural adhesive FM-73 (a product of American Cyanamid Corporation, supplied by courtesy of Mr R. Krieger). 'Solithane' is a well crosslinked polyurethane rubber having a long-term Young's modulus of approximately 3·45 MPa and is nearly incompressible. FM-73 is a rubber-toughened epoxy in its glassy state at room temperature and its scrimmed version, FM-73M, is used extensively in the aerospace industry.[35] The relative displacements of adherends normal and parallel to the bondline were independently controlled to within $0{\cdot}16\,\mu m$, in a microprocessor-controlled loading device that employed thermally driven displacement actuators.

Figure 8 shows a series of photographs, taken at different levels of bond-normal applied displacements, of a stationary crack in a 'Solithane' joint. It very quickly became apparent that three-dimensional effects were important. The width of the picture covers slightly more than a third of the specimen thickness and the crack, having the finger-like outline, is centred on the specimen midthickness. The dark area surrounding this finger indicates the region which is still bonded. Thus the crack does not extend all the way through the thickness but rather displays a finger, or tunnelling, mode of fracture. Under bond-parallel applied displacements, the crack front widened to the extent that it intersected the free surfaces of the specimen. It was also noted that cracks that were initially edge cracks tended to close up from the free edge. Although these three-dimensional effects were exaggerated by the incompressibility of the 'Solithane', crack front geometries involving FM-73 were very irregular.[34] Under the best of circumstances an 'Araldite 52' adhesive system[36] yielded

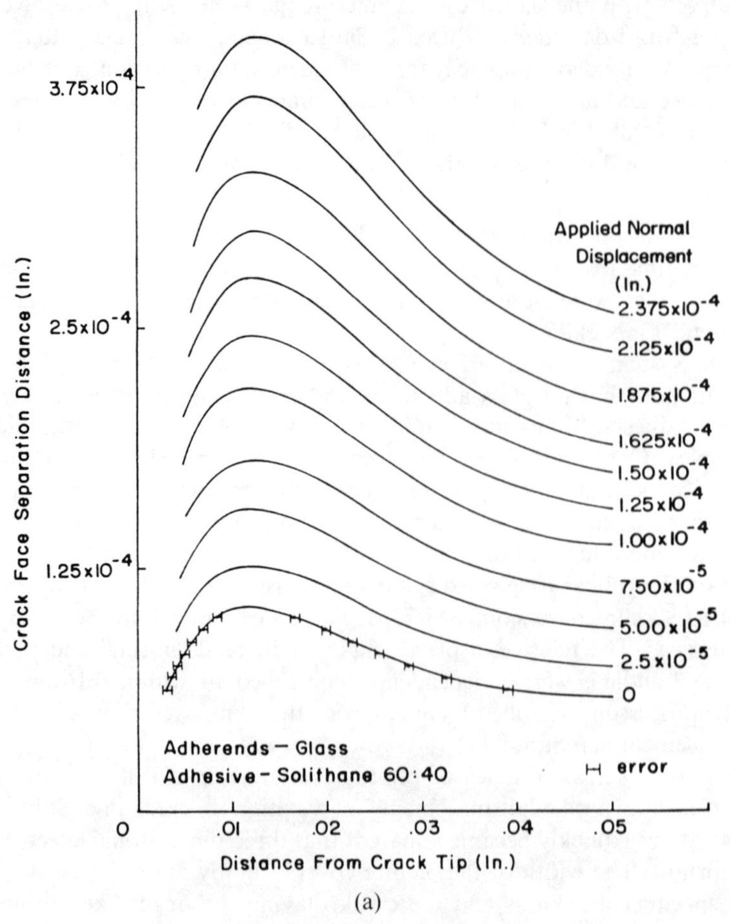

(a)

FIG. 9. Crack profiles in a glass/'Solithane' specimen: (a) bond-normal applied displacements (0·01 in ≡ 0·25 mm).

crack fronts which spanned the width of the specimen in a nearly semicircular fashion.

One of the underlying questions in all failure analyses is the validity of linearised deformation theories. This question was examined in a limited sense by determining the normal crack opening displacement (NCOD) resulting from relative displacements of the adherends either normal or parallel to the bondline. The NCOD were obtained by cross-plotting crack profile plots such as that shown in Fig. 9 for bond-normal loading. It was found that the relationship between the applied bond-normal

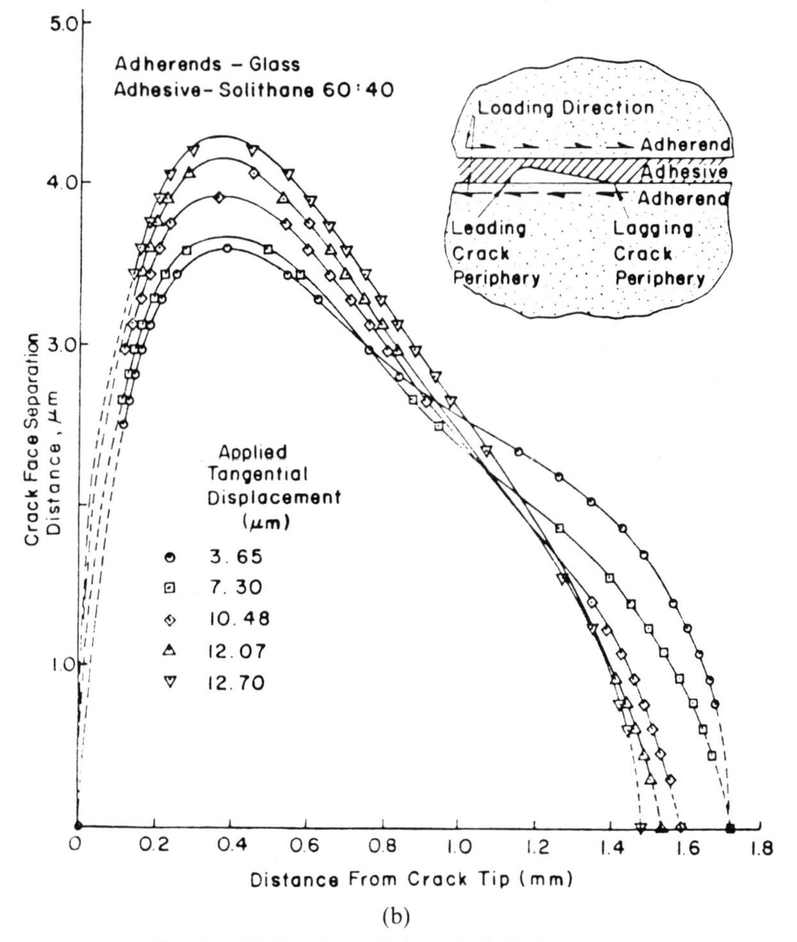

(b)

Fig. 9. (b) Bond-parallel applied displacements.

displacements and NCOD were at best linear for applied strains that were less than 0·25%. The non-linear contribution to the NCOD was not more than 10% at 0·5% strain. When purely bond-parallel displacements were applied, normal crack opening displacements certainly occurred (Fig. 9b). At the leading crack front crack opening occurred, whereas crack closure was consistently noted at the lagging crack front.[33] In the opening portion of the crack, the response to the bond-parallel loading was found to be highly non-linear with an initial independence of load level. The NCOD induced by the bond-parallel loading were less than those resulting from bond-normal displacements by a factor of about 20. In considering the

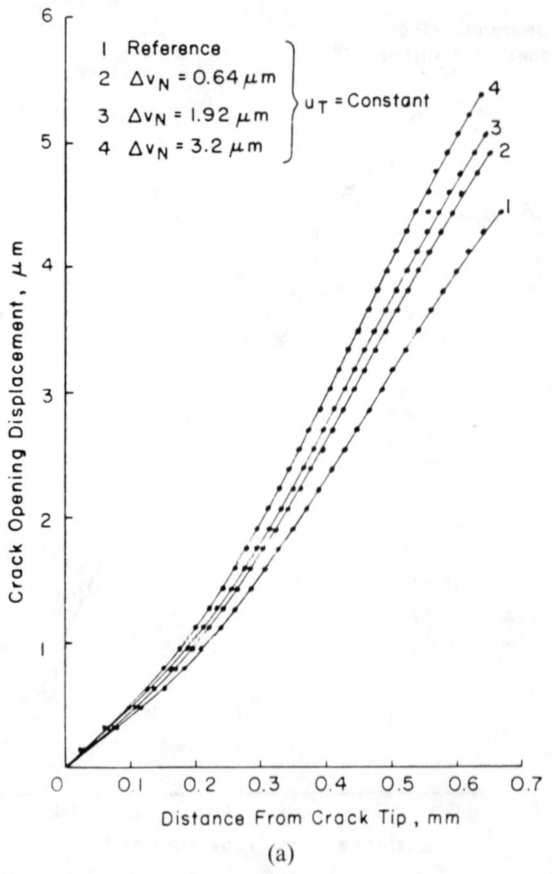

(a)

Fig. 10. Crack profiles in a glass/FM-73U specimen: (a) bond-normal applied displacements.

deformation response of a rigid adhesive (FM-73), it was found that the crack profiles near the crack front formed a much smaller angle with the interface (Fig. 10) than was the case in the 'Solithane' joint. By comparing Figs 10a and 10b, it can be seen that the bond-parallel loading resulted in NCOD that were of the same order as those produced by bond-normal loading.

There is a potential lesson to be drawn from this limited comparison. Recall from the introduction that current practice for failure analysis purposes is to assume a vanishingly thin bondline for bonded joints, without giving due consideration to the more detailed processes occurring in the bond. The two examples involving different bonding agents seem to point out substantial differences in crack tip response. This difference

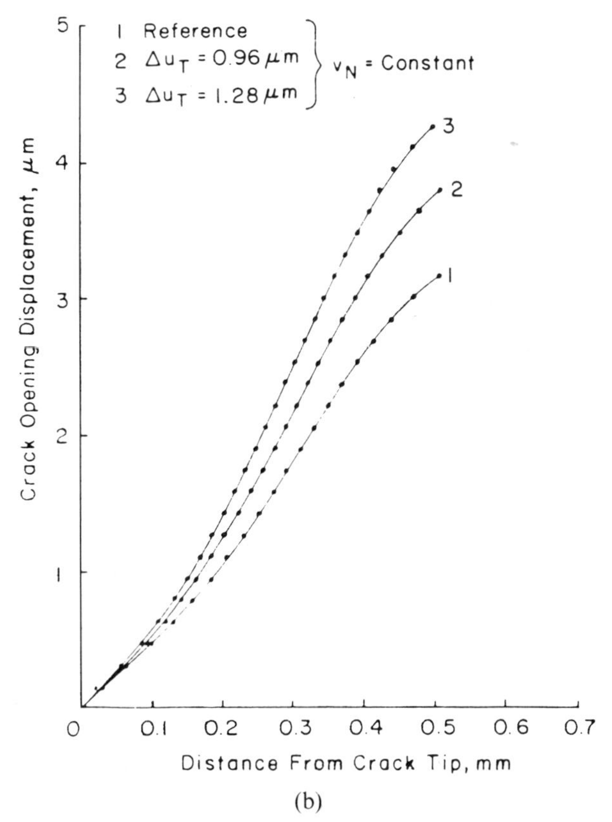

(b)

Fig. 10. (b) Bond-parallel applied displacements.

evidenced here in two systems could possibly occur in a single system, involving, for example, the more rigid FM-73 if the latter is then exposed to sustained loading at elevated temperatures in a moist environment.

4.3.5. Moiré of crack opening interferometry (COI)

An examination of Figs 9 and 10 indicates that in all cases the crack faces were not in contact under zero applied displacement. This initial crack opening is due to residual stresses that arise from shrinkage due to the polymerisation process as the adhesive cures, and also due to the mismatch in the thermal properties of the joint components in situations when high-temperature cures are employed. In order to obtain the NCOD due to an applied load, the crack profile at zero load must be subtracted from the profile at some subsequent load level. In ref. 18 the NCOD were obtained by a graphical subtraction of the profiles, a time-consuming process which is subject to the errors of cross-plotting. The subtraction can be accomplished much more directly in an optical sense if the interferograms corresponding to the two states are superimposed to produce a Moiré effect. The fringes that are formed are loci of constant NCOD. In other applications the Moiré effect was obtained by superimposing photographs of fringe patterns.[37,38] More recently, the Moiré of COI was obtained directly in real time[39] by making use of video image processing equipment to sübtract the fringe patterns electronically.

To see this, consider the distances h_1 and h_2 separating the crack faces in the initial and subsequent states. They are given, respectively, by

$$h_1 = \frac{m_1 \lambda}{2n \cos \phi}; \qquad m_1 = 0, 1, 2 \ldots \tag{4.2}$$

$$h_2 = \frac{m_2 \lambda}{2n \cos \phi}; \qquad m_2 = 0, 1, 2 \ldots \tag{4.3}$$

where m_1 and m_2 are the order of the fringe patterns at a particular location behind the crack fringe and the other variables are as defined before (eqn (4.1)).

When the interference patterns are superimposed a subtractive Moiré pattern is formed whose order, N, is given by

$$N = m_2 - m_1 = [2n(h_2 - h_1)\cos\phi]/\lambda \tag{4.4}$$

Rearranging eqn (4.4) for an air-filled crack and normal incidence gives the NCOD, Δv, as

$$\Delta v = h_2 - h_1 = N\lambda/2 \tag{4.5}$$

The subtractive Moiré fringes are therefore contours of constant normal crack opening displacement, with each successive contour corresponding to an increment of $\lambda/2$.

The technique was used[39] for the direct measurement of NCOD in a blister specimen made of a thick Plexiglas™ substrate to which a layer of 'Solithane' was cast. Pressure was introduced through a small, central hole in the Plexiglas to produce very regular, circular interfacial debonds. The axisymmetry of the arrangement, and lack of a second adherend, reduced the potential for the formation of dendrites or fingers. The regular geometry also meant that experimental results could be compared with existing two-dimensional, axisymmetric analyses of finite element codes so that the question of non-linearity could be examined in the absence of three-dimensional effects.

The Moiré of crack opening interferometry may be used to calculate fracture parameters from measurements that are made relatively close to the crack front. The measurements yield directly kinematic fracture parameters such as the crack opening displacements normal to the

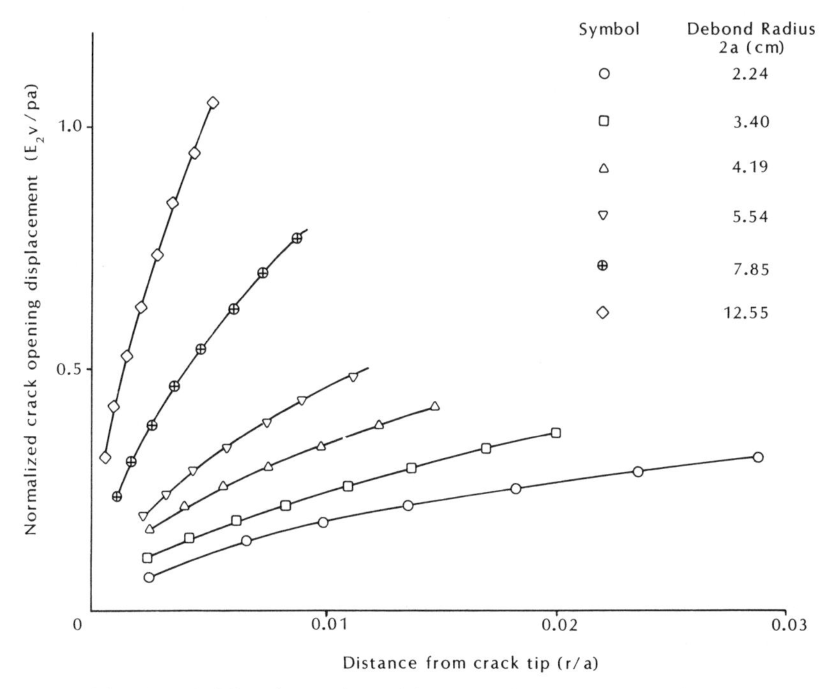

Fig. 11. NCOD for various debond radii in a blister specimen.

bondline or the crack opening angle. An advantage to considering these parameters as fracture parameters and criteria is that the determination of the criteria is independent of the representation of material behaviour in the crack front region. Although there is a degree of arbitrariness in the location at which criteria are determined, the full-field nature of the data (Fig. 11) may be used to examine this point.

In ref. 39, a range of debonds in a blister specimen were pressurised and the resulting NCOD in the crack front region were measured. It was found that the NCOD depended on the square root of the distance from the crack front, as predicted by linear elastic fracture mechanics. The opening mode energy release rate, G_I, was determined from the gradient of the NCOD from definitions of interfacial stress intensity factors and energy release rates.[40,41] It was found that the relative amount of G_I decreased with increasing debond length, thus confirming the mixed-mode nature of blister specimens that had been noted in the past.[41,42]

4.4. STATIC AND CYCLIC DEBONDING

Most of the work to be discussed in this section was taken from a series of reports by S. Mall and W. S. Johnson.[43–46] The specific areas to be covered are as follows.

(1) Assessment of the relative roles of opening mode strain energy release rate, G_I, and the shearing mode strain energy release rate, G_{II}, on both static and cyclic debonding.
(2) Evaluation of the threshold strain energy release rate as a design criterion.

4.4.1. Test specimens

The cracked lap shear (CLS) specimen, shown in Fig. 12, was used by Mall *et al.*[43] because it represents a simple structural joint subjected to in-plane loading. Both shear and peel stresses are represented in the bondline of this joint. The magnitude of each component of this mixed-mode loading can be modified by changing the relative thicknesses of strap and lap adherends.[15,48] For this study, the strap and lap adherends had 8 or 16 plies. These adherends had a quasi-isotropic lay-up of T300/5208 carbon/epoxy.

Several investigators such as Romanko, Liechti and Knauss[85] have used the CLS specimens first suggested by Brussat *et al.*[15] for testing

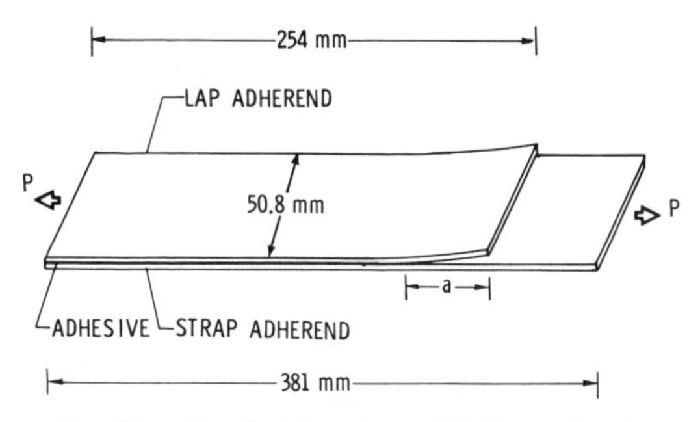

Fig. 12. Cracked lap shear (CLS) specimen.

bonded metallic joints. Wilkins[14] was the first to use the CLS specimen to test for the interlaminar fracture toughness of composites.

Two bonded systems were studied: carbon/epoxy (T300/5208) adherends bonded with EC 3445 adhesive and with FM-300 adhesive. EC 3445 is a thermosetting paste adhesive with a cure temperature of 121°C. Specimens with this adhesive were fabricated using a conventional secondary bonding procedure. However, specimens with FM-300 adhesive were fabricated by a co-curing procedure whereby adherends were cured and bonded simultaneously. FM-300 is a modified epoxy adhesive supported with a carrier cloth with a cure temperature of 177°C. The bonding process followed the manufacturer's recommended procedures for each adhesive. The nominal adhesive thickness was 0·10 and 0·25 mm for EC 3445 and FM-300 respectively. The adhesive material properties are given in Table 1.

The composite adherends consisted of quasi-isotropic lay-ups of $[0/45/-45/90°]_S$ and $[0/45/-45/90°]_{2S}$. The material properties of carbon/epoxy, presented in Table 2, were obtained from ref. 46. For each bonded system, two types of specimen were tested: (1) thin lap adherend of 8 plies bonded to thick strap adherend of 16 plies; and (2) thick lap

TABLE 1
Adhesive Material Properties

Adhesive	*Modulus (Gpa)*		*Poisson's ratio*
	E	*G*	*v*
EC 3445 (3M Company)	1·81	0·65	0·4
FM-300 (American Cyanamid Company)	2·32	0·83	0·4

TABLE 2
Carbon/Epoxy[a] Adherend Material Properties

Modulus[b] *(GPa)*			*Poisson's ratio*[b]	
E_1	E_2	G_{12}	ν_{12}	ν_{23}
131·0	13·0	6·4	0·34	0·34

[a] T300/5208 (NARMCO); fibre volume fraction is 0·63.
[b] The subscripts 1, 2, and 3 correspond to the longitudinal, transverse, and thickness directions, respectively, of a unidirectional ply.

adherend of 16 plies bonded to thin strap adherend of 8 plies. This arrangement provided four sets of CLS specimens.

The double cantilever beam (DCB) specimens, shown in Fig. 13, consisted of two bonded adherends, each having 14 unidirectional plies of the T300/5208, with an initial debond length of 38 mm. This debond was introduced by a 'Teflon' film of thickness equal to the adhesive bondline. Two 0·5 mm-thick aluminium end tabs were bonded to the DCB specimen, along with two 1·3 mm-thick aluminium reinforcing plates. The peeling load was applied through these tabs. Specimens were made with the EC 3445 and FM-300 adhesives as previously described.

4.4.2. Testing procedure

The test programme[43,44] included static and fatigue tests for both types of specimens. Its objective was to measure the critical strain energy release rate under the static loading, and to measure the debond growth rate

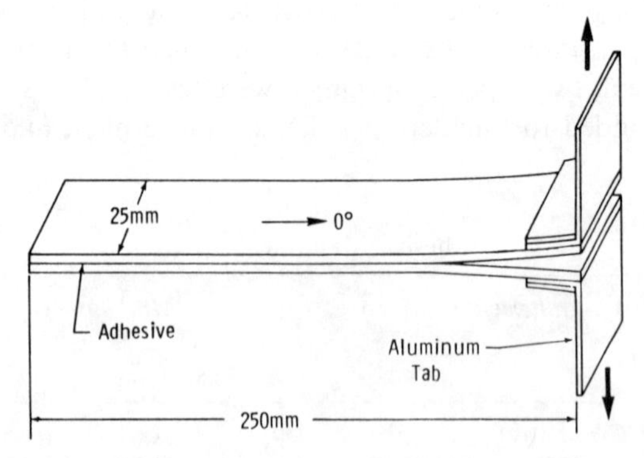

FIG. 13. Double cantilever beam (DCB) specimen.

under the cyclic loading. These are described separately below for each specimen.

4.4.2.1. Static tests of DCB specimen

All static tests of DCB specimens were performed in a displacement-controlled test machine. Both edges of the specimen were coated with a white brittle fluid, to aid in visually locating the debond tip. Fine visible marks were put on these edges, at 1 mm intervals, to aid in measuring the debond length. The debond length was measured visually on both sides with two microscopes having a magnification factor of 20. Prior to testing, either for static or fatigue loading, these specimens were fatigued to create a debond of at least 6 mm beyond the end of the 'Teflon' film. The static test involved the application of displacement at a slow crosshead speed (approximately 1·0 mm/min). The load corresponding to the applied displacement was also recorded. When the load reached the critical value, the debond grew. The onset of growth results in a deviation from linearity in the load versus crosshead displacement record. The applied displacement was then decreased until a zero load reading was observed. After each static test, the specimen was fatigued until the debond grew at least 6 mm further, thus forming a sharp crack for the next static test. A series of static tests was performed on each specimen, which provided compliance and critical load measurements at several debond lengths. These measurements provided the critical strain energy release rate as explained in Section 4.4.3.

4.4.2.2. Fatigue tests of DCB specimen

The fatigue tests of DCB specimens were conducted in a servo-hydraulic test machine at a cyclic frequency of 3 Hz. Two constant-amplitude testing modes were employed: (1) constant-amplitude cyclic load; and (2) constant-amplitude cyclic displacement. In both modes the ratio of minimum to maximum load (or displacement) in a fatigue cycle was 0·1. In displacement-controlled tests debond growth rates reduced as the debond propagated, whilst in the case of load-controlled tests debond growth rates increased as the debond propagated. Debond lengths, fatigue cycles, applied loads, and displacements were monitored continuously throughout each test. The measured relation between the debond length and fatigue cycle provided the debond growth rate, $\mathrm{d}a/\mathrm{d}N$. The strain energy release rate, G_I, was computed from the measured compliance and applied load, as explained in Section 4.4.3. Thus, a relation between G_I and $\mathrm{d}a/\mathrm{d}N$ was established for the cyclic debonding under Mode I loading.

4.4.2.3. Static tests of CLS specimen

Static tension tests on CLS specimens were conducted in a displacement-controlled mode. Prior to static testing this specimen was fatigued, and thus it had an initial sharp debond. During the test the axial load and displacement were recorded. The displacement was measured with two displacement transducers attached on opposite sides of the specimen. The applied load was increased slowly until the debond propagated. The critical load corresponding to unstable debond growth was measured and verified by the deviation from linearity in the recorded load–displacement curve. Only one such measurement could be obtained from each specimen, since debonds grew into the composite strap adherend. Static tests were conducted on all four sets of specimens (i.e. with two geometries and two adhesives).

4.4.2.4. Fatigue tests of CLS specimen

A detailed investigation of cyclic debonding under mixed-mode loading was conducted in ref. 1. In that study, the CLS specimen was tested under constant-amplitude cyclic load at 10 Hz frequency and stress ratio $R = 0{\cdot}1$. In ref. 2, fatigue tests of the CLS specimen were conducted at 3 Hz in order to compare mixed-mode results with Mode I results from DCB specimens which were also obtained at 3 Hz frequency. Only the eight-ply strap bonded to the 16-ply lap with the EC 3445 adhesive system was tested at 3 Hz.

Three techniques to measure the cyclic debond growth were evaluated. For each technique, the location of the debond front was measured periodically to calculate debond growth rates. The first method used a sheet of photoelastic material bonded to the lap adherend of the specimen, as discussed in ref. 46. Isochromatic fringes developed at the debond front as a result of the high strain gradient in that vicinity when subjected to load. These isochromatic fringes were observed through a polariser and were used to locate the debond front. The second method involved locating the debond front with an X-ray technique using a dye penetrant, zinc iodide. The third method involved measuring the compliance of the specimen and then calculating the debond length using a crack length compliance formula. The compliance of the specimen was measured with two displacement transducers attached on opposite sides of the specimen. All three methods provided good agreement for debond growth measurement, as shown in Fig. 14. In the present study the photoelastic technique was selected; an automated measurement system photographed the isochromatic fringes at predetermined intervals.

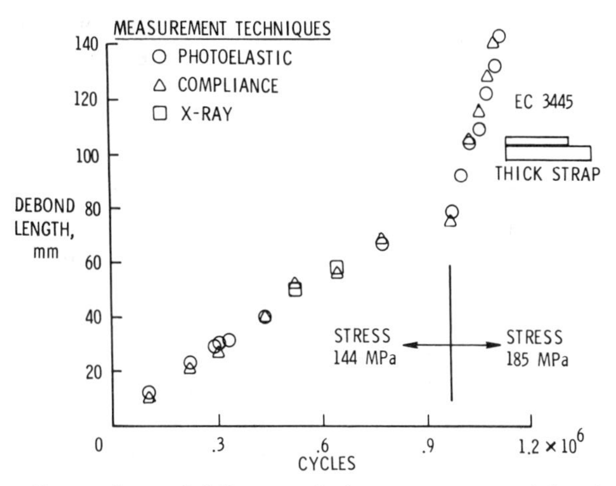

Fig. 14. Comparison of different techniques to measure debond growth.

The strain energy release rates (G_T (total), G_I and G_{II}) are usually uniform in the cracked lap shear specimen for a significant debond growth region. The debond data were measured over this region, which is discussed in Section 4.4.3.1. Tests were conducted at two, or more, constant-amplitude stress levels to get several values of debond growth rate (da/dN) from each specimen. At each stress level the debond was measured as it grew over about 3–4 cm, to ensure an accurate estimate of the debond growth rate over that region. Figure 15 shows typical debond

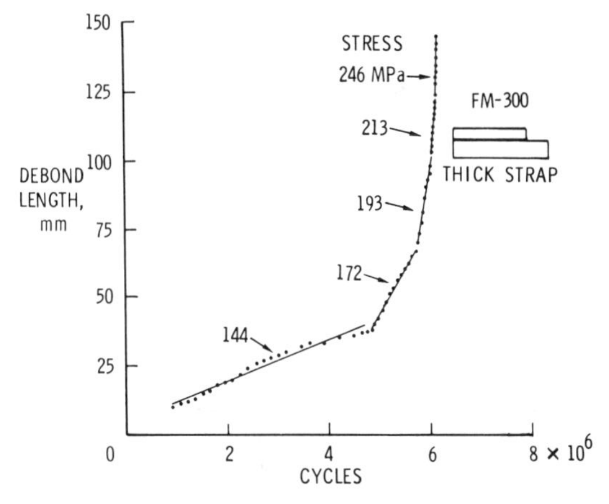

Fig. 15. Typical variation of debond length with fatigue cycles at different load levels.

data for different stress levels. In all cases debonding was initially non-linear, but became linear after the debond progressed a short distance (about 10 mm). For each stress level, the debond data were fitted with a straight line using regression analysis.

4.4.3. Specimen analysis

4.4.3.1. Cracked lap shear

Previous studies of fatigue damage mechanisms in adhesively bonded joints have shown that the strain energy release rate, defined from fracture mechanics principles, may be useful for correlating cyclic debond growth rate.[15,50] Therefore, all four sets of specimens were analysed with a finite element program, GAMNAS,[48] to calculate strain energy release rates. This two-dimensional analysis accounted for the geometric non-linearity associated with the large rotations in the unsymmetric cracked lap shear specimen. The importance of a geometrically non-linear analysis for the cracked lap shear specimen has been discussed in detail in ref. 6.

A typical finite element mesh, shown in Fig. 16, consisted of about 1600 isoparametric four-node elements and had about 3000 degrees of freedom. Each ply of the composite was modelled as one layer in the finite element model, except for the ply at the adhesive interface which was modelled in two or three layers. A multipoint constraint was applied to prevent rotation of the loaded end of the model (i.e., all the axial

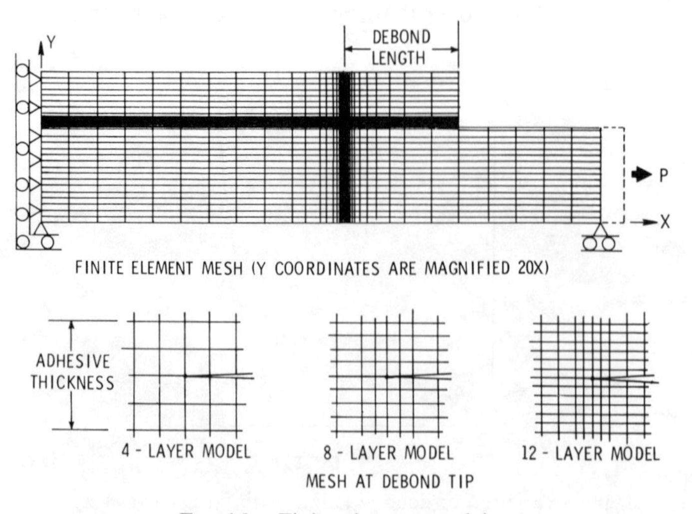

Fig. 16. Finite element model.

displacements along the ends are equal) to simulate actual grip loading of the specimen. A double cracked lap shear specimen with isotropic adherends was analysed using a quasi-three-dimensional analysis like that in ref. 10 to show that the present two-dimensional analysis should be based on the plane strain condition. The strain energy release rates G_T, G_I and G_{II} in the analysis were computed for the maximum load in the fatigue cycle, using a virtual crack closure technique.[52] The details of this procedure are given in ref. 51. The calculation of strain energy rates in the present analysis depends on three parameters: (1) debond location; (2) load level; and (3) debond length. These parameters will be discussed next.

Under cyclic loading the debond grew near the interface of the adhesive and strap adherend in all tests. It was very difficult to measure the exact location of the debond, but in general the debond grew within the one-quarter of the thickness of adhesive closest to the interface. The effect of the location of the debond within the adhesive in the cracked lap shear specimen was investigated using GAMNAS.[48] Figure 17 shows the variation of calculated G_T, G_I and G_{II} with debond location within the thickness of the EC 3445 adhesive. These results were calculated by modelling the adhesive with four, eight, and 12 layers of elements, as shown in Fig. 17, which clearly shows that G_T remains constant for all locations of the debond, while G_I has its maximum value near the adhesive/strap interface and G_{II} has its maximum value near the adhesive/

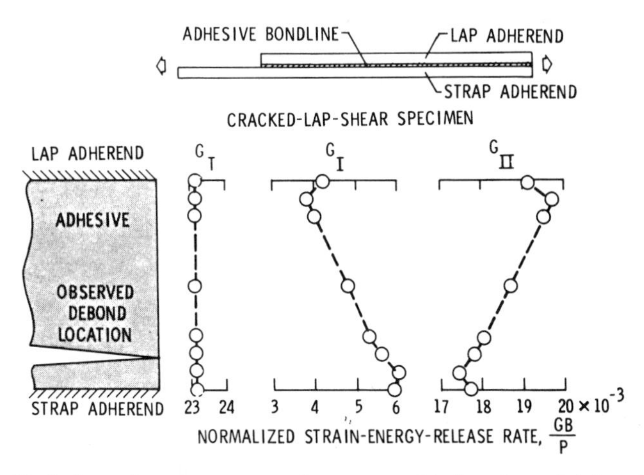

Fig. 17. Variation of strain energy release rate with location of debond within EC 3445 adhesive.

lap interface. The debond always initiated and grew in the region of highest G_I (near the adhesive/strap interface). This indicates that G_I has the greater influence on the debond location in the adhesive joint. This is consistent with the observations that adhesives are inherently weaker under peel loading than under shear loading.[53,54] Additionally, these results show that an accurate evaluation of G_T can be achieved by a four-layer model, while accurate evaluation of G_I and G_{II} requires a more refined model. To analyse the experimental debond growth rates, the debond location for subsequent calculations was selected by engineering judgement to be at one-sixth of the adhesive thickness away from the adhesive-strap interface. Also, the 12-layer model was used in these calculations.

Figure 18 shows the typical variation of strain energy release rates G_I and G_{II} with applied load on the cracked lap shear specimen obtained from geometric non-linear analyses. The G_I and G_{II} in non-linear analyses were found to be functions of the square of the applied load, to within 1%.

All four sets of specimens were then analysed to determine the variation of G_T, G_I and G_{II} with debond length. Figure 19 shows the typical dependence of G_T, G_I and G_{II} on debond length for both types of specimen with EC 3445 adhesive. For specimens with thick straps, G_T, G_I and G_{II} were constant up to 140 mm of debond length. Similar behaviour was found for co-cured specimens with thick straps. For specimens with thin straps, G_T, G_I and G_{II} were constant up to debond lengths of 115 mm and

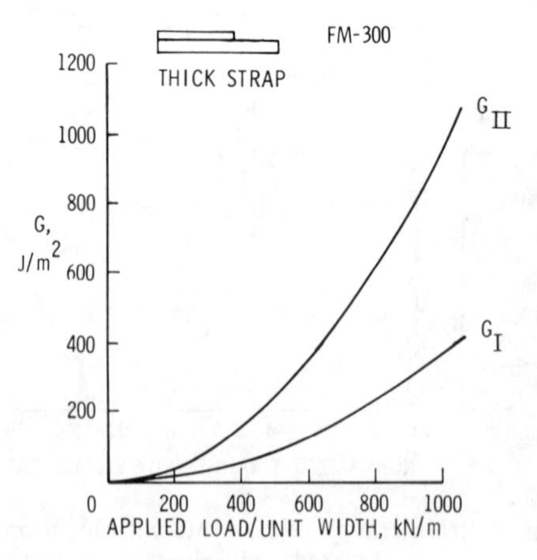

Fig. 18. Variation of strain energy release rate with applied load.

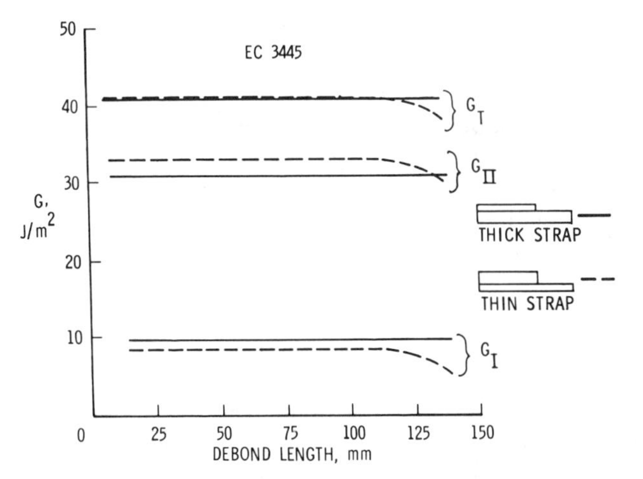

Fig. 19. Variation of strain energy release rates with debond length.

65 mm for secondary and co-cure bonded systems, respectively. The constant values of strain energy release rates for all four sets of specimens are provided in Table 3 for a specified stress level.

To compute a debond threshold strain energy release rate for a non-debonded CLS specimen, a small debond must be assumed to exist at the location of expected debond initiation, i.e. at the end of the lap adherend. For the current computations, a debond length of 1 mm was assumed. This debond length was approximately the minimum size of debond that could be found with the X-ray inspection technique used during the debond initiation tests. Computations using a debond length of 0·5 mm resulted in calculated initiation stresses about 3% lower than those for a 1 mm length.[45] Thus, G is not highly sensitive to the finite debond length.

TABLE 3
Strain Energy Release Rate

Specimen details		*Strain energy release rate (J/m^2) for the applied stress of 82·0 MPa*		
Adhesive	*Strap type*	G_I	G_{II}	G_I/G_{II}
EC 3445	Thick	9·75	31·08	0·31
(secondary bonding)	Thin	8·23	33·20	0·25
FM-300	Thick	11·21	29·60	0·38
(co-cure bonding)	Thin	10·66	32·40	0·33

4.4.3.2. Double cantilever beam

For the DCB specimen, the critical load, P_{cr}, and the experimentally measured compliance, C, for each debond length, were used with linear beam theory to compute the fracture toughness G_{Ic}. The details of this procedure are elaborated by Wilkins.[14] A brief description of the technique used by Wilkins, and others, is given below. Figure 20 shows the variation of compliance with debond length in a typical DCB specimen with EC 3445 adhesive. A compliance relation of

$$C = A_1 a^3 \tag{4.6}$$

was fitted through the experimental data points by the method of least squares, and is shown in Fig. 20 as a solid line. This relation, based on linear beam theory, fits very well with the experimental data. The constant A_1 in eqn (4.6) is $2/3EI$ where E is the extensional stiffness and I is the second moment of area of each adherend of the DCB specimen. The experimental values of A_1 are within $\pm 7\%$ of the linear beam theory value of $3{\cdot}77 \times 10^{-7}$.

Finite element analysis[48] was also used to analyse the DCB results. The adhesive was modelled with eight layers of elements. The analysis was conducted assuming plane strain conditions. The experimental values of compliance were within $\pm 5\%$ of those given by a geometrically linear finite element analysis. The computer compliances at various debond lengths were within 5% of the experimental values. These computed values are also shown in Fig. 20. Further, the geometrically non-linear

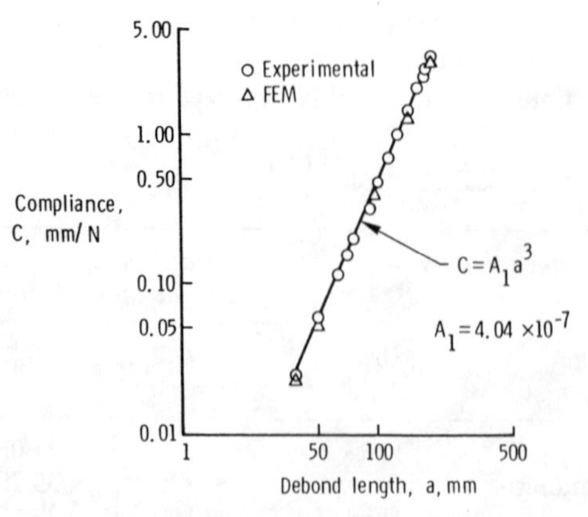

FIG. 20. Relation between compliance and debond length for a DCB specimen with EC 3445 adhesive.

analysis of this specimen did not show any significant change from the linear analysis. The maximum difference in the computed compliance from non-linear and linear analyses was 5% for the maximum debond length employed in the investigation (i.e. 200 mm), at its maximum or critical load. Thus, the compliance–debond length relation, expressed by eqn (4.6), represents the appropriate behaviour of the DCB specimen employed at present. All results from the DCB specimen are calculated using linear beam theory.

Figure 21 shows the measured critical load as a function of debond length for a typical specimen with EC3445 adhesive. Based on linear beam theory,[14] the relation between the critical load, P_{cr}, and the debond length, a, is

$$P_{cr} = A_2/a \tag{4.7}$$

A solid line shown in Fig. 21 with a slope of -1 was fitted to the experimental data by the method of least squares. Then, the averaged value of C_{Ic} for each specimen was computed from the relation:

$$C_{Ic} = \frac{P_{cr}^2}{2w}\frac{\partial C}{\partial a} = 3A_1A_2^2/(2w) \tag{4.8}$$

where w is the specimen width. A similar procedure was used to compute the strain energy release rate G_I associated with cyclic debonding where the critical load was replaced by the maximum load of the fatigue cycle.

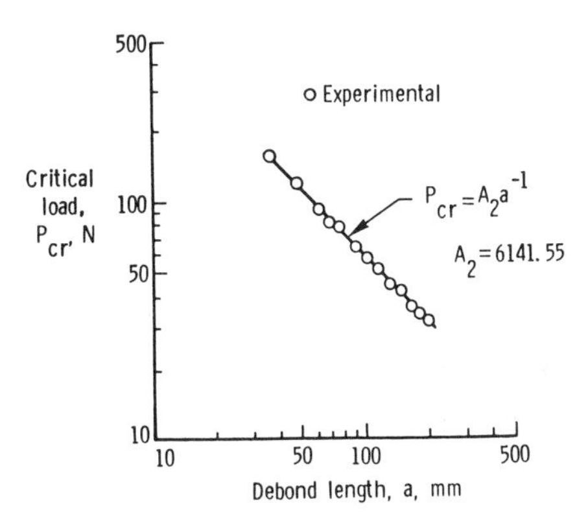

FIG. 21. Relation between critical load and debond length for a DCB specimen with EC 3445 adhesive.

4.4.4. Debond locations

All DCB and CLS specimens, with both adhesives EC 3445 and FM-300, failed by debond propagation during both static and fatigue tests. However, the debond grew in a different manner in each case. In the case of DCB specimens with EC 3445 adhesive, the debond grew in a cohesive manner during both static and fatigue tests. Here, the debond grew consistently in the middle portion of the adhesive layer. In DCB specimens with FM-300, the debond propagated in an irregular manner during both static and fatigue tests, involving cohesive, adhesive, or mixed cohesive–adhesive debonding. Typical debonded surfaces with these failure details are shown in Figs 22 and 23 for both adhesives.

The CLS specimens debonded in a cohesive manner during the fatigue tests for both adhesive systems. Although some adhesive remained on both the strap and lap adherends, significantly more adhesive remained on the lap adherend. Also, there was some 0° fibre pull-off from the strap adherend in most specimens. On close examination of the debonded surface the following conclusions regarding the cyclic debond can be drawn. The debond was basically of a cohesive nature, i.e. failure was within the adhesive with some 0° fibre pull-off from the strap and the debond was always closer to the strap than to the lap. A possible explanation for this debond characteristic was discussed in Section 4.4.3.

4.4.5. Static debonding behaviour

Figure 24 shows the critical strain energy release rate G_{Tc} and G_{Ic} obtained from static tests of CLS and DCB specimens, respectively. The total critical strain energy release rate G_{Tc}, from the CLS specimen, is in agreement with fracture toughness G_{Ic}, from the DCB specimen, in each case. This shows the total critical strain energy release is also the driving parameter for debond growth during static loading. The only exceptions are the adhesive failures in the FM-300 DCB specimens. These adhesive failure strengths are 40% lower than the cohesive failure strengths.

4.4.6. Cyclic debonding behaviour

An attempt was made to determine whether one of the components of strain energy release rate (G_T, G_I or G_{II}) dominates the cyclic debonding. The measured debond growth rates were, therefore, correlated with each of the calculated strain energy release rates G_I, G_{II} and G_T. These correlations are shown in Figs 25 and 26 for EC 3445 and FM-300, respectively. If one component of strain energy release rate had a dominant influence, it would correlate significantly better than the others when

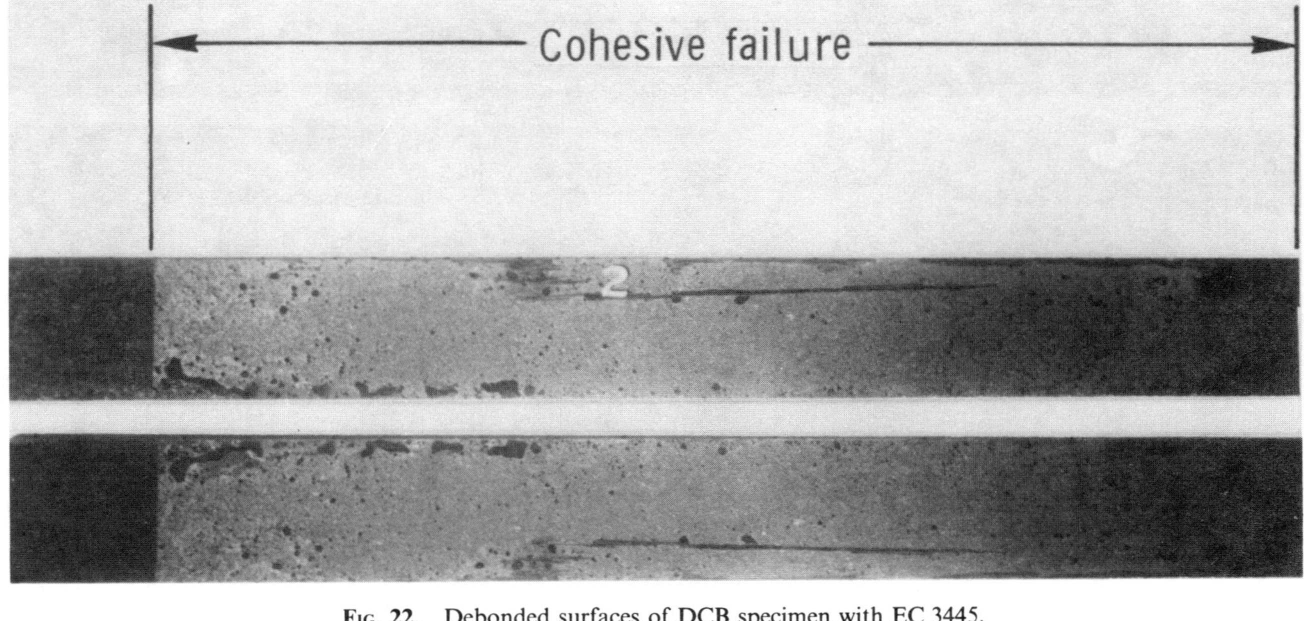

FIG. 22. Debonded surfaces of DCB specimen with EC 3445.

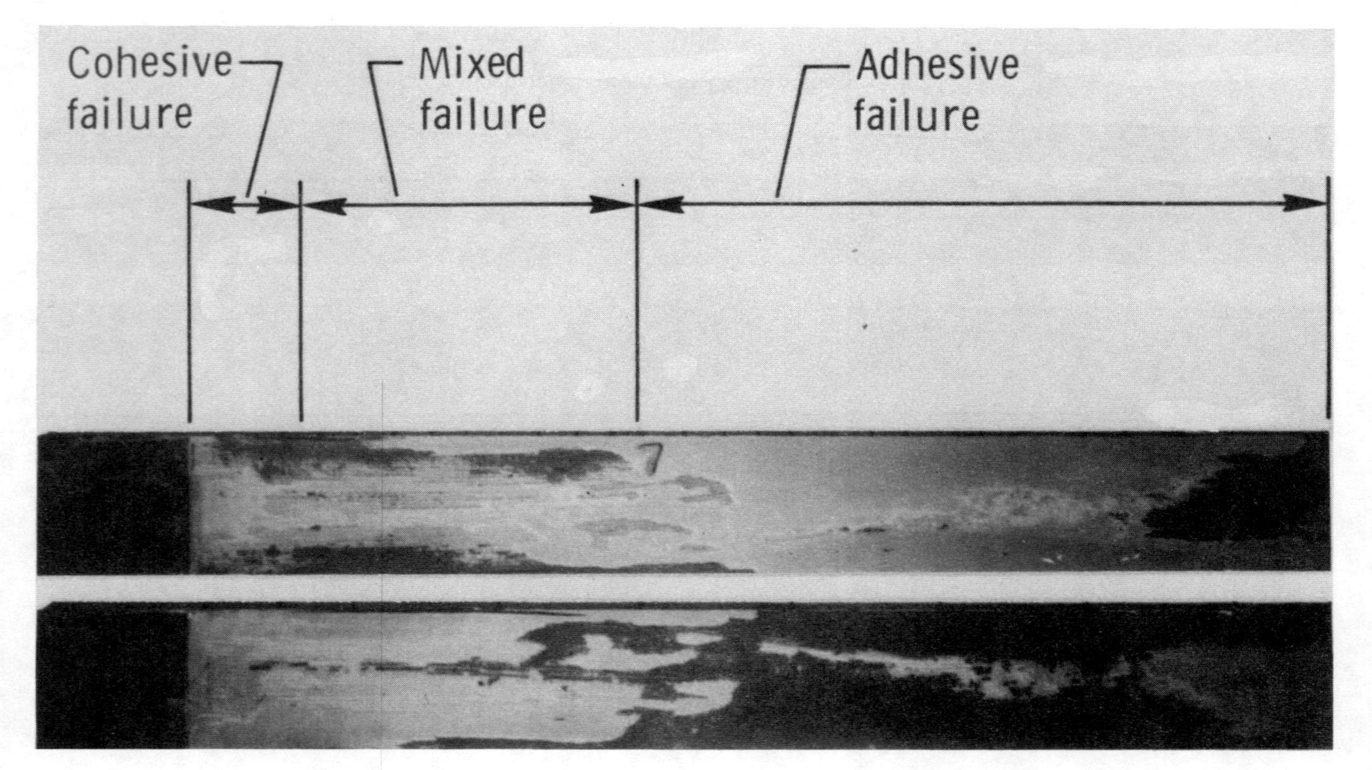

Fig. 23. Debonded surfaces of DCB specimen with FM-300 adhesive.

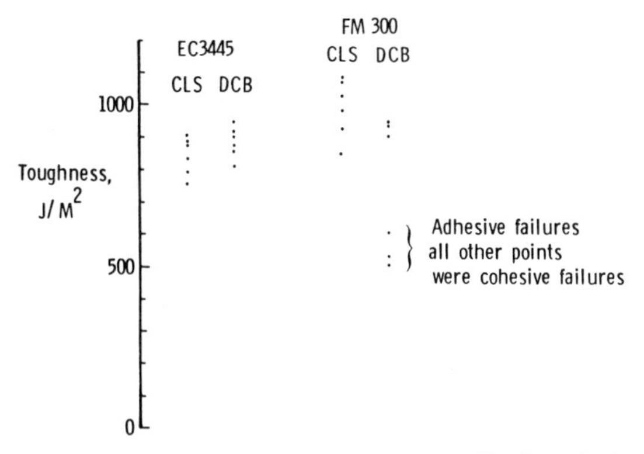

Fig. 24. Static toughness of FM-300 and EC 3445 adhesives derived from DCB and CLS specimens.

debond data from specimens with different $G_{\mathrm{I}}/G_{\mathrm{II}}$ ratios are compared.

An equation of the form

$$\frac{\mathrm{d}a}{\mathrm{d}N} = \mathrm{c}G^n \tag{4.9}$$

was fitted to the data in Figs 25 and 26 by using a least-squares regression analysis. The values of c and n, as well as the sum of errors Σr^2, are shown in the Figures. For each adhesive, the values of Σr^2 are about

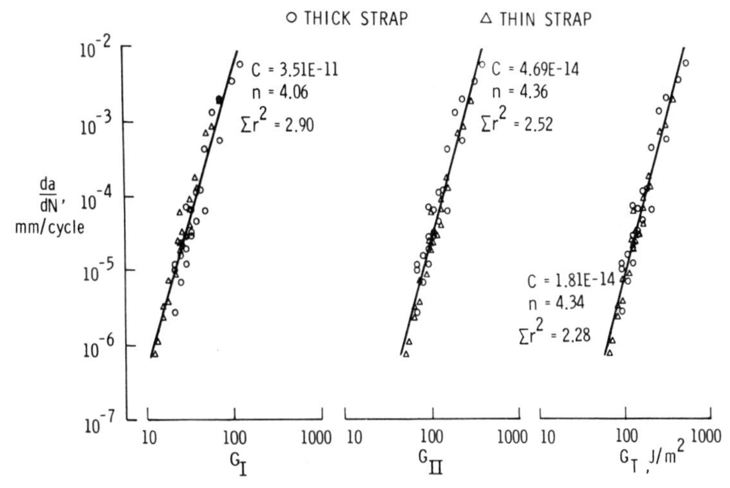

Fig. 25. Relation between strain energy release rates and debond growth rate for EC 3445 adhesive.

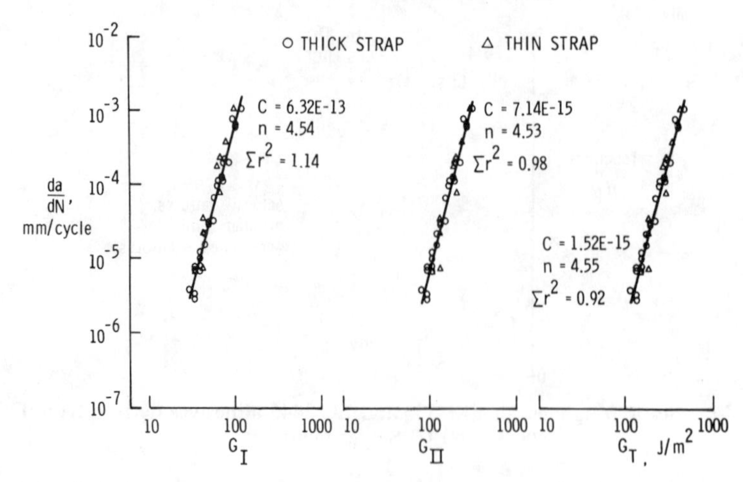

Fig. 26. Relation between strain energy release rates and debond growth rate for FM-300 adhesive.

the same for G_I, G_{II} and G_T. However, the Σr^2 term is lowest for G_T, indicating that G_T provided a somewhat better correlation than either G_I or G_{II}. This suggests that debond growth rate is a function of the combined effects of G_I and G_{II}. However, the G_I/G_{II} ratios for the test specimens were all within the rather narrow range of 0·25–0·38. As a result, it is not surprising that G_I, G_{II} and G_T all did a reasonably good job of correlating the data in Figs 25 and 26. Furthermore, each Figure shows that data from both specimen geometries are within an acceptable scatter band (similar to that observed in fatigue crack propagation in metals).[55] This indicates that specimen geometry did not influence the relationship between the debond growth rate and strain energy release rate. The relations between da/dN and G_T, for both adhesives, are compared in Fig. 27. This Figure shows that the debond rate with FM-300 is about 40% of that for EC 3445, for the same applied load and debond length.

Because of the log–log scale in Fig. 27, the curves relating da/dN and G have slopes equal to the n term in eqn (4.9). The values of n found in this investigation ranged from 4 to 4·5. This is quite high compared with typical values of n for fatigue crack growth in aluminium and steel alloys that range from 1·5 to 3.[55] These steep slopes mean that a small change in applied load would cause a large change in debond growth rate. Thus, the debond propagation in adhesive joints is more sensitive to errors in design loads than are typical cracks in metallic structures. Because of

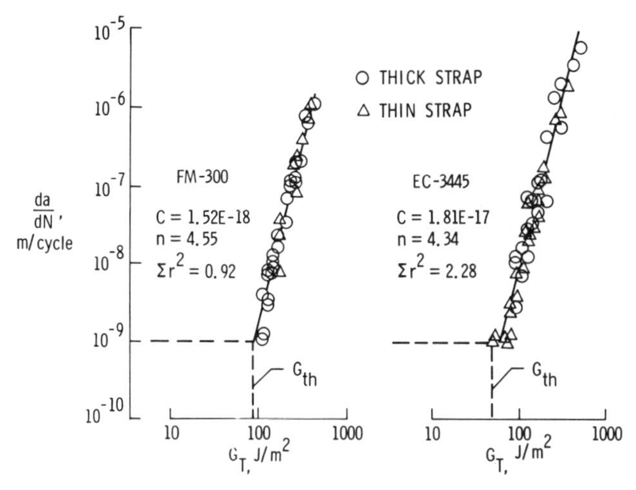

Fig. 27. Relation between total strain energy release rate and debond growth rate for both FM-300 and EC 3445 adhesives.[43]

these steep slopes it may be difficult to design bonded joints for finite life. Minor design alterations, or small analysis errors, could cause a much shorter life than the design value. A viable alternative would involve an infinite-life approach. For this purpose, the no-growth threshold, G_{th} (based on G_T for discussion purposes here), may be an important material property for bonded systems. If a 10^{-6} mm/cycle rate is arbitrarily assumed to be the no-growth threshold, the curves in Fig. 27 show that G_{th} values for EC 3445 and FM-300 are 38 J/m² and 87 J/m², respectively.

All fatigue tests with DCB specimens were conducted at 3 Hz. Figure 28 shows the comparison between the G_T versus da/dN relation for two cyclic frequencies, 10 Hz and 3 Hz, obtained from CLS specimens with EC 3445 adhesive. The solid line shown is a power-law relationship,

$$\frac{da}{dN} = cG_T^n \tag{4.10}$$

which was obtained from Fig. 27 by the method of least-squares fit to experimental data at 10 Hz. The data in Fig. 28 correspond to the 3 Hz cyclic test performed in the present study. The scatter in data is of the same order as that obtained at 10 Hz (which is not shown here for the sake of clarity). The relation between G_T and da/dN is, therefore, not affected by this change in frequency from 10 Hz to 3 Hz in a laboratory environment.

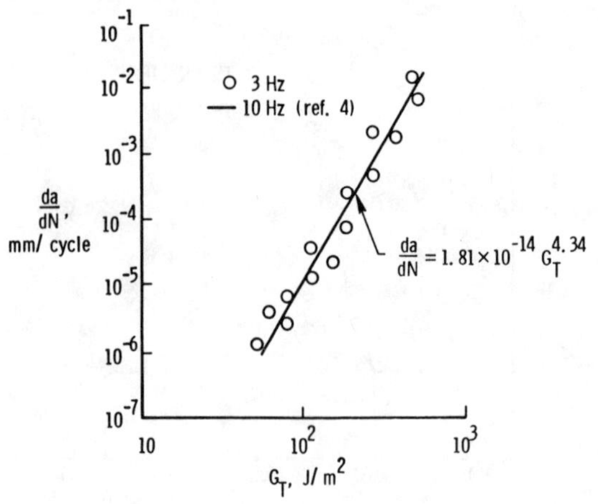

Fig. 28. Relation between total strain energy release rate and debond growth rate of EC 3445 adhesive at two cyclic frequencies using CLS specimens.

The measured debond growth rates from DCB specimens were correlated with the corresponding strain energy release rate G_I as shown in Fig. 29. As previously mentioned, the DCB specimens were tested with constant-amplitude cyclic load and constant-amplitude cyclic displacement. Data obtained from these two testing modes are shown in Fig. 29. The constant-load testing results in G increasing with debond length,

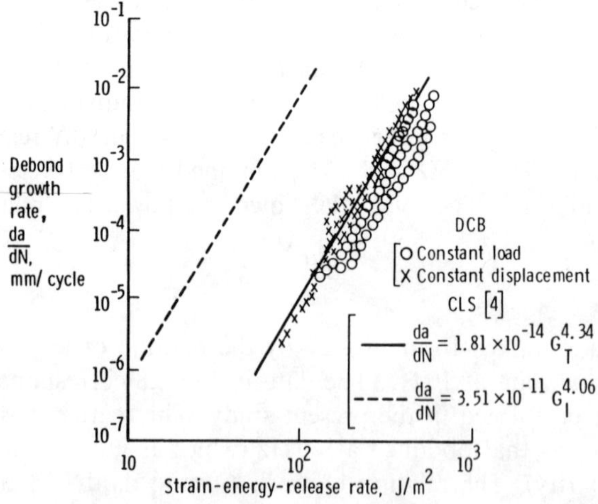

Fig. 29. Relation between strain energy release rates and debond growth rate of EC 3445 adhesive for DCB and CLS specimens.

whilst the constant-displacement results in G decreasing. Since the constant-displacement tests resulted in faster debond rates, the debond process appears to be influenced by the G gradient. Figure 29 also shows G_T versus da/dN and G_I versus da/dN relations from the CLS specimens under mixed-mode loading. The scatter in data from the DCB specimens was larger than from the CLS specimens (Fig. 27). The CLS data points are not shown here for the sake of clarity. The G_I versus da/dN data from the DCB specimen are in good agreement with the G_T versus da/dN relationship from the CLS specimen, represented by the solid line. On the other hand, the G_I versus da/dN relationship from the CLS specimen, represented by the dashed line, did not agree with the DCB specimen. This indicates that the cyclic debond growth is a function of total strain energy release rate.

A similar phenomenon was observed in the case of FM-300 adhesive. Figure 30 shows the comparison of the G_T versus da/dN relation from the DCB specimen. The data shown from the DCB specimen were obtained under constant-amplitude cyclic displacement. As previously mentioned, cyclic debonding occurred in a cohesive manner, adhesive manner, or a combination of both in the DCB specimens. In Fig. 30, data on the right-hand side correspond to cohesive failure, data on the left to adhesive failure, and data in between these correspond to mixed failure.

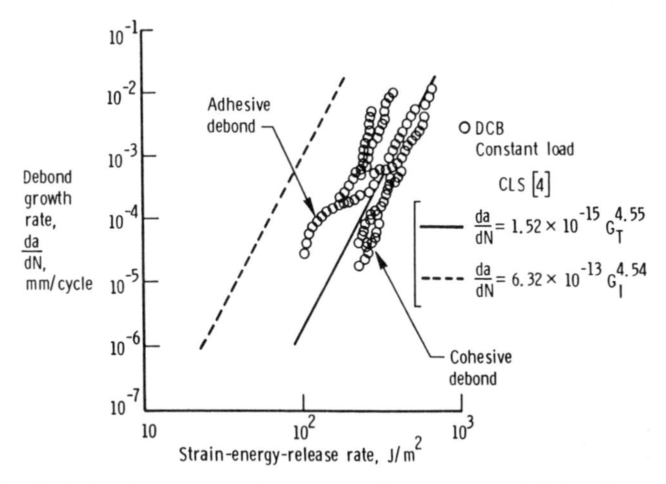

Fig. 30. Relation between strain energy release rates and debond growth rate of FM-300 adhesive for DCB and CLS specimens.

4.4.7. Potential applications for design

The G_T has been shown to correlate the debond growth rate data better than either G_I or G_{II}. The G_T versus da/dN data are plotted in Fig. 27 for FM-300 and EC 3445. On these log–log plots, the data are represented very well by a straight line given by the following equation:

$$\frac{da}{dN} CG_T^n \tag{4.11}$$

where n is the slope of the line in the plot. The value of n ranged from 4 to 4·5. These values are quite high when compared with typical values of n derived from applying eqn (4.10) to fatigue crack growth in aluminium and steel alloys, where n ranges from 1·5 to 3.[55] As noted above, steep slopes mean that small changes in applied load cause a large change in debond growth rate. Thus, the debond propagation rate in adhesive joints is more sensitive to errors in design loads than is typical crack growth rate in metallic structures. Because of this sensitivity, the design of bonded joints for finite life may be difficult. Minor design alterations or small analysis errors could cause large changes in actual and predicted lives. A viable alternative design procedure would be an infinite-life approach. The no-growth threshold, G_{th}, may be an important material property for bonded systems. The premise for this research is as follows:

> Given that the value of G_{th} has been correctly defined and found experimentally for a given adhesive, one should be able to predict the threshold (i.e. no debond growth) stress for any joint geometries using the same adhesive system.

The system in ref. 46 used the same cracked lap shear specimens already described, except that many of the lap adherends were machined to tapers, as shown in Fig. 31. All of these CLS specimens had 0° plies next to the bondline. Specimens with taper angles of 5°, 10°, 30° and 90° (untapered) were tested. Table 4 presents lay-ups and taper angles. Different tapers gave different values of G_I, G_{II} and G_T for the same applied load. This will be discussed further.

4.4.8. Debond initiation

The test programme was conducted to establish the minimum applied cyclic stress, in the strap adherend, that would cause debond initiation and growth in the adhesive bondline of the tapered cracked lap shear specimen. To this end, virgin specimens (no debonds) were tested at a given load range for 1 million cycles, then inspected using enhanced (zinc

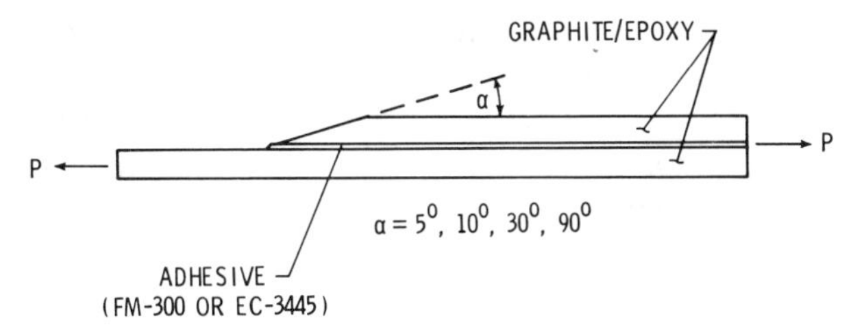

FIG. 31. Tapered cracked lap shear specimen.

iodide) radiography. The cyclic stress ratio, R, was 0·1 and the cyclic frequency was 10 Hz. If there was evidence of debond initiation, the test was stopped. Otherwise, the cyclic load level was raised by approximately 10%, keeping $R = 0{\cdot}1$. The specimen was tested for an additional 1 million cycles and then radiographed. These steps were repeated, progressively increasing the load range until the specimen showed signs of debonding. This procedure allowed more data points to be obtained per specimen.

Several tests are run on virgin specimens to verify that the increasing load did not influence the results. The virgin specimens debonded at stress levels at or very near the one found by the progressive technique. Thus, the prior low-level cycling was assumed to have no significant effect on the specimen.

After a fatigue-induced debond was found, the specimen was removed from the test machine. In many cases, the specimen was re-machined to

TABLE 4
Tapered CLS Specimens

Adhesive	*Taper angle, α (degrees)*	*Plies strap/ plies lap*	*Number of test specimens*	*Analytical results*	
				S_{th} *(MPa)*	G_I/G_{II}[a]
EC 3445	5	8/16	3	120	0
$G_{th} = 38\ J/m^2$	10	16/8	3	102	0
	30	16/8	4	88	0·18
	90	8/16	4	80	0·20
FM-300	5	16/8	3	164	0
$G_{th} = 87\ J/m^2$	10	16/8	4	149	0
	30	16/8	4	130	0·22
	90	16/8	3	121	0·26

[a] Approximate ratio. Mesh was not fine enough for convergence. Four elements through the adhesive thickness.

another lap adherend taper and re-tested. The tip of the new taper was always at least 8 mm ahead of the previous debond tip.

The premise that a material property, G_{th}, could be determined and used to predict debond threshold stresses in arbitrary joint geometries was verified in the study by the following procedure.

(1) The minimum cyclic load that initiated debonds in tests of the untapered adherend ($\alpha = 90°$) specimens was used in the GAMNAS program to calculate G_T. This G_T was taken to be equal to G_{th}.

(2) For each of the tapered adherend geometries, the GAMNAS program was used to calculate the stress required in the strap adherend to create a value of G_T equal to G_{th}.

(3) The minimum cyclic stresses required to initiate debonds in tests of the tapered adherend specimens were compared with the predicted stresses.

To compute a debond threshold stress, a small debond must be assumed to exist at the location of expected debond initiation, i.e. at the end of the strap adherend. For the current computations, a debond length of 1 mm was assumed. This debond length was typical of the lengths found by the X-ray inspection technique during the debond initiation tests. Computations using a debond length of 0·5 mm resulted in predicted initiation stresses about 3% lower than those for a 1 mm length.

In addition to G_T, a rough estimate of the G_I/G_{II} ratio was calculated for each specimen type. All analytical results are presented in Table 4.

The minimum cyclic stresses that initiated debonds in the tests of the tapered adherend specimens and the analytical predictions of those stresses are plotted in Fig. 32. The solid symbols in the figure represent specimens that initiated and grew debonds within 1 million cycles and no debonding was evident from the radiographs. The lines in the Figure are the predicted debond threshold stresses as a function of taper angle presented in Table 4.

The data show a significant improvement in debond resistance for taper angles below 10°. The 5° taper results in a threshold stress level about 50% higher than that for no taper ($\alpha = 90°$). Figure 32 also shows clearly that joints fabricated with FM-300 adhesive can be subjected to 50% higher loads than EC 3445 without debonding. The relative difference between the two adhesives would probably vary with test environment.

The predicted debond initiation threshold stresses are in excellent agreement with the experimental data for each adhesive for the specimen geometries studied. These results are particularly revealing in that thres-

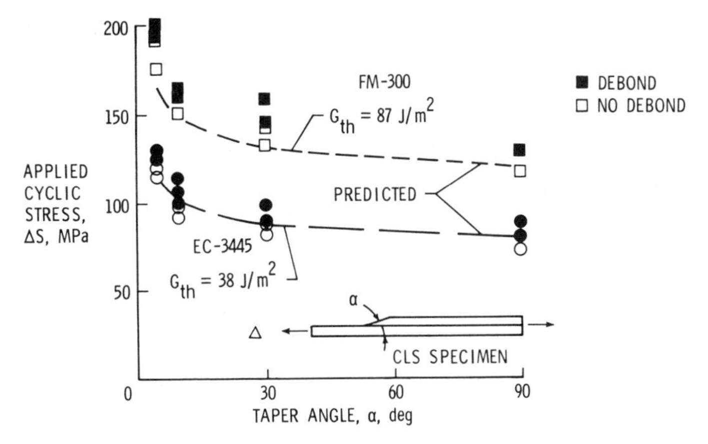

Fig. 32. Comparison with experimental data of predicted maximum stress for no debonding.

hold values based on total strain energy release rate predicted the test results so well. At a 90° taper, the G_I/G_{II} ratio is approximately 0·20 and 0·26 for EC 3445 and FM-300 adhesives, respectively. For the 5° and 10° taper, the G_I/G_{II} ratio is practically zero for both adhesives. According to Hart-Smith,[57] a 5° taper should eliminate all of the peel stresses and the specimen should not debond. This study indicates that a 5° taper does indeed eliminate peel stresses for all practical purposes, but the taper hardly guarantees that no debonding will occur. Clearly, G_I alone cannot be used to explain the trend of the data in Fig. 32.

The debond initiation tests resulted in a G_{th} very close to the G_T associated with a debond propagation rate of 10^{-6} mm/cycle ($3{\cdot}94 \times 10^{-8}$ in/cycle). The authors feel that 10^{-6} mm/cycle is a sufficiently low crack growth rate upon which to base a threshold value—particularly if G_{th} is taken on the conservative side of the data, as shown in Fig. 27. The two approaches ($da/dN = 10^{-6}$ mm/cycle and debond initiation) agreed closely despite the absence of a clear 'knee' in the da/dN vs G data commonly associated with threshold. However, before a clear correlation between debond initiation and the crack growth threshold can be established, more crack growth rate data are needed at very low rates to see if a 'knee' really exists.

Total strain energy release rate, G_T, appears to be the driving factor for debonding of these rather tough structural adhesives (EC 3445 and FM-300). Liechti and Knauss[33] have also suggested, for adhesive joints, that G_T may be the driver for debond extension based on observations of a polyurethane elastomer.

The proposed design technique can be applied to actual structures as follows. First, from basic laboratory coupons, such as the untapered (i.e. $\alpha = 90°$) CLS specimen, tested in the usage environment (e.g. temperature, humidity, and cyclic frequency), the G_{th} value for the adhesive system of interest can be determined. This, of course, requires a proper analysis of the specimen to determine G_T.

Second, an initial flaw (debond) size must be assumed. This would normally be either (1) the largest size of debond due to manufacture that might not be found during inspection, or (2) the largest debond that might result from operational damage.

Finally, a proposed structural joint geometry can be sized to ensure no debonding. This requires an iterative process by which the geometry and loading are analysed to ensure that G_T at the assumed debond tip is less than or equal to G_{th}.

4.5. LONG-TERM ENVIRONMENTAL EFFECTS

4.5.1. General remarks

In view of the sensitivity of polymers to time, temperature and moisture, it is important to enquire into the behaviour of the polymeric adhesives and the matrix materials of the adherends in adhesively bonded joints under such conditions. Often the weak link in joints fabricated with metallic adherends is the specially prepared oxide surface layer which deteriorates with time and moisture. Since such specially prepared oxide layers do not exist in bonded joints made of composite adherends, this phenomenon will not be present.

Durability is a major concern with the use of polymeric materials for any application. Throughout recorded history mankind has been gaining familiarity and confidence with bronze, iron, and a host of more modern structural metals. The public, as well as some design engineers, have a lingering fear that generic ‘plastic’ is not a material capable of long-term endurance. Indeed, polymeric adhesives for any application may carry some of these negative connotations.

The long-term integrity of bonded joints implies both chemical and mechanical durability of the adhesive and the bond in the presence of varying temperature, moisture and other environmental factors. Joint failure may involve different mechanisms, depending on the materials involved and the environment to which they are subjected. Long-term bond integrity implies many things, including: slow stable crack propagation; the lack of excessive environmental degradation of the adhesive, oxide layers, or interphase regions; and the avoidance of excessive creep and viscoelastic delayed failures.

As with bonded joints of metal adherends, the adhesives used to bond FRP are polymeric materials and, as such, are subject to time-dependent behaviour which can be greatly accelerated by higher operating temperatures, increased stresses, or the absorption of plasticisers such as moisture, fuel, or oil. Viscoelastic effects can result in creep deformation, stress relaxation, and/or delayed failures when subjected to long-duration loading. Reduced crosslinking of a polymer also results in excessive time dependence. Great care should be exercised in the cure process to ensure that adequate crosslinking occurs. When designing with fibre-reinforced plastics, adhesives, or any polymeric material, it is essential to recognise the potential for viscoelastic failures of the product. The design engineer must recognise the interdependence of time, temperature, stress, and environment in producing these effects. Some industries have taken the attitude that viscoelasticity is an unknown which cannot be dealt with successfully. As such they have sought to avoid viscoelasticity rather than deal with it on a rational basis. The quest for light-weight, efficient structures is forcing the use of materials, stress levels and operating temperatures which can lead to viscoelastic behaviour. Rapidly applied or shock loads make polymers stiffer and more brittle. Because stresses in bonded joints are often highly concentrated at the ends, plastic yielding of the adhesive is necessary to achieve high strength. At high loading rates, plastic yielding is reduced and brittle failures can occur. Efficient design with polymeric materials necessitates an understanding of these time dependent phenomena.

Bonded FRP structures are a dual viscoelastic problem. Because the matrix of most fibre-reinforced plastics is polymeric, composite materials are subject to viscoelastic behaviour. A number of investigators have reported on the time-dependent nature of composites and the reader is referred, for example, to refs 58–62. Creep of these materials can be substantial, particularly when subjected to high loads, elevated temperatures, or moisture absorption. The Time Temperature Superposition Principle has been successfully applied to permit the shifting of compliance data at various temperatures to predict a response master curve which is valid over many decades in time.[63] Figure 33 illustrates the non-linear nature of the creep compliance for a 10° off-axis unidirectional tension

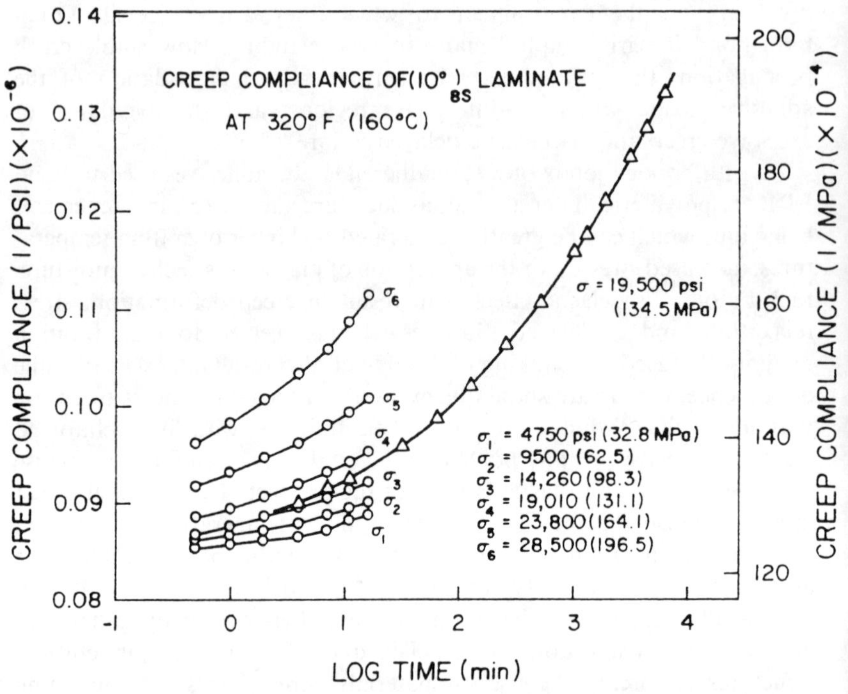

FIG. 33. Non-linear creep compliance of 10° off-axis carbon/epoxy specimens at 160°C.

specimen of T300/934 at 160°C. Because these materials tend to be quite non-linear, techniques such as Time Stress Superposition,[64] the Findley approach[65,66] and the Schapery single-integral technique[67,68] have all been used to predict the stress dependence. The effects of temperature and stress level on the time to delayed failure have also been incorporated[64,71] into the models using procedures such as those due to Zhurkov,[69] and Bruller.[70] Figure 34 illustrates typical delayed failure results for $(90/60/-60/90°)_{2s}$ specimens of T300/934 at three different temperatures. Although these data are for a matrix-dominated laminate, similar failures have been observed in practical, fibre-dominated laminates.[72] Delayed failures should be considered for any long-duration loading situation, especially when elevated temperatures and/or exposure to moisture are involved. Numerical procedures have been developed to predict the compliance and delayed failure phenomena in laminated composites based on a relatively small amount of short-term testing of unidirectional

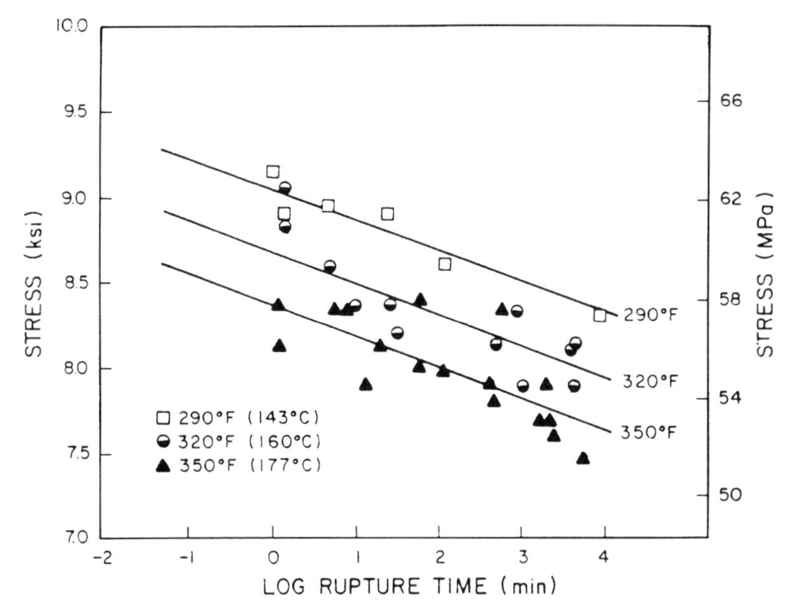

Fig. 34. Typical delayed failure data for laminated carbon/epoxy specimens at three temperature levels. Lines are least-squares fits of the data.

material.[73] These procedures provide a rational approach to predicting time-dependent response in FRP.

It is important to note that this time dependence for most high-performance composites is directly linked to the matrix behaviour. Of particular concern here is how the stress state induced in the composite by an adhesive joint may lead to delayed failure. Laminated composites tend to be extremely weak in the out-of-plane direction for both tensile (peel) and shearing loads. In overlap joint configurations, peel stresses and shear stresses tend to be highly concentrated at the adherend ends. For metal adherends, these stresses may be critical when compared with allowables in the adhesive, but are normally much smaller than allowable stresses for the adherend. This is not the case for laminated composites. In fact, residual stress effects may already impose critical out-of-plane stresses at the free edge. As such, the joint stresses may contribute to a pre-existing critical stresss state within the laminate. Over time these high stress levels, combined with certain environmental conditions, could lead to delayed failures of composite adherends. Adams and Wake[17] have investigated the effect of adhesive spew on increasing joint strength with metal and composite adherends. Spew may play an even more important role in composite joints than in metal joints because of the support and more uniform loading it gives to the highly stressed free edge. Adams and Wake[17] have shown more than a three-fold increase in the strength of unidirectional carbon/epoxy bonded to steel by reverse-tapering the steel and using a 30° spew fillet angle. This changed the failure mode from a transverse fracture in the composite to an adhesive failure. In the case of co-cured joints, where the adhesive and matrix are the same material, one expects the out-of-plane stiffness in the composite to be two to three times higher than that of the neat resin because of the fibre inclusions. On the other hand, the out-of-plane strength for the composite may be less than that of the neat resin because of the stress concentrations and weak interfaces with fibres. Thus interlaminar failures are more likely than adhesive failures.

The adhesives used to bond FRP materials are polymeric and may even consist of excess matrix resin which flows to the bond surface during the cure process. Since these adhesives are also time-dependent, the combination of a viscoelastic adherend and a viscoelastic adhesive provides an interesting configuration which has not been fully studied. Improved techniques are needed better to characterise the constitutive relations of adhesives and adherends. The appropriateness of bulk adhesive properties for modelling the *in situ* behaviour of an adhesive remains a topic of discussion. A portion of the discrepancies may be described by the constraint placed on the adhesive by the adherends, the intermixing or influence of the adherend and adhesive, and by differences in curing procedures and the resulting thermal stresses.[22] Despite the problems associated with the use of bulk properties, these properties continue to be widely used because they are relatively easy to obtain, and accurate determination of *in situ* properties remains an extremely difficult task.

Brinson[74] has used bulk adhesive viscoelastic properties to study the time-dependent nature of bonded joints. He has shown that the viscoelastic response of adhesives such as FM-73, FM-300, and Metlbond 1113 is non-linear. Because typical joints involve very high stresses near the overlap ends, even relatively low load levels may result in significantly non-linear behaviour. Brinson and his coworkers have applied several models to bulk tensile data for adhesives, including a modified Bingham model and the Schapery single-integral technique. Typical results for ramp-loaded tension coupons of Metlbond 1113 in Fig. 35 clearly show a significant time dependence.

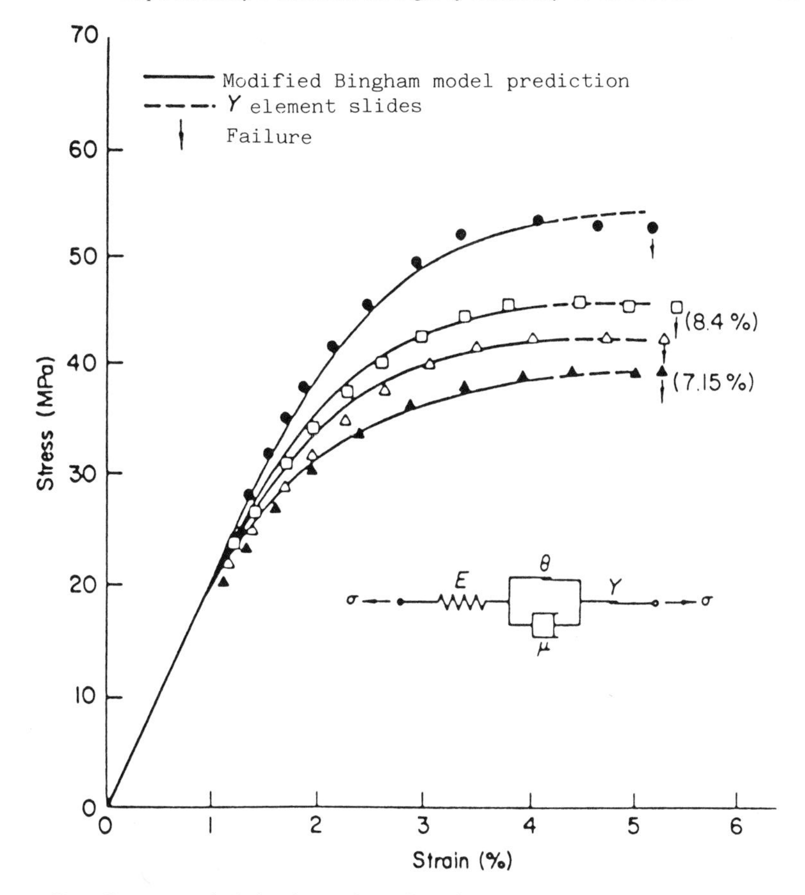

FIG. 35. Stress–strain behaviour of Metlbond 1113-2. Fit with a Bingham model. Strain rates (%/s):[70] ▲, $7{\cdot}00 \times 10^{-4}$; △, $6{\cdot}75 \times 10^{-3}$; □, $7{\cdot}05 \times 10^{-2}$; ○, $6{\cdot}35 \times 10^{-1}$. Brinson, H. F., The viscoelastic constitutive modelling of adhesives, *Composites*, October 1982, Butterworth and Co. (Publishers) Ltd.

The Schapery single-integral method[75] is an attractive technique for modelling the non-linear viscoelastic response of adhesives. Henriksen[76] has utilised this approach to develop a non-linear viscoelastic finite-element program. He has used this procedure to analyse a butt joint bonded with a viscoelastic adhesive. Another technique which shows great promise is the modified form of the Leaderman equation proposed by Knauss and Emri.[77] This approach incorporates a shift factor based on the free volume of the material. This permits temperature, moisture

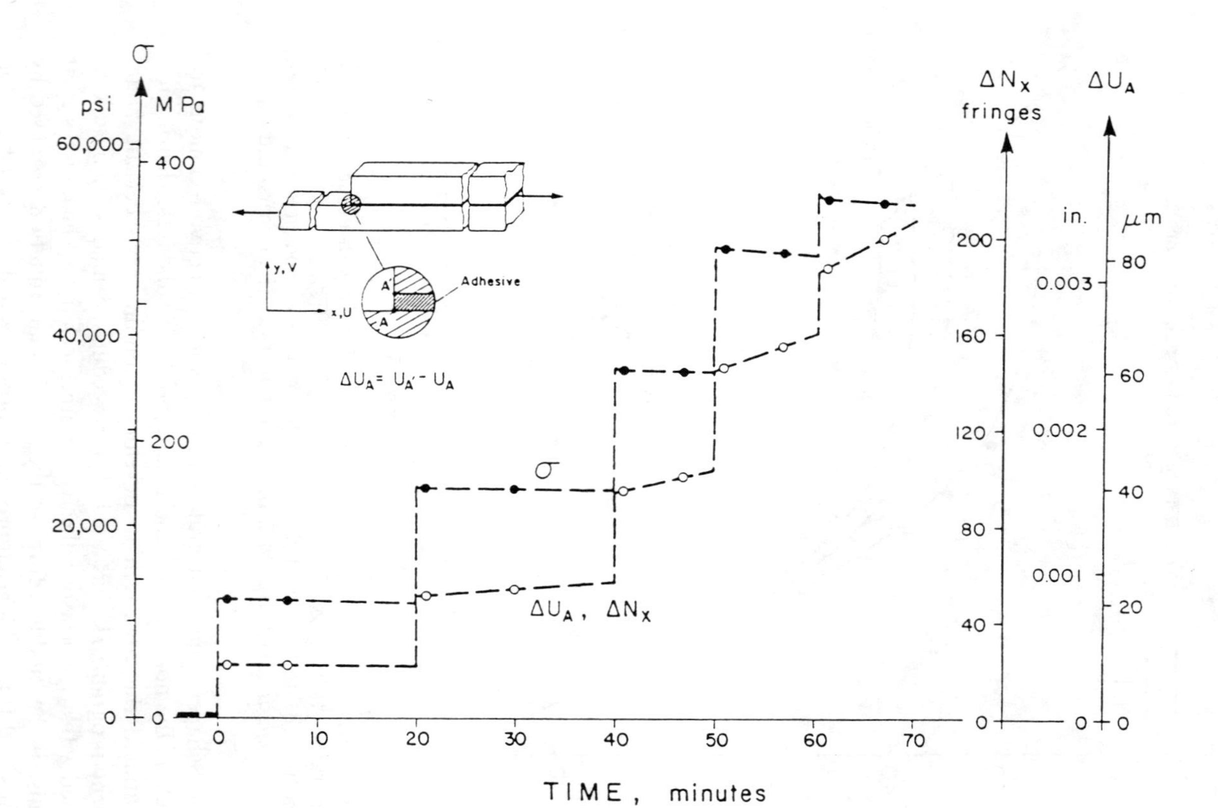

Fig. 36. Shear stress and displacement history for the thick adherend lap joint specimen.

absorption and stress-induced dilation all to be combined in a rational manner to predict the acceleration of the viscoelastic process. Becker *et al.*[10] have used this approach in developing the finite-element program VISTA which is being used by several investigators to predict the response of bonded joints. VISTA also permits the analysis of moisture diffusion into a joint.

Post *et al.*[26] have utilised Moiré interferometry to measure experimentally time-dependent deformation in a thick adherend joint. These results for FM-73M epoxy with aluminium adherends are illustrated in Fig. 36. The specimen was loaded in a relaxation mode at several successively higher displacements. As would be expected the load, and hence the average shear stress, which is plotted versus time, is seen to decay. Interestingly, however, the shear strains measured optically are seen to increase significantly at higher displacements. The flexibilities of the specimen and of the loading frame were sufficient to convert the apparent relaxation loading mode on the joint to more of a creep-type load within the viscoelastic bond.[26]

4.5.2. Moisture effects

When the composite adherend is thin in comparison with the bondline length, moisture migration to the adhesive may be much faster than in metal joints because of the shorter path, assuming the diffusivities for the adhesive and composite are of the same order of magnitude. Matrix damage in the composite may even provide direct access of moisture to the adhesive layer by capillary wicking. The absorption of moisture and other plasticisers tends to accelerate time-dependent processes by lowering the glass transition temperature.[78] Figure 37[83] illustrates the significant decrease in the glass transition temperature which occurs with increasing moisture content for a carbon/epoxy system. A similar effect occurs in most adhesives,[79] resulting in a significant shift of the master curve for compliance or modulus. Moisture absorption also induces swelling of the matrix and adhesive, and can result in differential swelling stresses which can damage the bond.

In properly prepared metal joints the failure is essentially cohesive within the adhesive layer. Testing of specimens subjected to environmental exposure, however, often reveals that the failure locus is at or very near the adhesive/adherend interface. Although cohesive/adhesive failures are uncommon in structural joints of composites, the implication from metal experience is that moisture may induce failure modes which are different from those in dry joints.

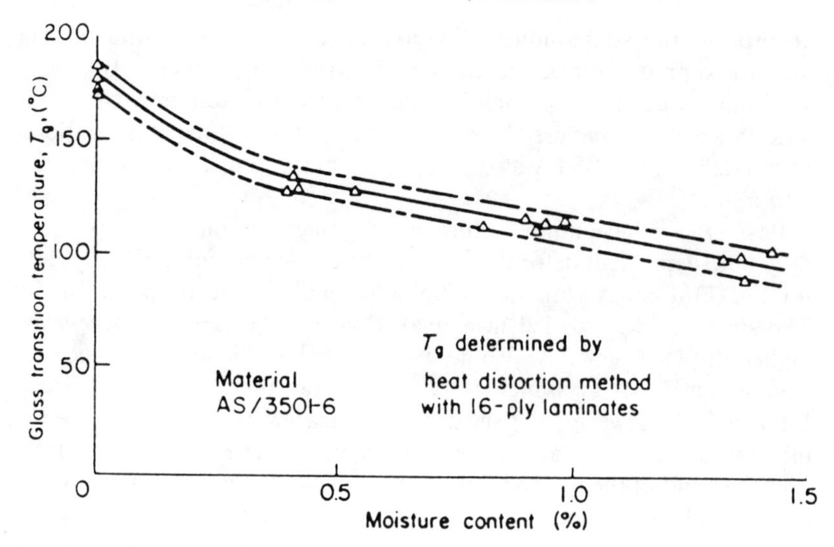

Fig. 37. Effect of moisture on glass transition temperature of AS/3501-6 carbon/epoxy.[83] Labor, J. D., *Depot Level Repair for Composite Structures,* NADA-79171-60, Northrop Corporation, Aircraft Division, October 1981.

In addition to mechanical effects, moisture can promote chemical reactions in the adhesive, the composite adherend, or the interphase region. Some adhesives are much more susceptible to chemical degradation induced by moisture than others. The plasticisation of a polymer by water is normally recoverable when the material is redried. Butt and Cotter[80] report, for example, that for a commercial nylon/epoxy adhesive exposed to 43°C at 97% RH, for times ranging from 142 to 2040 h, the complex dynamic tensile modulus decreased to 18% of the dry value. Upon redrying, the moduli returned to within a few per cent of the original moduli. This represented a reduction in the glass transition temperature of approximately 40°C. Interestingly, when aluminium adherends were bonded with this adhesive, and exposed to a similar environment, the loss of strength was not regained upon redrying. Hydrolysis of bonds weakens joint strength, but some of this strength can be regained when dried. If corrosion products are formed, however, strength loss is permanent.

One should note that in addition to chemical reactions and plasticisation, moisture absorption and desorption can result in crazing and cracking of the adhesive and/or polymeric matrix, due in part to swelling and shrinkage stresses. These processes are, in general, irreversible. These cracks can serve as initiators for fatigue cracks to develop.

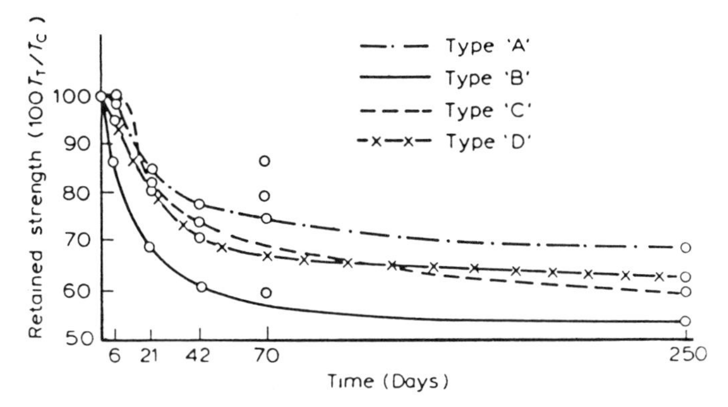

FIG. 38. Effect of 70°C and 90% RH on tensile lap shear joint strength of glass/epoxy specimens bonded with several adhesive systems.[21] A–D refer to different unidentified adhesives. All are two-part acrylic adhesives. Bowditch, M. R. and Stannard, K. J., Adhesive bonding of GRP, *Composites*, July 1982, Butterworth and Co. (Publishers) Ltd. Controller, HMSO, London, UK, 1982.©

Bowditch and Stannard[21] have studied the durability of several two-part acrylic adhesives for bonding glass-reinforced plastics to themselves and to aluminium. Unstressed lap shear specimens were exposed to 70°C at 90% RH and then tested to failure after various exposure times. Typical results are shown in Fig. 38. They noted that failures after relatively short exposures and very long exposures tended to be cohesive within the composite adherends. For three of the four adhesives, however, failures at intermediate exposures tended to be within the adhesive. Their hypothesis to explain this is schematically illustrated in Fig. 39 where the failure is controlled by the weakest link, but the adhesive and composite degrade at different rates. Substantial reductions in retained strength are observed, although each of the four adhesive systems studied seemed to exhibit an endurance limit in the sense that strength reductions all reached an asymptote. It should be noted, however, that this strength retention asymptote cannot be interpreted in the sense of a true endurance limit. Indeed, in tests performed by Minford[81] with metal adherends exposed while under stress, no endurance limits were observed for some systems, as shown in Fig. 40.

Minford[81] has shown that stressed joints tend to degrade faster when exposed to moisture than do unstressed joints. Several mechanisms are possible for this including increased diffusion rate in loaded specimens, stress levels inducing cracks and debonds which permit capillary wicking of the environment into the joint, and the higher energy state of the

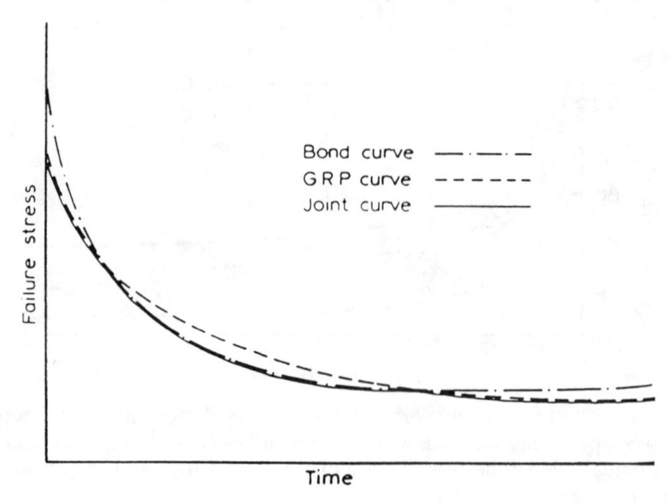

Fig. 39. Schematic representation of the relative degradation rates of adherend and bond line.[21] Bowditch, M. R. and Stannard, K. J., Adhesive bonding of GRP, *Composites*, July 1982, Butterworth and Co. (Publishers) Ltd. Controller, HMSO, London, UK, 1982.©

molecules which accelerates degradation of the bond. To facilitate stress tests during environmental exposure, the durability test has become widely used by producers and users of adhesives for evaluating the long-term integrity of adhesives. Some concern has been raised regarding the applicability of delayed failure data from short overlap (13 mm) single lap joints to actual production joints which may utilise long overlap lengths. Whilst the possibility for delayed failures is significantly reduced by the forgiving cushion of long overlaps which are typical of many production joints, concern is still raised that a crack could grow under the relaxation conditions.

Because environment and moisture migration play such an important role in reducing bond strength, durability predictions must often take diffusion into account. Fick's Law has been widely used to predict moisture absorption, but there are limitations. Adamson[82] has found that temperature changes can be used to pump much more moisture into epoxy materials than would be predicted by equilibrium saturation levels. Matrix cracking can also cause substantial deviations from Fickian predictions. In making predictions of diffusion rates, one should also recall that diffusivity normally varies with temperature according to an activation energy criterion.

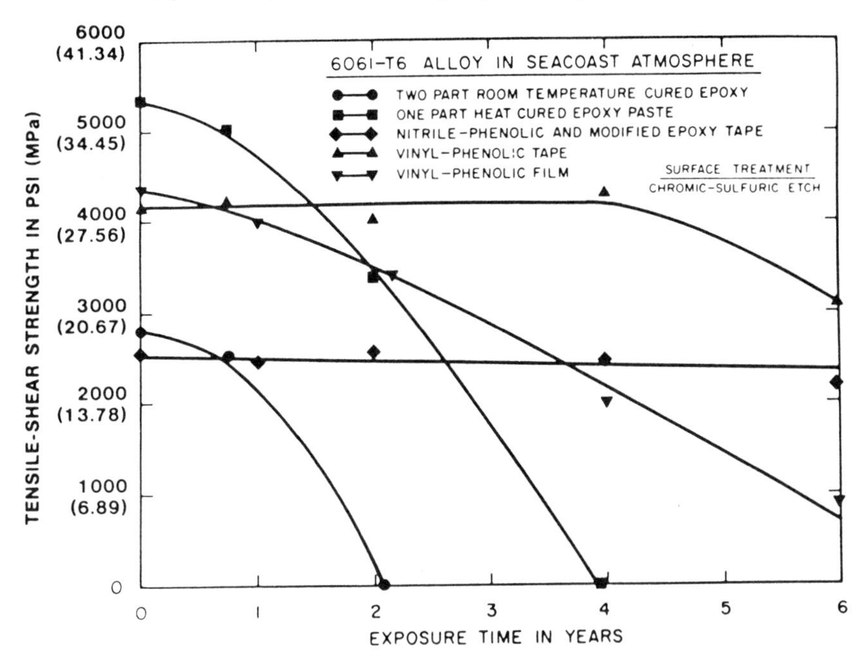

Fig. 40. Effect of adhesive on the durability of etched 6061-T6 aluminium alloy joints exposed to a marine environment. Minford, J. D., Adhesives, *Durability of Structural Adhesives* (Ed. A. J. Kinloch), Applied Science Publishers, London, 1983.

4.5.3. Effects on cyclic crack growth

In ref. 49 the thick adherend model lap joint was used in a study of temperature and moisture effects. The specimen geometry (Fig. 41) was chosen because, upon loading, it has a relatively low peel component, corresponding to joint geometries of a more practical design which have large overlap length-to-bondline thickness ratios. The adherends were bare aluminium alloy 7075-T651 which were alkaline-cleaned, acid-etched and anodised using the Boeing phosphoric acid process. Following water rinse and oven drying, the anodised parts were coated with Bloomingdale BR-127 primer. The adhesive employed throughout the work was FM-73M (manufactured by American Cyanamid, Havre de Grace, MD, USA), a 121°C cure tape having a nominal thickness of 0·20 mm with a matte 'Dacron' carrier.

The joints were subjected to cyclic loads of constant amplitude in various environments. The gap deflection was measured using a clip gauge

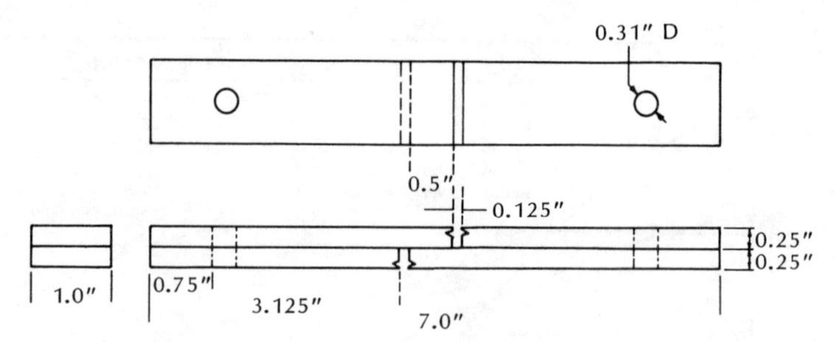

Fig. 41. Thick adherend model single lap shear joint (1 in ≡ 2·5 cm).

so that the load–deflection history could be plotted as a function of the number of cycles. The load–deflection history was characterised by a progressive amount of deformation with cycling (cyclic creep), in which each cycle gave rise to a hysteresis loop, indicating that part of the deformation energy was recoverable. A third feature was that the slope of the major axis of the loops decreased with the number of cycles. The slope was taken to be an average stiffness of the joint. The ratio of the stiffness after a given number of cycles K_N to the initial stiffness K_i of the

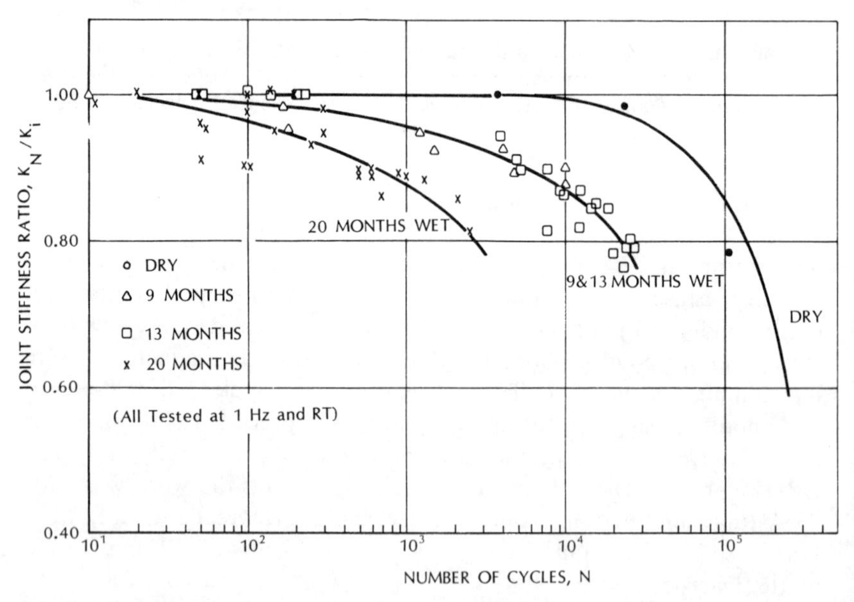

Fig. 42. Joint stiffness ratios vs number of cycles at 1 Hz and 24°C for different moisture conditions at 4·448 kN maximum load.[28]

joint (measured after about five 'shakedown' cycles) was compared with finite element analyses of joints containing various debond lengths. The comparison indicated that the reduction in joint stiffness could be associated with crack growth. In view of the possibility that temperature and moisture could cause a decrease in stiffness, tests were conducted involving sacrificial specimens in which the extent of crack growth was measured directly. The tests indicated that the reduction in stiffness ratio was due entirely to crack growth, which in turn allowed the stiffness ratios to be used as measures of crack growth.

The effects of various environments are shown[49] in Figs 42 and 43 where the joint stiffness ratio is plotted against the number of cycles. In Fig. 42 the effect of moisture is shown by comparing data sets for dry, and 9-, 13- and 20-month moisture conditioned joints. Increasing moisture content very clearly accelerates the fatigue crack growth. Increasing temperature had the same effect (Fig. 43).

In all cases the cracks initiated as adhesive cracks from the high stress region caused by the 90° corner between the adhesive and adherend, along the load transfer edge of the adhesive. The interface cracks grew relatively

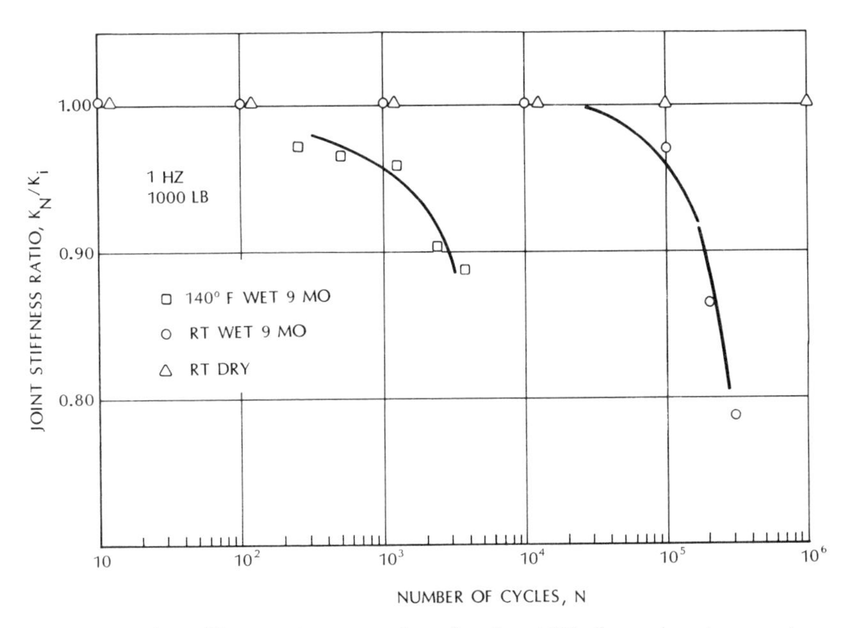

Fig. 43. Joint stiffness ratios vs number of cycles at 1 Hz for various temperatures and moisture conditions at 4·448 kN maximum load.[28]

slowly until cohesive crack growth occurred in a catastrophic manner. The two regions of adhesive and cohesive failure were most distinct in the cases of dry environments, high frequency and room temperature. High moisture contents were found to give rise to very small amounts of adhesive crack growth. The same topography was evident in dry joints subjected to low-frequency cycling (0·17 Hz) and high temperature (60°C). A similar dependence on cyclic frequency has also been observed in other studies.[83]

The reason for the varying amounts of interfacial crack growth can be attributed to a decrease in adhesive modulus that accompanies high temperature, high moisture contents and low frequencies. The decrease in modulus decreases the severity of the stress concentration at the load transfer edge and increases the energy release rates associated with cohesive crack growth, both making cohesive growth more likely. These tendencies are compounded by the fact that the scrim plane, which defined the location of cohesive crack growth, was weakened by moisture ingress between the (untreated) scrim fibres and the adhesive.

The effects of cyclic frequency and temperature on crack growth rates were also examined[84] in cracked lap shear specimens having a similar design to that originally proposed in ref. 15. The geometries are shown in Fig. 44 and are designated CLS_1 and CLS_2, corresponding to specimens having thick and thin strap adherends, respectively. The specimens were

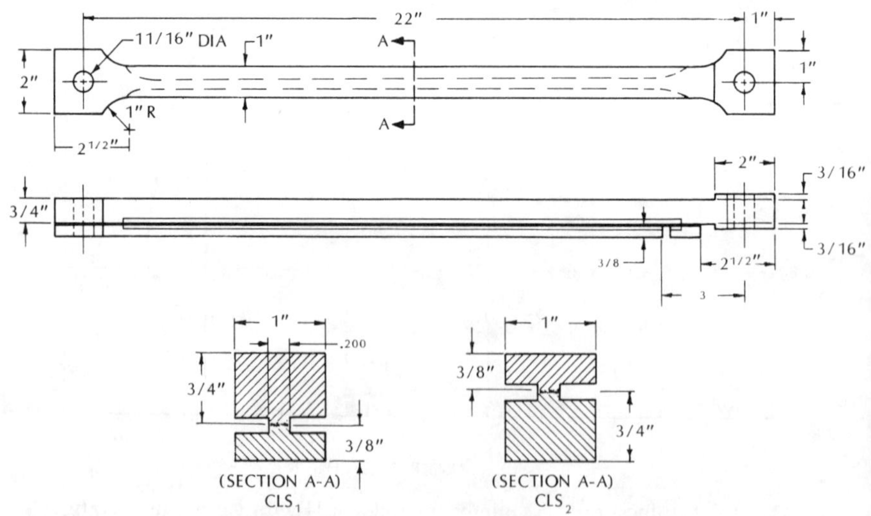

FIG. 44. Cracked lap shear test geometry (1 in ≡ 2·5 cm).[15]

side-notched to prevent failure in the adherends. The surface preparations, adhesive and cure schedules were the same as those that were employed in the fabrication of the thick adherend lap shear joints. Loads were introduced to the specimens through a clevis grip arrangement that allowed for rotation to occur at the specimen ends. The specimens were precracked under gradually decreasing loads, from an initial high load, to avoid potential retardation effects. Fatigue cycling was conducted under constant load amplitude with a load ratio of 0·1. Load amplitudes were increased every 0·038–0·051 mm to provide a range of crack growth rates for each specimen. Crack lengths were measured ultrasonically at periodic intervals. Tests were conducted at three temperatures (-54°C, room temperature and 60°C) and three frequencies (0·3, 3·0 and 10 Hz). Crack growth rates at a given crack length were correlated with the change in total energy release rate over one cycle, ΔG, obtained from finite element analyses. In the CLS_1 specimen, which has a nominal ratio of $G_1/G_2 = 0{\cdot}13$, the tests at 3 Hz indicated that for a given ΔG crack growth rates at -54°C and room temperature were the same. However, increasing the temperature gave rise to a noticeable increase in crack growth rates. The range of frequencies examined gave rise to very little change in crack growth rates. There did not appear to be any temperature or frequency effects in the CLS_2 specimens ($G_1/G_2 = 0{\cdot}35$) whose da/dN vs ΔG curves were characterised by very steep slopes.

4.5.4. Time-dependent crack growth

Another type of slow, subcritical crack growth that can occur is the time-dependent growth due to the viscoelastic nature of the polymeric adhesive. For structural adhesives in their glassy state at room temperature, this phenomenon is most likely to occur over long periods of time and/or hot wet conditions. As in the case of fatigue crack growth, the question arises as to which fracture parameter controls the crack growth rates under the mixed-mode conditions that are likely to prevail. This question was examined[33] in terms of the locally measured crack flank displacements for the case of interfacial crack growth, in the glass–'Solithane' joint that was described in Sections 4.3.3 and 4.3.4.

Crack propagation rates were measured when the crack speed had reached a steady-state value following incrementation of boundary displacements, at which time the crack profile was also recorded. Thus, for each crack growth rate that was recorded, both the boundary displacements and the local crack opening displacements were determined. Two sets of experiments were conducted: in the first, bond-normal displacements were applied, keeping the bond-parallel displacements fixed at zero; the conditions were reversed in the second series of experiments.

The data relating the crack velocities to the applied boundary displacements are shown in Fig. 45. Note that the boundary displacements parallel to the bondline were an order of magnitude greater than those normal to the bondline to achieve the same crack growth rate. The results are in keeping with the general practice of designing bonded joints to take shear rather than normal loads. Also, since the studies on static cracks indicated that the shear loads gave rise to NCOD that were an order of magnitude smaller than those that were produced by similar levels of normal applied displacements, it seemed likely that it is the NCOD that controls mixed-mode interfacial crack growth. The crack opening interferometry measurements were used to examine this possibility.

The normal crack opening displacements at a distance of 2·5 μm (approximately 0·5% of the bondline thickness) under bond-normal and bond-tangential applied displacements are plotted against debond rates in Fig. 46. The data resulting from the two types of tests were not unified by this critical NCOD criterion; they were separated by a factor of about 2·5 on the ordinate which constitutes an insufficient improvement over the separation in Fig. 45. The critical profile gradient (or crack opening angle) criterion was distinguished from the NCOD criterion by a difference,

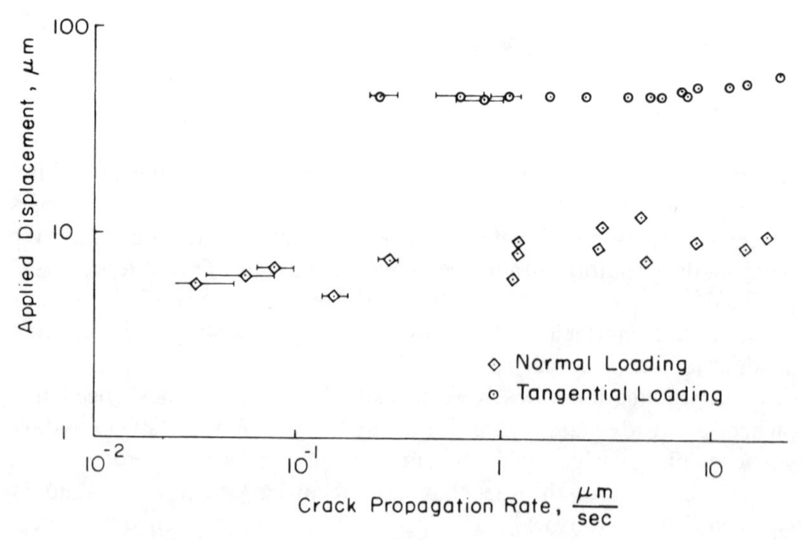

Fig. 45. Applied displacements vs crack propagation rate.

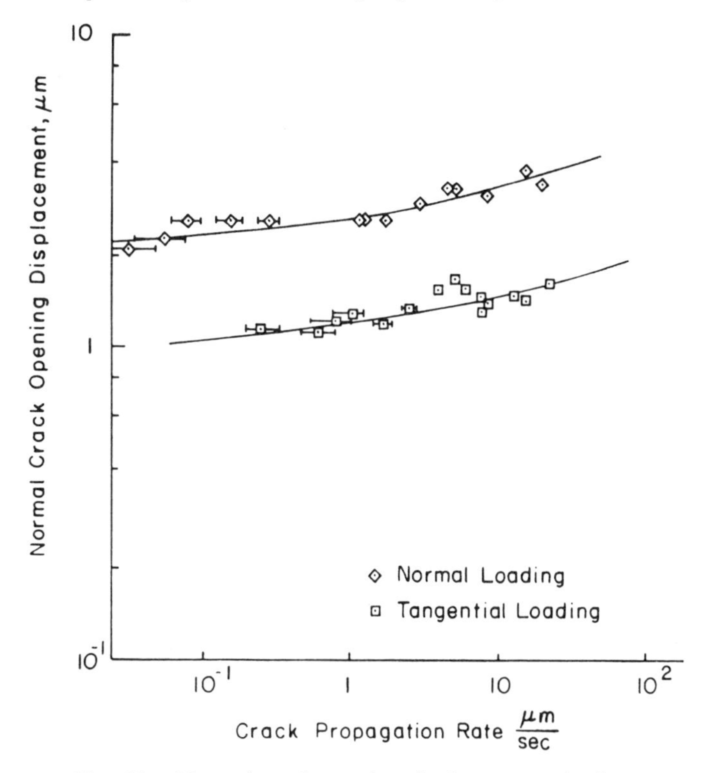

Fig. 46. Normal crack opening displacement criterion.

essentially a constant, which represented the distance from the crack front to the point of interest. Correspondingly, the data plotted virtually as those of Fig. 45, except for a shift in the ordinate.

Better agreement between the two data sets was achieved when the vectorial crack opening displacement (VCOD), made up by the vectorial addition of the NCOD and TCOD, was considered. Because of experimental limitations at the time, the TCOD could not be measured directly and were therefore estimated on the basis of linearised theory.[85] The result of these computations is shown in Fig. 47 where the two sets of crack propagation data differ by only 30%. This is an expected error that is reasonable in the light of the non-linear aspects of the problem that were discussed earlier. It is thus likely that interfacial crack growth rates are governed by the vectorial crack opening displacement. It was also noted that, in the limit of infinitesimal deformations, this displacement parameter

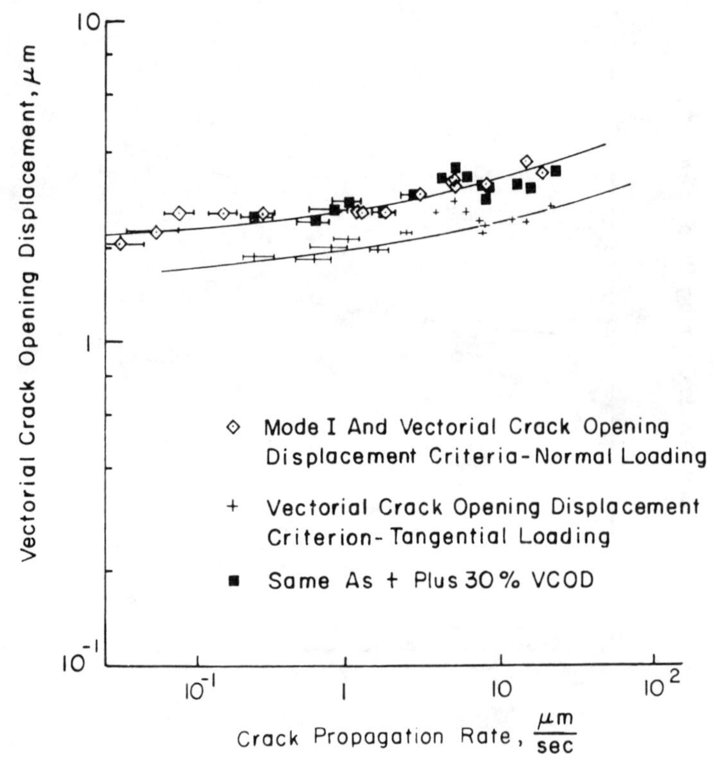

Fig. 47. Vectorial crack opening displacement criterion.

is directly proportional to the square root of the total energy release rate of linear (bond) fracture mechanics.

4.6. FAILURE MODES

In adhesively bonded composite joints, it is widely recognised that the adherend matrix material is more brittle than the state-of-the-art structural adhesives (Fig. 9, ref. 86). This leads to many situations (particularly in joints exhibiting more peel) where the adherends will delaminate before the adhesive fails. However, there are many situations in which debonding in the adhesive layer occurs.

Failure modes in bonded composite joints can be significantly different

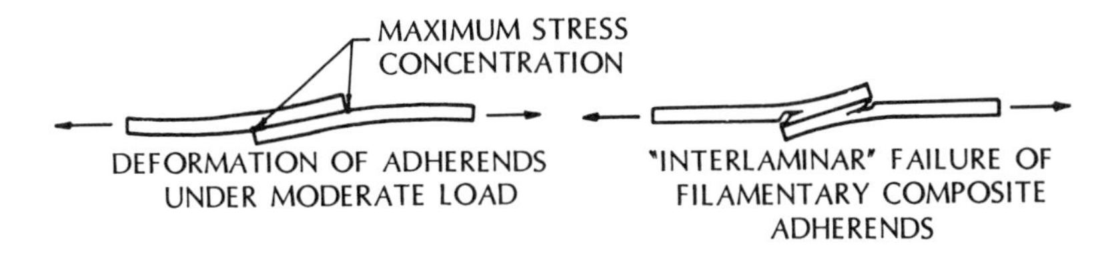

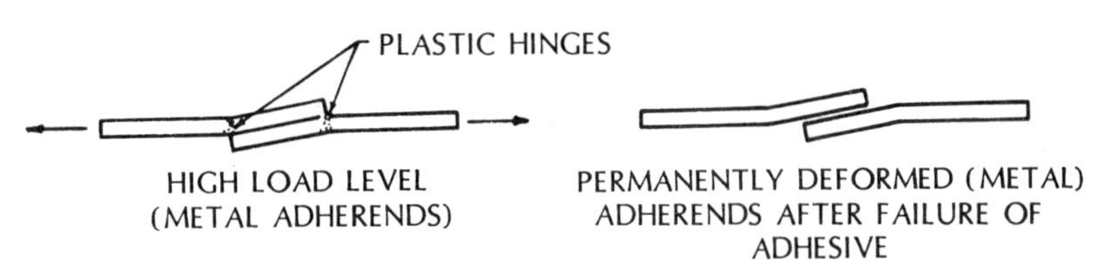

Fig. 48. Single lap bonded joints with eccentric load path; brittle and ductile adherends.[88]

from those in joints with metal adherends. A good example is the single lap shear (SLS) specimen. This specimen generates significant peel stresses in the adhesive and adherends at the end of the lap when pulled in tension. Since the adhesive is generally tougher than the matrix material in composites, the composite adherend fails by delamination. Figure 48 is from Hart-Smith.[88] Because the SLS specimen has a short over-lap and is unsymmetric, it produces peel stresses that are much higher than would be found in specimens with longer overlaps, such as the CLS specimen. (Failure modes of CLS specimens will be discussed later.) Therefore data generated for strength of bonded composite joints using the SLS specimen can be very misleading and practically useless for two reasons.

(1) The failure usually occurs in the composite and not in the adhesive bondline.
(2) Most practical bonded joints of composites would not be similar to the single lap. Therefore different stress states and probably different failure modes, would exist.

Guess *et al.*[13] have made an interesting study of the single lap shear specimen. Although their work dealt with metal adherends only, it should be read by anyone using the SLS specimen.

There are a number of similarities in joints fabricated from isotropic materials, such as metals, and those in which either one or both of the adherends consist of fibre-reinforced plastics. Several notable exceptions, however, manifest themselves in failure modes which are unique to FRP bonded joints. These are realised by recalling that laminated composites tend to be extremely weak in the out-of-plane directions for both tensile (peel) and shearing loads. Also, unless the laminate is quasi-isotropic, the in-plane strength varies with orientation. Other factors which also affect the stress distributions include the relative stiffness of the composite in bending, tension and interlaminar shear. The stacking sequence strongly affects the bending stiffness and also affects the decay of the interlaminar stress induced by the adhesive shear force.

In addition to the cohesive/adhesive failures in the adhesive layer, and net tensile failures of the adherend which are typical in joints with isotropic adherends, two other modes of failure are possible. For lap shear geometries, transverse cracking can occur when there is sufficient Poisson contraction mismatch. These occurrences are most prevalent in unidirectional material with non-identical adherends. A much more serious problem is the interlaminar failures which result from flaw growth between plies.

If one employs the simple shear lag model, by assuming that plane sections within an isotropic adherend remain plane, the shear stress distribution within each adherend is seen to decay in a linear fashion to load the entire thickness of the adherend. On the other hand, for layered materials, the interlaminar shear stress does not decay in a uniform manner. Instead, there are rather discrete drops in shear stress as high modulus plies are encountered. This implies that the stacking sequence will play a major role in the possibility and location of interlaminar failures.

A consequence of the low interlaminar strength is that composite adherends are normally thinner than those permissible in well balanced joints of isotropic materials. Thus one is able to visualise the basis for the generalisation that composite joints normally fail in the adherend rather than in the bond. In practical structures, scarf or step lap bonded joints provide a distribution of the load through the thickness of the composite adherend to avoid highly loaded ply interfaces, as well as providing a more uniform shear stress distribution along the bondline.[21]

Another feature of composite adherends is often the brittle nature of these materials. For single lap configurations, the induced bending moments can substantially increase the outer ply stresses. If failure of a load-bearing ply or group of plies bonded to the adhesive fails, the entire load must be carried by interlaminar shear to the adjacent plies. This shear stress may induce failure in an interlaminar fashion.

The previously discussed cyclic experimental tests using the CLS specimen had 0° fibres next to the adhesive. As a result of this, the static and cyclic debonding occurred within the adhesive, usually as a cohesive failure. Johnson and Mall[45] conducted tests using the same material systems discussed so far in this section, except for variations in the lay-ups of the composite adherends. These variations resulted in either 0°, 45°, or 90° plies being next to the bondline as indicated in Table 5. The purpose of this testing was to evaluate the influence of the interface ply orientation on the fatigue damage mode of the joint.

In all specimens with 0° and 45° interface plies, the fatigue damage initiated with cyclic debonding within the adhesive for both EC 3445 and FM-300. Thereafter, in all specimens with 0° interface plies, the debond grew in a cohesive manner within the adhesive region. A detailed investigation of the mechanics of this cyclic debond growth has already been discussed. For 45° interface plies, the debond grew in a combination of cyclic debonding in the adhesive and intraply failure in the ±45° plies of the strap adherend. The damage continued to grow in this mode for about 5–10 mm until it grew through the ±45° plies and reached the 0° ply. Thereafter, fatigue failure grew in the form of cyclic delamination between the −45° and 0° plies. Photographs of these modes are shown

TABLE 5
Lay-Ups Tested and Results

Adhesive	*Ply at interface*	*Lay-up strap/ Lay-up lap*	*Average minimum cyclic stress MPa*	G_{th}, *J/m²*
EC 3445	0°	$[0/\pm45/90]_s/[0/\pm45/90]_{2s}$	78	38
	45°	$[\pm45/0/90]_{2s}/[\pm45/0/90]_{2s}$	70	42
	90°	$[90/\pm45/0]_{2s}/[90/\pm45/0]_{2s}$	62	
FM-300	0°	$[0/\pm45/90]_{2s}/[0/\pm45/90]_s$	123	87
	45°	$[\pm45/0/90]_{2s}/[\pm45/0/90]_{2s}$	106	96
	90°	$[90/\pm45/0]_{2s}/[90/\pm45/0]_{2s}$	62	

Fig. 49. Edge view of fatigue damage in the EC 3445 adhesive CLS specimen with $[\pm 45/0/90]_{2S}$ lay-up.

in Figs 49 and 50 for EC 3445 adhesive. Figure 49 shows this failure in a side view while Fig. 50 shows a radiograph in a front view.

For both adhesive systems in all specimens with 90° interface plies, fatigue failure initiated with transverse cracking in the 90° ply of the strap adherend, localised at the end of the lap adherend. The fatigue failure then grew as combined delamination and intraply failure between ±45° plies for about 5–10 mm until it reached the first 0° ply in the strap adherend. Thereafter, fatigue damage grew in the form of delamination between the −45° and 0° plies. This damage is shown in Figs 51 and 52 for a typical FM-300 specimen. Figure 51 shows the side view and Fig. 52 shows a radiograph of the front view of the fatigue damage. The damage modes are summarised in Table 6.

TABLE 6
Damage Modes for EC 3445 and FM-300 Adhesives in Cracked Lap Shear Specimens

Interface orientation	*Initiation*	*Early growth*	*Later growth*
0°	Adhesive debonding	Within adhesive	Within adhesive
45°	Adhesive debonding	Combination of in-adhesive and between ±45° plies	Delamination between −45° and 0° plies
90°	Cracks in 90° ply in strap adherend	Delamination and ply cracking in 90° and ±45° plies	Delamination between −45° and 0° plies

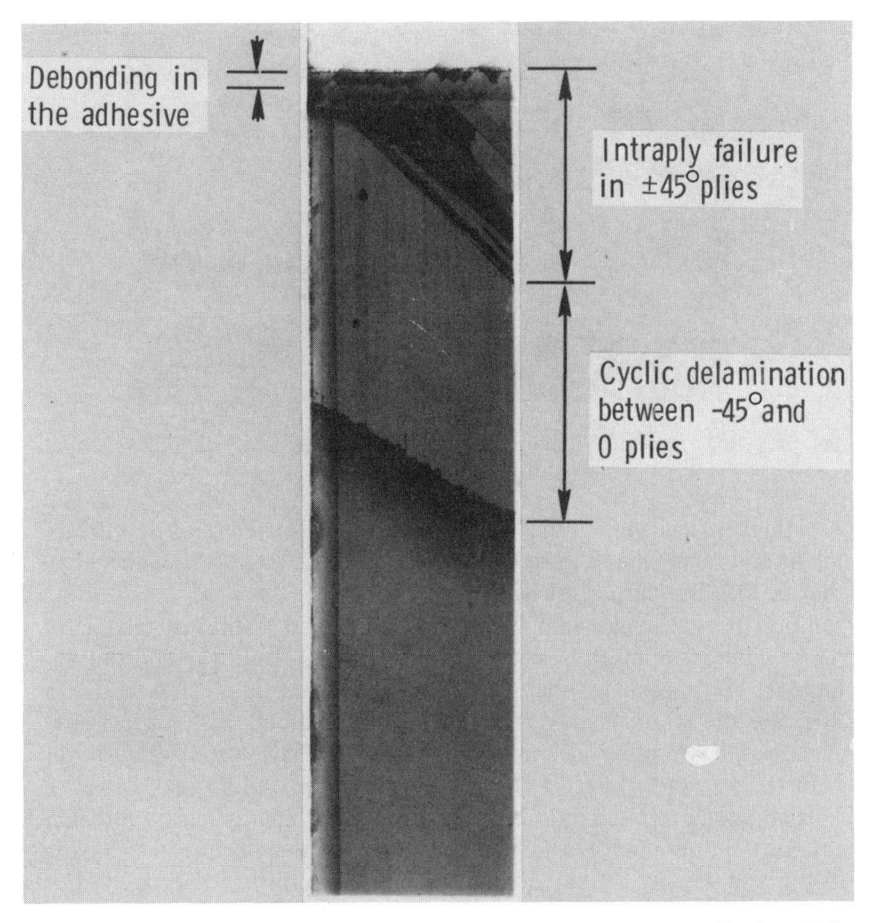

FIG. 50. Radiograph showing fatigue damage in the EC 3445 adhesive CLS specimen with $[\pm 45/0/90]_{2S}$ lay-up.

The minimum cyclic stresses that initiated adhesive debonding or adherend damage within 1 million cycles are plotted in Fig. 53. These test results are also presented in Table 5. Since the number of plies was the same for the specimens with 90° and 45° plies at the interface, the stress levels at damage initiation may be compared. The stress levels for damage initiation in the 90° interface plies are the same, for both adhesives, as damage initiated in the strap adherends for these specimens. For EC 3445 specimens, the 45° interface plies allowed a no-damage stress level

Fig. 51. Edge view of fatigue damage in the FM-300 adhesive CLS specimen with $[90/\pm 45/0]_{2S}$ lay-up.

13% higher than that of the 90° interface plies. For FM-300 specimens, the 45° interface plies allowed a no-damage stress level 71% higher than that of the 90° interface plies.

Since the specimens with 0° interface plies had a different geometry (i.e. a different number of plies in the adherends) from the 90° and 45° interface ply specimens, they cannot be compared on a stress basis. In this case the strain energy release rates were calculated, as previously explained, based upon the minimum cyclic stress at damage initiation for both the 45° and 0° specimens (values shown in Fig. 53 and Table 5). This approach resulted in a threshold total strain energy release rate, G_{th} equal to 38 J/m^2 and 42 J/m^2 for bonded systems with EC 3445 having 0° ply and 45° ply at the adhesive/adherend interface, respectively. For FM-300 the values of G_{th} were 87 J/m^2 and 96 J/m^2 for 0° and 45° plies, respectively. This variation in G_{th} obtained from specimens with 0° or 45° plies at the interface (for both adhesive systems) was within the observed experimental scatter band. Therefore, joints with 0° or 45° plies at the interface had practically equal strain energy release rate thresholds. This was expected since the fatigue failure mechanism was the same for both cases. Further, the same is expected for all joints with any ply orientation at the interface as long as cyclic debonding in the adhesive is the cause of fatigue damage initiation.

These results are significant because they indicate that 0° fibres do not have to be in the direction of maximum stress to achieve a maximum no-damage threshold stress level. This assumes that safe-life design is used

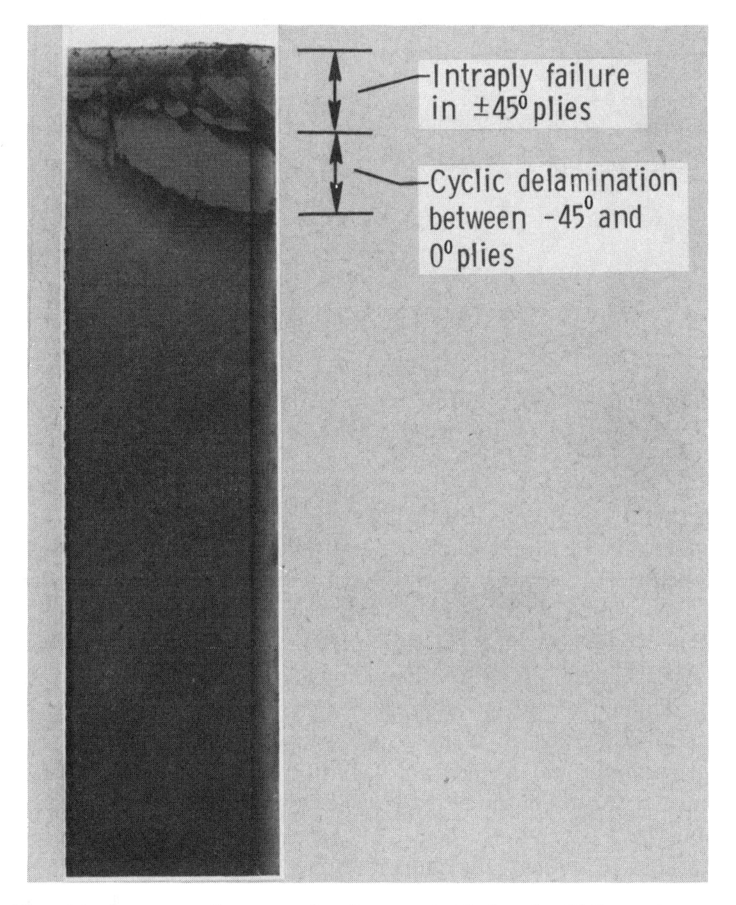

Fig. 52. Maximum cyclic stress for damage initiation in CLS specimens with various lay-ups ($R = 0{\cdot}1$).

based on maximum design stress levels below those required to cause G_{th} in the adhesive at an assumed initial defect.[46] Care must be exercised to assure that 90° (or near-90°) plies are not adjacent to the bondline in the direction of principal stress, or else delamination in the composite adherend will result unless lower design stress levels are used.

The effect of the interface ply orientation on the resulting G value of the adhesive was less than 3%. The stiffness and thickness of the total adherend were the primary factors. Therefore, essentially the same minimum cyclic stress for damage initiation would be expected for 0° or

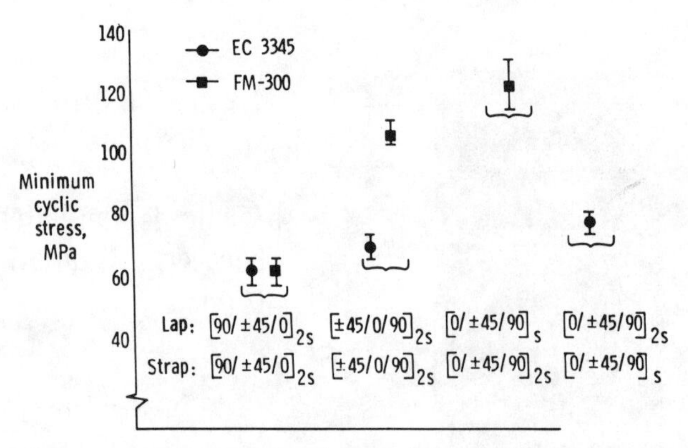

Fig. 53. Minimum cyclic stress for damage initiation in CLS specimens with various lay-ups ($R = 0{\cdot}1$).

45° interface plies if the same specimen geometry was used (i.e. the same number of plies in a quasi-isotropic laminate).

Apparently the strength of the 90° ply was so low that ply cracks developed below that stress required to create G_{th} in the adhesive. The static strength of a 90° T300/5208 lamina is listed as approximately 40 MPa in ref. 56, whilst the strength of a 0° lamina is approximately 1455 MPa. These strength values indicate how relatively easy it is to initiate damage in a 90° ply compared with a 0° ply. Once damage started in the strap adherend, the damage propagated in the form of ply cracking, intraply damage, and delamination until a 0° ply was reached. At that point the damage continued to spread by delamination between the 0° ply and the adjacent ply closest to the adhesive bondline, once again illustrating it is difficult for damage to grow past a 0° ply.

The debond initiation threshold values, G_{th} obtained in the present study are very close to the values of total strain energy release rate associated with the debond growth rate of 10^{-6} mm/cycle shown in Fig. 27. Thus, the G_{th} associated with a low crack growth rate (i.e. corresponding to 10^{-6} mm/cycle) can be used to predict or assess the durability of bonded joints in composite structures where fatigue would occur by debond initiation (as in the case of 0° or 45° interface plies).

4.7. SUMMARY

A number of factors have been considered in the experimental determination of the strength of adhesively bonded joints. Because joint failure is a debonding process that involves crack growth, the concept of strength has been defined in terms of fracture mechanics parameters. In cases where pre-existing flaws are present as manufacturing defects, bond strength may be defined in terms of the static fracture toughness for critical or catastrophic debonding, or a threshold toughness if cyclic or fatigue loading is anticipated. Time-dependent debonding due to viscoelastic and environmental effects may be characterised in terms of correlations of debond growth rates with a fracture parameter. At this stage, debond initiation in the absence of pre-existing flaws is accounted for by considering the effect of a small (1 mm) debond at potential initiation sites. Bond strength is then determined in the same way as was the case for pre-existing flaws.

The three specimens that have been suggested for the determination of fracture properties reflect the mixed-mode nature of debonding in adhesively bonded joints. Where purely Mode I effects are anticipated, the double cantilever beam specimen may be used. Various ratios of Mode I to Mode II conditions are provided by varying the relative stiffnesses of the adherends in the cracked lap shear family of specimens. Pure Mode II effects may be accounted for by making use of the end-notched flexure specimen. It is recommended that the fracture properties be determined on the basis of all three specimens because debonding in some (more brittle) adhesives may be governed by Mode I effects, whereas the total energy release rate seems to control debonding in tougher adhesive systems. At the present time, environmental effects are mainly addressed through the use of single lap shear specimens. Strength is characterised by time to failure, a measure which is useful for ranking purposes (all other factors being equal), but is not practical for rational design purposes. Much more useful data can be obtained (under Mode I conditions) by making use of the wedge test, especially if debond growth rates are correlated with Mode I fracture parameters. Mixed-mode effects may be examined by varying the relative stiffnesses of the adherends.

In some cases, fracture parameters in fracture property specimens may be determined on the basis of beam theory analyses. However, in other specimen geometries and structural applications, finite element stress analyses will be required for the determination of fracture parameters. Simple linear analyses are often sufficient but non-linear effects can arise, mainly as a result of large deflections of the joints, in which case geometrically non-linear analyses are called for. To date, material non-linear effects have been relatively small and linear elastic fracture mechanics concepts have been sufficient. This situation can, however, be expected to change as tougher adhesives are used and more fundamental studies of debond initiation are conducted.

In addition to the potential for non-linear effects, uncertainties in stress analyses may arise due to three-dimensional effects and mesh size sensitivities. Solution techniques should therefore at least be checked against existing solutions. Ideally, comparisons should also be made with measurements of joint deflections. In the simplest case, comparisons can be made on the basis of pointwise measurements of the relative displacements of adherends. More detailed comparisons may be effected by making use of optical interference techniques which have provided a number of insights into three-dimensional, non-linear and viscoelastic effects in cracked and uncracked adhesively bonded joints. Assumptions made in analytical and numerical models should also be checked with fractographic observations. Such observations yield valuable information as to failure modes, and should therefore form the basis of any subsequent analysis.

REFERENCES

1. Adams, R. D., in *Developments in Adhesives—2* (Ed. A. J. Kinloch), 1981, Applied Science Publishers, London, UK, p. 45.
2. Greenwood, L., Boag, T. R. and McLaren, A. S., in *Adhesion: Fundamentals and Practice* (Ed. A. S. McLaren), 1969, Butterworths, London, UK, p. 273.
3. Hart-Smith, L. J., in *Developments in Adhesives—2* (Ed. A. J. Kinloch), 1981, Applied Science Publishers, London, UK, p. 1.
4. Adams, R. D., Coppendale, J. and Pepiatt, N. A., in *Adhesion—2* (Ed. K. W. Allen), 1978, Applied Science Publishers, London, UK, p. 105.
5. Ripling, E. J., Corten, S. and Mostovoy, S., *J. Adhesion*, 1971, **3**, 107.
6. Mostovoy, S., Ripling, E. J. and Busch, C. F., *J. Adhesion*, 1971, **3**, 125.
7. Bascom, W. D., Timmons, C. O. and Jones, R. L., *J. Mater. Sci.*, 1975, **10**, 1037.
8. Wang, S. S., Mandell, J. F. and McGarry, F. S., *Int. J. Fracture*, 1978, **14**, 39.
9. Dattaguru, B., Everett, R. A., Jr, Whitcomb, J. D. and Johnson, W. S., *J. Eng. Mater. and Tech.*, 1984, **106**, 59.
10. Becker, E. B., Chambers, R. S., Collins, L. R., Knauss, W. G., Liechti, K. M. and Romanko, J., AFWAL-TR-84-4057, 1984, Air Force Wright Aeronautical Laboratories, Dayton, OH, USA.
11. McGavill, W. T. and Bell, J. P., *J. Adhesion*, 1974, **6**, 193.
12. Frazier, T. B., *Proc. Nat. SAMPE Tech. Conf.*, 1970, **2**, 71.
13. Guess, T. R., Allred, R. E. and Gerstle, F. P., Jr, *J. Testing and Evaluation*, 1977, **5**, 84.
14. Wilkins, D. J., NAV-GD-0037 (DTIC ADA-112474), Naval Systems Air Command, USA, 1981.

15. Brussat, T. R., Chiu, S. T. and Mostovoy, S., AFML-TR-1963, Air Force Materials Laboratory, Dayton, OH, USA, 1977.
16. Russell, A. J. and Street, K. N., in *Delamination and Debonding* (Ed. W. Johnson), 1985, American Society for Testing and Materials, ASTM STP 876.
17. Adams, R. D. and Wake, W. C., *Structural Adhesive Joints in Engineering*, 1984, Elsevier Applied Science Publishers, London, UK.
18. Brown, S. R., NADC-82270-60, 1983, Naval Air Development Center, USA.
19. Parker, B. M. and Waghorne, R. M., *Composites,* 1982, **13**, 280.
20. Sage, G. N. and Tiu, W. P., *Composites,* 1982, **13**, 228.
21. Bowditch, M. R. and Stannard, J. J., *Composites,* 1982, **13**, 298.
22. Anderson, G. P., Bennett, S. J. and DeVries, K. L., *Analysis and Testing of Adhesive Bonds,* 1977, Academic Press, New York, USA.
23. Anderson, G. P., Publication No. 84446, 1985, Morton Thiokol Inc.
24. Krieger, R. B., paper *Proc. 22nd Nat. SAMPE Symposium,* San Diego, CA, USA, April 1977.
25. Tuttle, M. E., Barthelemy, B. M. and Brinson, H. F., *Experimental Techniques,* 1984, **8**, 31.
26. Post, D., Czarnek, R., Wood, J., Joh, D. and Lubowinski, S., NASA Contractor Report 172474, 1984.
27. Post, D., *Optical Engineering*, 1985, **24**, 663.
28. Sommer, E., *Engng. Frac. Mech.,* 1970, **1**, 705.
29. Packman, P. F., in *Experimental Techniques in Fracture Mechanics* (Ed. A. Kobayashi), 1975, SESA Monograph Series, p. 59, Society for Experimental Mechanics, Connecticut, USA.
30. Patoniak, F. J., Grandt, A. F., Montulli, L. T. and Packman, P. F., *Engng. Frac. Mech.,* 1974, **6**, 663.
31. Crosley, P. B., Mostovoy, S. and Ripling, E. J., *Engng. Frac. Mech.,* 1971, **3**, 421.
32. Liechti, K. M. and Knauss, W. G., *Experimental Mechanics,* 1982, **22**, 262.
33. Liechti, K. M. and Knauss, W. G., *Experimental Mechanics,* 1982, **22**, 383.
34. Liechti, K. M. and Knauss, W. G., in *Advances in Aerospace Structures and Materials* (Ed. S. S. Wang and W. J. Renton), 1981, ASME ADOI, p. 51.
35. Romanko, J. and Knauss, W. G., *J. Adhesion,* 1980, **10**, 269.
36. Liechti, K. M., 1980, Ph.D. Dissertation, California Institute of Technology, Pasadena, CA, USA.
37. Sargent, J. P. and Ashbee, K. E. G., *J. Adhesion*, 1980, **11**, 175.
38. Fourney, M. E., *Proc. 4th Brazilian Congress of Mechanical Engineering,* 1977.
39. Liechti, K. M., *Experimental Mechanics,* 1984, **25**, 255.
40. Erdogan, F. and Gupta, G. D., *Int. J. Solids and Structures,* 1971, **7**, 1089.
41. Arin. K. and Erdogan, F., *Int. J. Eng. Sci.,* 1972, **10**, 115.
42. Anderson, G. P., De Vries, K. L. and Williams, M. L., *Int. J. Fracture*, 1974, **10**, 33.
43. Mall, S., Johnson, W. S. and Everett, R. A., Jr, *Adhesive Joints* (Ed. K. L. Mittal), 1984, Plenum Press, New York, USA, pp. 639–58.
44. Mall, S. and Johnson, W. S., ASTM STP 893, J. M. Whitney (Ed.), 1985, ASTM, Philadelphia, PA, USA.

45. Johnson, W. S. and Mall, S., TM 86443, 1985, NASA, Washington DC, USA.
46. Johnson, W. S. and Mall, S., in *Delamination and Debonding of Materials* (Ed. W. S. Johnson), 1985, ASTM STP 876, Philadelphia, PA, USA.
47. Shivakumar, K. N. and Crews, J. H., Jr, *Long-Term Behavior of Composites* (Ed. T. K. O'Brien), 1983, ASTM STP 813, ASTM, Philadelphia, PA, USA, p. 5.
48. Dattaguru, B., Everett, R. A., Jr, Whitcomb, J. D. and Johnson, W. S., *J. Engng Mater. Technol.*, 1984, **106,** 59.
49. Romanko, J. and Knauss, W. G., AFWAL-TR-80-4037, 1980, Air Force Wright Aeronautical Laboratories, Dayton, OH, USA.
50. Roderick, G. L., Everett, R. A., Jr and Crews, J. H., Jr, in *Fatigue of Composite Materials*, 1975, ASTM STP 569, ASTM, Philadelphia, PA, USA, p. 295.
51. Raju, I. S. and Crews, J. H., Jr, *Computers and Structures*, 1981, **14**, 21.
52. Rybicki, E. F. and Kanninen, M. F., *Engng Frac. Mech.*, 1977, **9**, 931.
53. Matthews, F. L., Kilty, P. F. and Godwin, E. W., *Composites*, 1982, **13**, 29.
54. Hart-Smith, L. J., CR-2218, 1974, NASA, Washington, DC, USA.
55. Anon., *Damage Tolerant Design Handbook*, 1982, Battelle Metals and Ceramics Information Center, Columbus, OH, USA.
56. Anon., *DOD/NASA Advanced Composite Design Guide,* Volume IV-A: Material, 1983, Air Force Wright Aeronautical Laboratories, Dayton, OH, USA.
57. Hart-Smith, L. J. and Bunin, B. L., Douglas Paper 7299, McDonnell Douglas Corp. Presented at *Sixth Conference on Fibrous Composites in Structural Design, New Orleans,* LA, 24–27 January 1983.
58. Dillard, D. A., Morris, D. H. and Brinson, H. F., ASTM STP 787, 1982, p. 357.
59. Wu, E. M. and Ruhmann, D. C., ASTM STP 580, 1975, p. 263.
60. Crossman, F. W. and Flaggs, D. L., Lockheed Missiles and Space Center—D633086, 1978.
61. Schaffer, B. G. and Adams, D. F., *J. Appl. Mech.*, 1981, **48**, 859.
62. Heller, R. A., Thakker, A. B. and Arthur, C. E., ASTM STP 580, 1975, p. 298.
63. Yeow, Y. T., Ph.D. Dissertation, 1978, Virginia Polytechnic Institute and State University, Blacksburg, VA, USA.
64. Griffith, W. I., Morris, D. H. and Brinson, H. F., VPI-E-80-15, 1980, Virginia Polytechnic Institute and State University Report, Blacksburg, VA, USA.
65. Findley, W. N. and Peterson, D. B., *ASTM Proc.*, 1958, **58**, 84.
66. Dillard, D. A. and Brinson, H. F., *Proc. 1982 Joint Conf. on Experimental Mechanics,* 1982, Oahu, Hawaii, p. 102.
67. Lou, Y. C. and Schapery, R. A., *J. Composite Materials,* 1971, **5**, 208.
68. Tuttle, M. E. and Brinson, H. F., VPI-E-84-9, 1984, Virginia Polytechnic Institute and State University Report, Blacksburg, VA, USA.
69. Zhurkov, S. N., *Int. J. Frac. Mech.*, 1965, **1**, 311.
70. Bruller, O. S., *Polym. Engng Sci.*, 1978, **18**, 42.
71. Hiel, C., Cardon, A. H. and Brinson, H. F., NASA Contractor Report 3772, 1984.

72. Dillard, D. A., Morris, D. H. and Brinson, H. F., *Proc. SESA Spring Meeting,* Dearborn, Michigan, 1981, p. 151.
73. Dillard, D. A. and Brinson, H. F., ASTM STP 813, 1983, p. 23.
74. Brinson, H. F., *Composites,* 1982, **13**, 377.
75. Schapery, R. A., *Polym. Engng Sci.,* 1969, **9**, 295.
76. Henriksen, M., *Computers and Structures*, 1984, **18**, 133.
77. Knauss, W. G. and Emri, I. J., *Computers and Structures*, 1981, **13,** 123.
78. Beckwith, S. W. and Wallace, B. D., *SAMPE Quarterly*, 1983, **14**, 38.
79. Myhre, S. H., Labor, J. D. and Aker, S. C., *Composites*, 1982, **13**, 289.
80. Butt, R. I. and Cotter, J. L., *J. Adhesion,* 1976, **8**, 11.
81. Minford, J. D., in *Durability of Structural Adhesives* (Ed. A. J. Kinloch), 1983, Applied Science Publishers, London, p. 135.
82. Adamson, M. J., *J. Mater. Sci.,* 1980, **15**, 1736.
83. Labor, J. D., NADA-79171-60, 1981, Northrop Corporation, Aircraft Division, USA.
84. Marceau, J. A., McMillan, J. C. and Scardino, W. M., *Proc. 22nd Nat. SAMPE Symposium,* San Diego, CA, USA, April 1977.
85. Romanko, J., Liechti, K. M. and Knauss, W. G., AFWAL-TR-82-4139, 1982, Air Force Wright Aeronautical Laboratories, Dayton, OH, USA.
86. Rice, J. R. and Shih, G. C., *J. Appl. Mech.,* 1965, **32**, 418.
87. Johnson, W. S. and Mangalgiri, P. D., TM 87571, 1985, NASA, Washington DC, USA.
88. Hart-Smith, L. J., Contract F33615-80-C-5092, Quarterly Progress Report No. 8, 1982, Air Force Wright Aeronautical Laboratory, Dayton, OH, USA.

Chapter 5

Theoretical Stress Analysis of Adhesively Bonded Joints

R. D. ADAMS

Department of Mechanical Engineering, University of Bristol, UK

5.1. INTRODUCTION

A major advantage of adhesive bonding is that it enables dissimilar materials to be joined, even when one or both of these is non-metallic. A major application of bonding is therefore where composite materials are concerned. Composites take many forms: in aerospace applications, these materials usually consist of highly aligned layers of carbon or glass fibres, each oriented to accommodate the expected loads. Some aerospace composites are woven or stitched so that the fibres are not perfectly aligned. High-quality chemical plant may be made from satin weave glass fibre-reinforced polyester or epoxy resin, whilst lower grade composites usually consist of random glass fibres in polyester. Thus, composites can be highly anisotropic in respect of both stiffness and strength and, although a unidirectional composite may be very strong and stiff in the fibre direction, its transverse and shear properties may be very much poorer. Bolts and rivets can sometimes be used with composites, but it is then often necessary to have load-spreading inserts *bonded* into the structure. Adhesive bonding is attractive as it reduces the localised stresses encountered when using bolts and rivets. A further form of bonding is the so-called co-curing technique in which the composite is bonded without the use of an adhesive. To do this, the composite is prepared in its uncured (pre-impregnated fibre) form and heat and pressure are applied. As the composite cures, the excess matrix material (usually a high-performance laminating epoxy) is squeezed out in a liquid form and contacts the adjacent component. When curing is complete, the bond is made with corresponding saving in production costs.

The techniques of analysis are essentially the same as when isotropic adherends are used, although due attention must be paid to the low longitudinal shear stiffness of unidirectional composites. As Demarkles[1] showed, even with metallic adherends in which the shear modulus is of the order of 25–30% of Young's modulus, it is necessary to take account of adherend shears. With unidirectional composites, this modulus ratio may be as low as 2%, and so the adherend shears become extremely important. The use of lamination techniques in which fibres are placed at different angles to the plate axis leads to reduced longitudinal and increased in-plane shear moduli. However, the *lateral* modulus (i.e. through the thickness of the adherend) remains low, being only two or three times that of the matrix material. In addition, the lateral *strength* is low, usually being of the same order or less than that of the matrix. The composite matrix has to meet a variety of requirements, only one of which is strength. Also, it will have been formulated from the same basic family of materials as the adhesive, which has been largely chosen for its strength and ductility. Thus, if the joint experiences lateral (peel) loading, there is a strong likelihood that the composite will fail in through-thickness tension before the adhesive fails. Adhesive peel stresses should therefore be minimised where composite adherends are used, lest this leads to adherend failure.

Matthews *et al.*[2] have published a review on the strength of joints in fibre-reinforced plastics. They considered both theoretical and experimental results, although mainly the former. The reader is referred to this excellent article. In essence, they concluded that it is necessary to consider non-linear adhesive behaviour if joint strength is to be predicted. Joint strength is improved as the adherend stiffness increases and the adhesive stiffness decreases, and a ductile adhesive is always preferable to a brittle one.

A rational basis of structural engineering design must be based on the ability to predict the loads and hence stresses which are likely to be encountered in practice. The loading system is prescribed by the function, and the skill of an engineer is to use the best available materials and design techniques to arrive at a suitable and cost-effective solution. However, advances in technology, together with new and more demanding environments, have resulted in a strong emphasis being placed in modern engineering on the need to quantify the structural loads and stresses.

However, let us be under no illusions. It is no easier to predict the strength of a joint from first principles than it is to predict the fatigue life of a steel bar by knowing its crystal structure and the strength of the crystals. Any theoretical predictions must be justified by experiment. There has to be a limit, or rather a collection of limits, set for a given situation. For instance, for a joint which is often used, testing and experience may be better than any analysis. However, an expensive, one-off design in a critical component would repay a careful analysis before manufacture commences.

As has been pointed out above, it is necessary to carry out some form of mathematical analysis so that strength predictions can be made and experimental data interpreted.

The method most attractive to a mathematician is to set up a series of differential equations to describe the state of stress and strain in a joint. Then, by using stress functions or other methods, closed-form algebraic solutions may be obtained. In the simple case where the adhesive and the adherends are all elastic, it may be possible to devise a solution for given boundary conditions. These methods become increasingly more difficult to use as non-linearities arise. Causes of non-linearity vary from joint rotation to material plasticity. However, by making various simplifications, it is possible to produce solutions which can readily be used to vary geometric and material parameters to cover a range of cases. Thus, once the solution has been obtained, a parametric study may be carried out with great efficiency, *provided the limits of the simplifications made are borne in mind.*

Modern digital computers have led to the use of a variety of numerical techniques for solving problems of mechanics, fluids, thermodynamics and so on. In one of these methods applied to adhesive joints, the structure is split into a series of small parts (finite elements), each of which obeys the prescribed material behaviour and each of which interacts regularly with its neighbours in terms of force continuity and displacement compatibility. The finite element technique (FET) is very powerful and can be efficiently used in that small elements need only be employed where there are large stress gradients. However, the big problem is that each solution applies only to a given set of parameters. If it is necessary to vary joint length, adhesive thickness, adhesive modulus, and so on, a new computer run is required each time. Whilst it is possible to use mesh generation programs to alter the geometry, and whilst changing moduli is very simple, the cost in computing time for a parametric study is not insignificant. Because very large stress gradients, approaching singularities, can exist in typical joints, it is necessary to use very fine meshes. Thus, the computing power required is large and the cost of a parametric study unattractive, particularly when non-linear behaviour is included. However, for investigations into the mechanics of real joints, there is no real substitute for finite element techniques. Also, where one-off joints are concerned, it is probably much more efficient than trying to evolve a closed-form solution.

This chapter deals with the stresses in lap joints, particularly when these are made with composite materials. Firstly, closed-form analyses, such as those by Volkersen,[3] and Goland and Reissner[4] are considered, then the results obtained from finite element analyses are discussed.

5.2. LINEAR ELASTIC ANALYSIS

5.2.1. Volkersen's analysis

Two sheets joined together with an overlap form one of the most common joints encountered in practice. Joints of this type made from thin (usually aluminium) sheet 25 mm wide, 12 mm long (1 in × 0·5 in) and 1·6 mm thick have long been used for quality control. The joint is easy to make and the results are sensitive to both adhesive quality and adherend surface preparation.

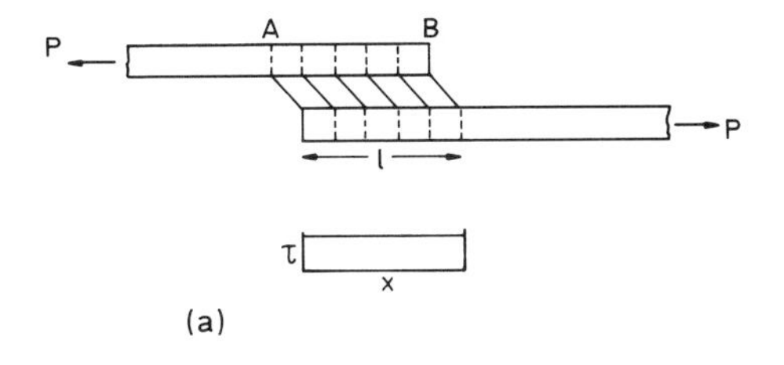

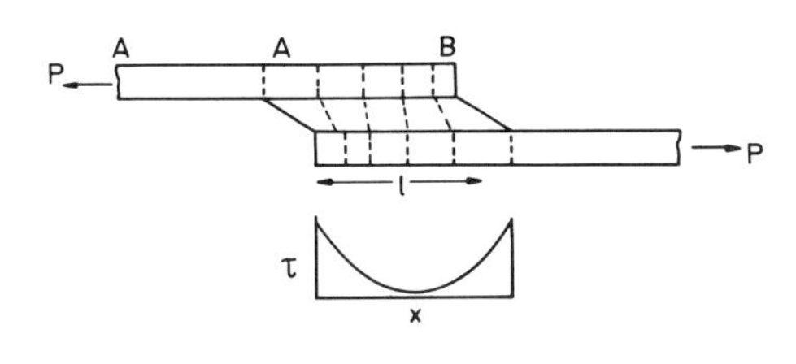

Fig. 1. Exaggerated deformations in loaded single lap joint: (a) with rigid adherends; (b) with elastic adherends.

The simplest analysis considers the adherends to be rigid and the adhesive to deform only in shear. This is shown in Fig. 1(a). If the width of the joint is b, the length l, and the load P, then the shear stress τ is given by:

$$\tau = \frac{P}{bl} \tag{5.1}$$

In Fig. 1(b) a similar joint is shown but the adherends are now elastic. For the upper adherend, the tensile stress is a maximum at A and falls to zero at B. Thus, the tensile strain at A is larger than that at B and this strain must progressively reduce over the length l. The converse is true for the lower adherend. Thus, assuming continuity of the adhesive/adherend interface, the uniformly sheared parallelograms of adhesive shown in Fig. 1(a) become distorted to the shapes given in Fig. 1(b). This leads to differential shear, which is the problem Volkersen[3] analysed. It is instructive to study Volkersen's analysis first and then to add to it the subsequent, more refined analyses, and to see what effect they have.

In Volkersen's 'shear lag' analysis it is assumed that the adhesive deforms only in shear, while the adherend deforms only in tension.

By setting up the differential equations of linear elasticity, Volkersen showed that the ratio of the shear stress τ_x at any position x along the joint, measured from one edge, to the average applied shear stress ($\tau_m = P/bl$) is given by:

$$\frac{\tau_x}{\tau_m} = \frac{\omega}{2} \frac{\cosh \omega X}{\sinh \omega/2} + \left(\frac{\psi - 1}{\psi + 1}\right) \frac{\omega}{2} \frac{\sinh \omega X}{\cosh \omega/2} \tag{5.2}$$

where

$$\omega^2 = (1 + \psi)\phi$$

$$\psi = t_1/t_2$$

$$\phi = \frac{Gl^2}{Et_1t_3}$$

$$X = x/l$$

G = shear modulus of the adhesive

E = Young's modulus of the adherends

t_1, t_2 = thickness of the adherends

t_3 = thickness of the adhesive

l = length of the joint

If the adherends are of equal thickness, $t_1 = t_2 = t$, $\psi = 1$ and $\omega = \sqrt{(2\phi)}$.

The maximum adhesive shear stress occurs at the ends of the point and is

$$\frac{\tau_{max}}{\tau_m} = \sqrt{\frac{\phi}{2}} \coth \sqrt{\frac{\phi}{2}} \tag{5.3}$$

It is worth noting that for large overlap lengths the absolute value of the maximum shear stress, τ_{max}, is proportional to:

$$\sqrt{\left(\frac{G}{2Ett_3}\right)}$$

which is independent of overlap length.

The theory developed by Volkersen takes no account of two important factors. Firstly, the two forces P in Fig. 1 are not collinear. There will therefore be a bending moment applied to the joint in addition to the in-plane tension. Secondly, the adherends bend, allowing the joint to rotate. The rotation alters the direction of the load line in the region of the overlap, giving rise to a geometrically non-linear problem.

5.2.2. Goland and Reissner's analysis

Goland and Reissner[4] took bending into account by using a factor, k, which relates the bending moment on the adherend at the end of the

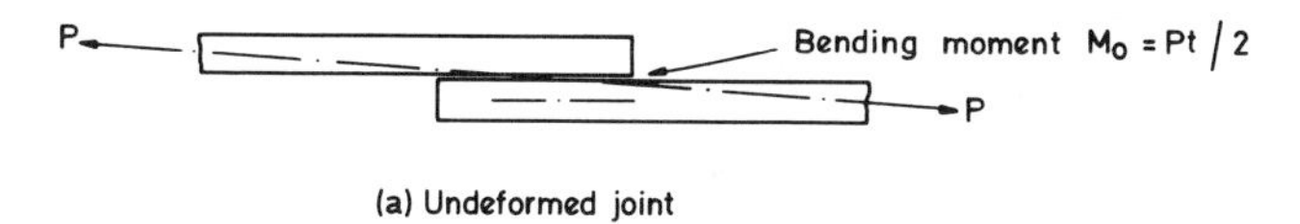

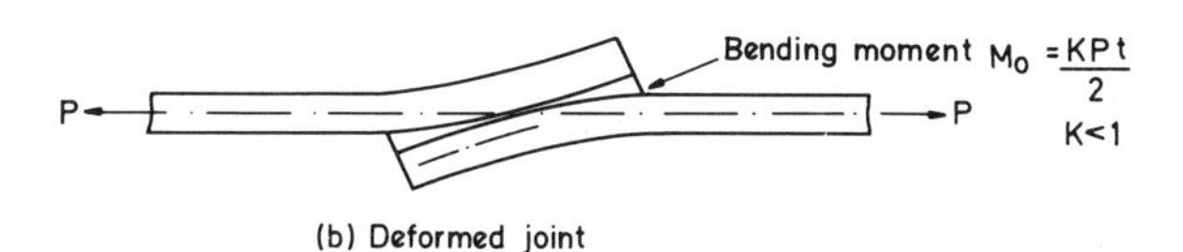

Fig. 2. A geometrical representation of the Goland and Reissner bending moment factor.

overlap, M_0, to the in-plane loading, by the relationship:

$$M_0 = kP\frac{t}{2} \tag{5.4}$$

where P is the applied load and t is the adherend thickness.

If the load on the joint is very small, no rotation of the overlap takes place, and the line of action of the load is as shown in Fig. 2(a), passing close to the edge of the adherends at the ends of the overlap. In this case, therefore, $M_0 \simeq Pt/2$ and $k \simeq 1{\cdot}0$. As the load is increased the overlap rotates, bringing the line of action of the load closer to the centreline of the adherends, as shown in Fig. 2(b), and thus reducing the value of the bending moment factor.

In their second theoretical approximation, Goland and Reissner treat the adhesive layer as an infinite number of shear springs and an infinite number of tension/compression springs in the y direction i.e. through-thickness direction (a description first explicitly used by Cornell[5]). Soluble differential equations, assuming plane strain, to describe both the shear stress (τ_{3x}) and the normal stress (σ_{3y}) distributions, can be derived (the subscript 3 denotes the adhesive).

Kutscha and Hofer[6] indicated that the solution for the normal stress, σ_{3y}, was incorrectly written in the original paper. The correct result was also given by Sneddon[7] and is:

$$\sigma_{3y} = \frac{\sigma t^2}{C^2 R_3}\left[\left(R_2\lambda^2\frac{k}{2} - \lambda k' \cosh\lambda\cos\lambda\right)\cosh\frac{\lambda x}{C}\cos\frac{\lambda x}{C} + \left(R_1\lambda^2\frac{k}{2} - \lambda k' \sinh\lambda\sin\lambda\right)\sinh\frac{\lambda x}{C}\sin\frac{\lambda x}{C}\right] \tag{5.5}$$

where

$$C = l/2$$

$$\lambda = \frac{C}{t}\left(\frac{6E_3 t}{E t_3}\right)^{1/4}$$

$$k' = k\frac{C}{t}\left(3(1-\nu^2)\frac{\sigma}{E}\right)^{1/2}$$

$$R_1 = \cosh\lambda\sin\lambda + \sinh\lambda\cos\lambda$$

$$R_2 = \sinh\lambda\cos\lambda - \cosh\lambda\sin\lambda$$

$$R_3 = (\sinh 2\lambda + \sin 2\lambda)/2$$

$$\nu = \text{Poisson's ratio}$$

Kutscha and Hofer noted that both the load on the joint and its width are not explicitly factorable from the functions for shear and normal stress (i.e. the applied adherend stress is not factorable). This is because of the change in value of the bending moment factor as the load is increased.

5.2.3. Effect of bending in a double lap joint

Although there is no net bending moment on a symmetrical double lap joint, as there is with a single lap joint, because the load is applied through the adhesive to the adherend plates away from their neutral axes, the double lap joint experiences internal bending, as shown diagrammatically in Fig. 3 (see also Fig. 2). In a symmetrical double lap, the centre adherend experiences no net bending moment, but the outer adherends bend, giving rise to tensile stresses across the adhesive layer at the end of

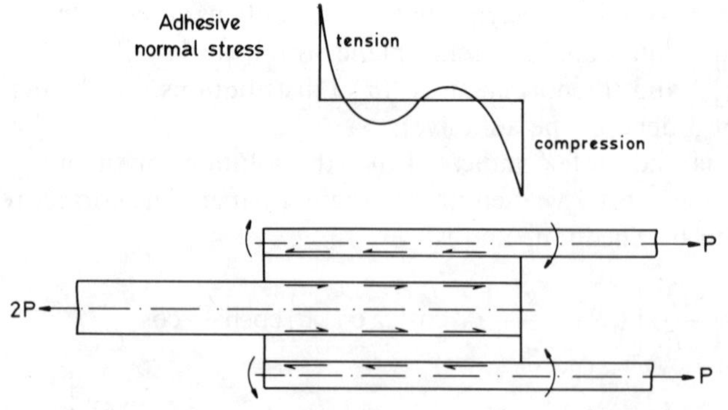

FIG. 3. Bending moments and induced stresses in the outer adherends of a double lap joint.

the overlap where they are not loaded and compressive stresses at the end where they are loaded, as shown in Fig. 3.

5.2.4. Later work

The classical work on Volkersen,[3] and Goland and Reissner[4] and other earlier workers was limited because the peel and shear stresses were assumed constant across the adhesive thickness, the shear was a maximum (and not zero) at the overlap end and the shear deformation of the adherends was neglected. Because the end face of the adhesive is a free surface, there can be no shear stress on it. Thus, by the law of complementary shears, the τ_x shear stress at the joint end must also be zero.

Recently, several authors, notably Renton and Vinson[8] and Allman[9] have produced analyses where the adherends have been modelled to account for bending, shear and normal stresses. They have also set the adhesive shear stress (τ_{3x}) to zero at the overlap ends. In addition, Allman has allowed for a linear variation of the peel stress across the adhesive thickness, although his adhesive shear stress is assumed constant through the thickness.

5.2.5. Transverse stresses

Adams and Peppiatt[10] showed that there exist significant stresses across the *width* of an adhesive joint. By photoelastic techniques, Hahn[11] showed that the shear stresses in the adhesive were highest at the corners. In this experiment, the adherends were allowed to bend, and the high stresses at the corner were thought to be caused by the anticlastic bending of the adherends.

Adams and Peppiatt[10] considered the existence of shear stresses in the adhesive layer and direct stresses in the adherends acting across the width at right-angles to the direction of the applied load, these stresses being caused by Poisson's ratio strains in the adherends. Adams and Peppiatt neglected the effects of bending (so that their results are more applicable to double rather than single lap joints) and treated the adhesive as an infinite number of shear springs. Thus, tearing and peeling stresses, together with longitudinal normal stresses in the adhesive, were also ignored. However, they took into consideration adherend shears, using the approach developed by Demarkles.[1] This was necessary since, for many practical joints, Goland and Reissner's criterion for neglecting adherend shear strains,

$$\frac{t_1 G_3}{t_3 G_1} \leqslant 0{\cdot}1$$

is not applicable. Here, t_1 and t_3 are the thicknesses of the adherend and adhesive respectively, while G_1 and G_3 are the corresponding shear moduli. For composites, G is usually very much smaller than for metallic adherends. This is especially true for unidirectional materials. For a typical adhesive joint with unidirectional adherends, $t_1 = 1{\cdot}5$ mm, $t_3 = 0{\cdot}25$ mm, $G_1 = 4{\cdot}0$ GPa, and $G_3 = 1{\cdot}0$ GPa, so

$$\frac{t_1 G_3}{t_3 G_1} = 1{\cdot}5$$

which is an order of magnitude larger than Goland and Reissner's limit.

Adams and Peppiatt[10] set up a series of differential equations for the longitudinal (x) and transverse (z) direct (σ) and shear (τ) stresses in a lap joint. They showed that

$$\frac{\partial^2 \sigma_{1x}}{\partial x^2} = K_a \sigma_{1x} - K_b \sigma_{1z} + C_a \tag{5.6}$$

and

$$\frac{\partial^2 \sigma_{1z}}{\partial z^2} = K_a \sigma_{1z} - K_b \sigma_{1x} + C_b \tag{5.7}$$

where

$$K_a = \frac{2G_1G_2G_3(E_1t_1 + E_2t_2)}{E_1t_1E_2t_2(t_1G_2G_3 + t_2G_1G_3 + 2t_3G_1G_2)}$$

$$K_b = \frac{2G_1G_2G_3(\nu_2E_1t_1 + \nu_1E_2t_2)}{E_1t_1E_2t_2(t_1G_2G_3 + t_2G_1G_3 + 2t_3G_1G_2)}$$

$$C_a = \frac{-2PG_1G_2G_3}{bt_1t_2E_2(t_1G_2G_3 + t_2G_1G_3 + 2t_3G_1G_2)}$$

$$C_b = \frac{2\nu_2PG_1G_2G_3}{bt_1t_2E_2(t_1G_2G_3 + t_2G_1G_3 + 2t_3G_1G_2)}$$

G = shear modulus

E = Young's modulus

ν = Poisson's ratio

t = thickness

P = applied load

The subscripts 1, 2 denote adherends and 3 indicates adhesive. The boundary conditions of this system are:

$$\text{At } x=0 \qquad \sigma_{1x}=\frac{P}{bt_1} \qquad \sigma_{2x}=0$$

$$\text{At } x=l \qquad \sigma_{2x}=\frac{P}{bt_2} \qquad \sigma_{1x}=0$$

$$\text{At } z=\pm\frac{b}{2} \qquad \sigma_{1z}=0 \qquad \sigma_{2x}=0$$

b being the joint width.

Unfortunately, there is no closed analytical solution for the above pair of simultaneous differential equations: a solution in series form may be derived but it would be cumbersome. However, an approximate analytical solution, which is exact at the boundaries, or a finite difference solution may be obtained. For the approximate analytical solution it is necessary to assume that σ_{1x} is constant with z and to neglect the term $K_b\sigma_{1z}$ which is small compared with $K_a\sigma_{1x}$. The second-order equation in σ_{1x} (eqn (5.6)) can then be written as:

$$\frac{\partial^2\sigma_{1x}}{\partial x^2}=K_a\sigma_{1x}+C_a \tag{5.8}$$

This is essentially the equation obtained by Volkersen,[3] Sazhin[12] and Demarkles,[1] the solution being:

$$\sigma_{1x}=\frac{P}{bt_1}\left[1-\psi(1-\cosh\alpha x)-\frac{(1-\psi(1-\cosh\alpha l))\sinh\alpha x}{\sinh\alpha l}\right] \tag{5.9}$$

where

$$\alpha=\sqrt{K_a} \qquad \psi=\frac{E_2t_2}{E_1t_1+E_2t_2}$$

Also,

$$\tau_{3x}=\frac{P\alpha}{b}\left[\frac{(1-\psi(1-\cosh\alpha l))\cosh\alpha x}{\sinh\alpha l}-\psi\sinh\alpha x\right] \tag{5.10}$$

These solutions for σ_{1x} and τ_{3x} are exact for the boundaries $z=\pm b/2$ because here $\sigma_{1z}=0$. We therefore have from eqn (5.7) that

$$\frac{\partial^2\sigma_{1z}}{\partial z^2}=\alpha^2\sigma_{1z}+C_b \tag{5.11}$$

giving the following solution for the normal transverse adherend stresses:

$$\sigma_{1z} = t_2(\nu_1\sigma_{1x}E_2 - \nu_2\sigma_{2x}E_1)\frac{\left[\cosh\left(\frac{\alpha b}{2}\right) - \cosh\alpha z\right]}{(t_1E_1 + t_2E_2)\cosh\left(\frac{\alpha b}{2}\right)} \tag{5.12}$$

and for the transverse shear stresses in the adhesive:

$$\tau_{3z} = \frac{t_1t_2\alpha(\nu_1\sigma_{1x}E_2 - \nu_2\sigma_{2x}E_1)\sinh\alpha z}{(t_1E_1 + t_2E_2)\cosh\left(\frac{\alpha b}{2}\right)} \tag{5.13}$$

When $x = 0$, $\sigma_{1x} = P/bt_1$ and $\sigma_{2x} = 0$; thus the equations for σ_{1z} and τ_{3z} are exact at the boundaries $x = 0$, and $x = l$.

The complete partial differential equations were also solved by a finite difference method, and the results are compared with the approximate analytical solution in Figs 4 and 5.

The results for a 25·4 mm × 25·4 mm (1 in × 1 in) lap joint show that the form of the adherend tensile stress and the adhesive shear stress (in

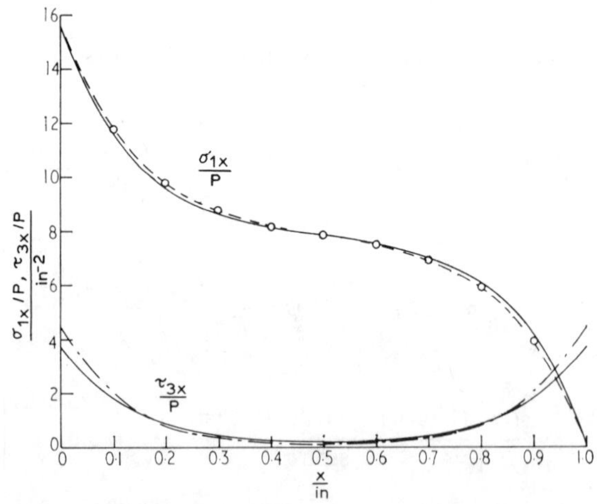

Fig. 4. Tensile stress in adherend and shear stress in adhesive (each per unit load) plotted against x for a 1 in × 1 in (25 mm × 25 mm) single lap joint (from Adams and Peppiatt):[10] ——, approximate analytical solution (exact at $z = \pm b/2$); ○, finite difference solution at joint centreline ($z = 0$); —·—·, approximate analytical solution neglecting adherend shears.

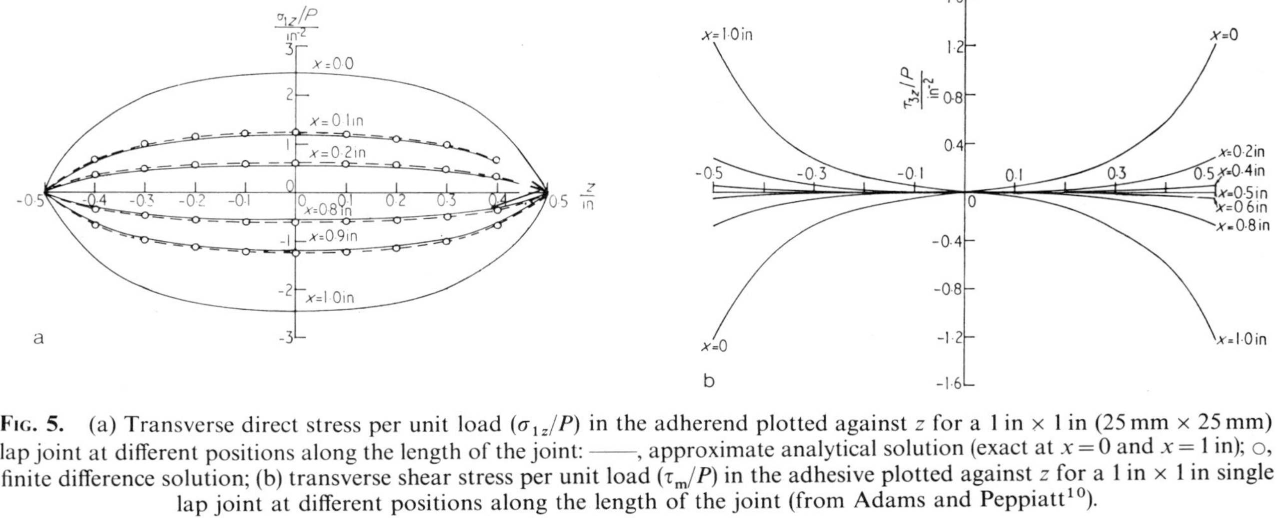

Fig. 5. (a) Transverse direct stress per unit load (σ_{1z}/P) in the adherend plotted against z for a 1 in × 1 in (25 mm × 25 mm) lap joint at different positions along the length of the joint: ——, approximate analytical solution (exact at $x = 0$ and $x = 1$ in); ○, finite difference solution; (b) transverse shear stress per unit load (τ_m/P) in the adhesive plotted against z for a 1 in × 1 in single lap joint at different positions along the length of the joint (from Adams and Peppiatt[10]).

the longitudinal direction) were much as would be expected from a Volkersen[3] type solution (Fig. 4). However, the transverse stresses show direct (tension or compression) and shear stress maxima at the ends of the joint (see Figs 5(a) and (b)).

Thus it is possible, by using closed-form analyses of varying complexity, to predict the stresses in simple lap joints. (This approach is termed continuum mechanics.) In many instances, such solutions may be deemed acceptable. However, two problems still remain to be solved if it is required to predict the *strength* of *real* joints. These may be summarised as *end effects* and *material non-linearity* (adhesive and adherend plasticity).

5.3. THE SINGLE LAP JOINT—END EFFECTS

One common result from all the closed-form analyses, whether they be simple or complicated, is that the maximum adhesive stresses always occur near the end of the bondline.

The closed-form algebraic lap joint analyses which have been discussed so far have all assumed that the adhesive layer ends in a square edge as is shown in Fig. 6(a). Coker,[13] experimentally (using photoelasticity), and Inglis,[14] analytically, have shown that a rectangular plate with shear loading on two opposite sides experiences high tensile and compressive stresses at its corners, the magnitude of these being about four times the applied shear stress, and the direction being at right angles to the sides on which the shear load is applied. These lateral direct stresses arise because the direct and shear stresses acting on the free surface must be zero. If the adhesive layer is assumed to have a square edge, it would be expected that similar tensile and compressive stresses must occur in the corners of this layer, because of the free surface.

Real structural adhesive joints do not have a square edge but have a fillet of adhesive spew (see Fig. 6(b)). The assumption that the adhesive layer has a square edge is thus unlikely to be realistic. Mylonas[15] has investigated the stresses induced at the end of an adhesive layer for a number of adhesive edge shapes using photoelastic techniques and has shown that the position of the maximum stress is dependent on the edge shape.

It was shown above that the existing closed-form solutions predict that the highest stresses should be near the ends of the joint. However, they do not take into account the influence on these stresses of the spew fillet which is formed at the ends, and so it is in just these regions of maximum

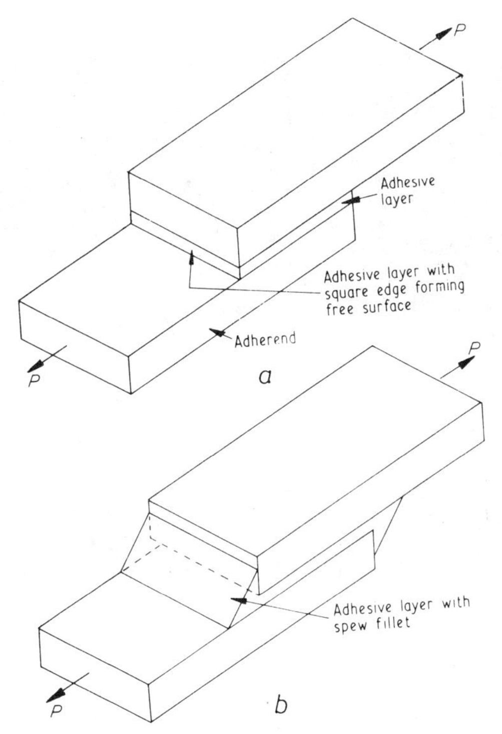

Fig. 6. Diagrammatic lap joints to show adhesive layers with (a) square edge; (b) spew fillet (from Adams and Peppiatt[16]).

stress and where failure is bound to occur that the assumed boundary conditions of the previous theories are the least representative of reality.

Because it is difficult to carry out experimental stress analysis on the adhesive layer in a typical joint, Adams and Peppiatt[16] constructed a model consisting of silicone rubber cast between two 'rigid' steel adherends. This model was based on the earlier experimental stress analyses by Adams *et al.*[17] using hard rubber for the adherends and soft (foam) rubber for the adhesive. The model is shown in its deformed state in Fig. 7(a), and good agreement is shown between the predicted and actual deflections. Figure 7(b) shows the principal stress pattern obtained by a finite element analysis. The length and direction of the lines represent respectively the magnitude and direction of the principal stresses at the centroid of each finite element. A bar at the end of the line implies a negative principal stress, i.e. compressive. It is evident that the presence

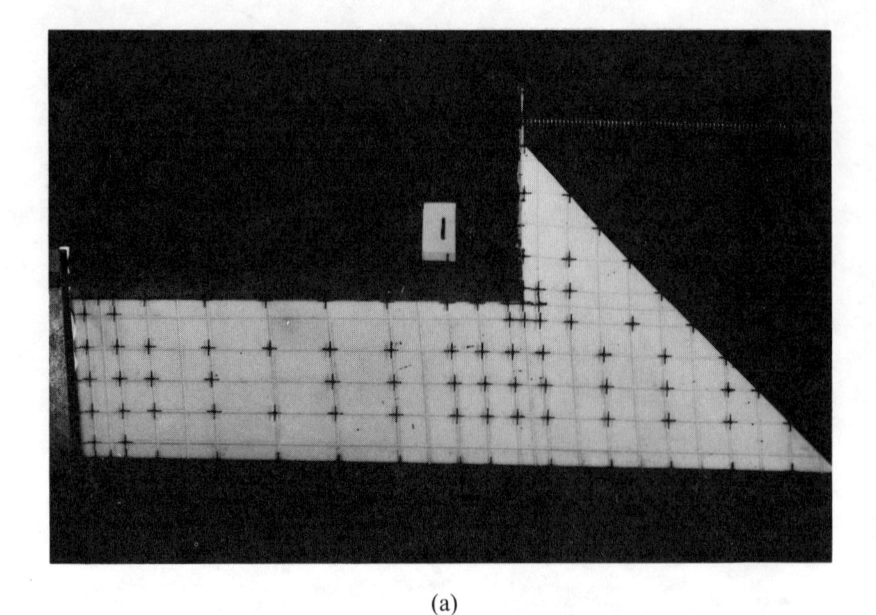

(a)

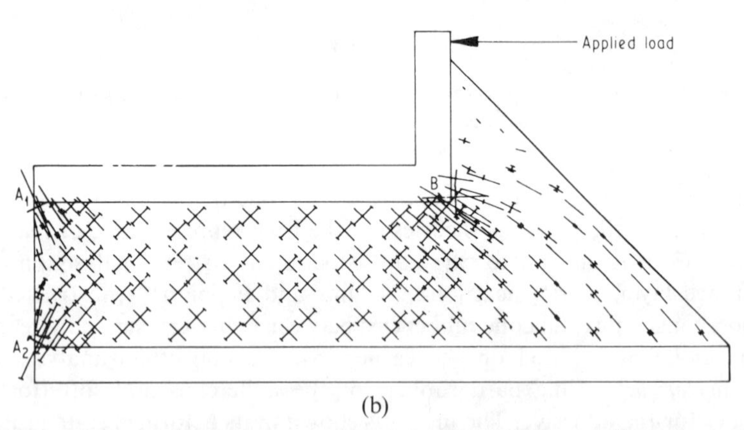

(b)

FIG. 7. (a) Comparison between calculated and experimental displacements of the silicone rubber model. (The black crosses are the finite element predictions of the intersections of the grid lines of the model.) (b) Principal stress pattern for silicone rubber model showing end effects (from Adams and Peppiatt[16]).

of the fillet causes the stress pattern to differ significantly from the pattern at the end with no fillet. At the points A_1 and A_2, the high tensile and compressive stresses predicted by Inglis[14] are shown, the absolute magnitude of the largest elemental principal stresses being at least 3·6 times the shear stress in the rubber between the plates. This is of similar size to the value predicted by Inglis, who says that the normal stress is more than four times as large as the applied shear stress. It should be noted that the rubber away from the ends of the steel plates is in pure shear, as is shown by the equal and opposite principal stresses in the elements in this region. The stresses in the fillet are predominantly tensile, the maximum principal tensile stress being at least 3·5 times the shear stress in the rubber between the plates.

However, it can be clearly seen that the presence of the fillet causes very large changes in the adhesive stress distribution in the highly stressed region at the joint end. Investigation of the stresses is *only* possible if finite element methods are used. No other modelling technique can *accurately* represent the stresses and hence be used to predict joint strength.

5.4 THE SINGLE LAP JOINT—ELASTO–PLASTIC ANALYSIS

5.4.1. General

Complicated mathematics is required if the stress situation in a single adhesive lap joint is to be determined precisely. However, end effects, particularly where a spew fillet is involved, are the most difficult to model accurately, whilst this is the most critical region since failure almost always occurs here. But even if this modelling could be done, we have to consider what happens to the stress distribution when the adhesive can *yield*. Some of the new adhesives are so strong that metal adherends may also be caused to yield. Physically, two opposite effects occur when this happens. In Fig. 1(b) the effect of adherend differential straining was shown to cause adhesive shear stress peaks towards the bondline end. If the adherends yield at their loaded end, the differential straining is enhanced and so the adhesive stresses will be increased, leading to premature failure. However, if the adherends are stressed to the yield point, they will more easily rotate under the effect of the non-collinear applied loads. This causes the Goland and Reissner joint factor k to decrease more than if the adherends remained elastic, thus reducing the stresses. Thus, not only adhesive plasticity but where appropriate adherend plasticity also needs to be considered.

There have been two basic approaches to studying the stresses in lap joints when plasticity occurs. The first of these to be considered is based on the use of continuum mechanics, while the second uses finite element techniques.

5.4.2. Continuum methods

The main advocate of continuum mechanics is L. J. Hart-Smith, who has produced an enormous amount of work on this subject for the PABST programme (Primary Adhesively Bonded Structure Technology) under contract to the USAF Flight Dynamics Laboratory. This method is a development of the shear lag analysis of Volkersen[3] and the two theories of Goland and Reissner.[4] While Hart-Smith has published many papers on the analysis of joints (see, for example, refs 18–24) the reader is referred particularly to his excellent review.[25] The design philosophy behind Hart-Smith's work is that the adhesive should never be the weak link. Thus, if peel stresses are likely to occur, they should be alleviated by tapering the adherends (scarfing) or by locally thickening the adhesive layer. Hart-Smith's continuum mechanics approach has the advantage over the finite element technique in that it allows a parametric investigation concerning the effects of glueline thickness, joint length, and so on, together with adherend and adhesive mechanical properties, to be carried out at low cost.

Since Hart-Smith has contributed a chapter (Chapter 7) to the design of joints in this text, only a brief account of his work will be given here, but it is included for completeness.

The first difficulty is how to characterise the adhesive. Hart-Smith chose an elastic–plastic model (see Fig. 8) such that the total area under the stress–strain curve was equal to that under the true stress–strain curve. If the maximum stress is less than yield, the true elastic curve may be used while, for a peak stress intermediate between yield and failure, a different, and more accurate, model is chosen. The bilinear model is closer to the true adhesive characteristic over its entire range of loads, so that a single model can be used for calculating the stresses without having first to calculate these to establish which intermediate elastic–plastic model should be used! It is argued that the model or models to be adopted for a given situation will be a compromise between precision and convenience.

Hart-Smith has developed computer programs for analysing various joints, double lap and single lap, with equal or dissimilar adherends, for

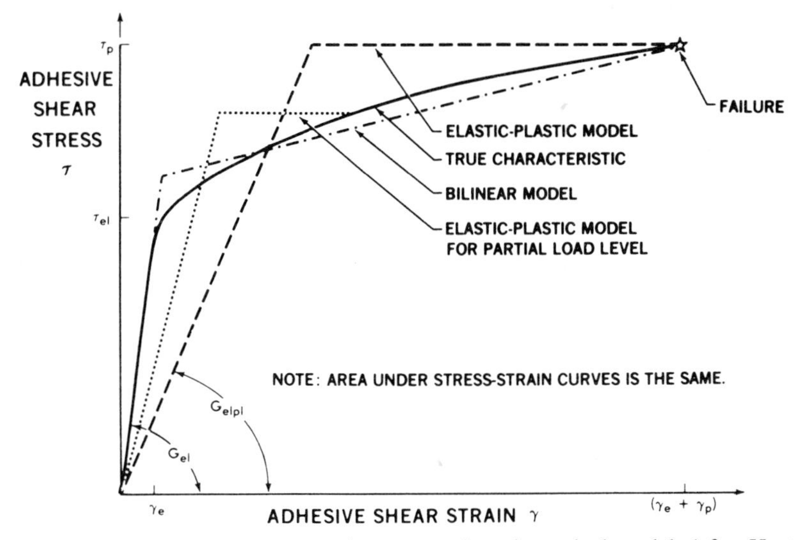

Fig. 8. Adhesive shear stress–strain curve and mathematical models (after Hart-Smith[25]).

parallel, stepped, scarf and double straps. Similar programs are available from ESDU (Engineering Sciences Data Unit, 251–9 Regent Street, London, UK) for elastic and elastic–plastic calculations. The ESDU programs were developed from the work of P. Grant[26,27] of British Aerospace and are again based on the work by Volkersen[3] and Goland and Reissner.[4] In the ESDU/Grant method, the adhesive is modelled as shown in Fig. 9 with the following conditions:

$$\text{If } \gamma < \gamma_e \qquad \text{then } \tau = \gamma G_e$$

$$\text{If } \gamma > \gamma_e \qquad \text{then } \tau = \tau_e + \left(\frac{\alpha\beta}{\alpha + \beta}\right)$$

where $\alpha = \gamma G_e - \tau_e$ and $\beta = \tau_{max} - \tau_e$.

For single lap joints, where peel stresses are significant, especially at the ends of the overlap, the PABST philosophy is to reduce them by tapering the adherends and by increasing the overlap. For instance, the ratio of overlap length to adherend thickness in a typical laboratory lap shear test (0·5in long × 0·063 in thick; 12·5 mm × 1·6 mm) is about 8:1 and such joints usually experience adherend yielding prior to bond failure. In the PABST programme, a ratio of 80:1 was chosen: weight for weight,

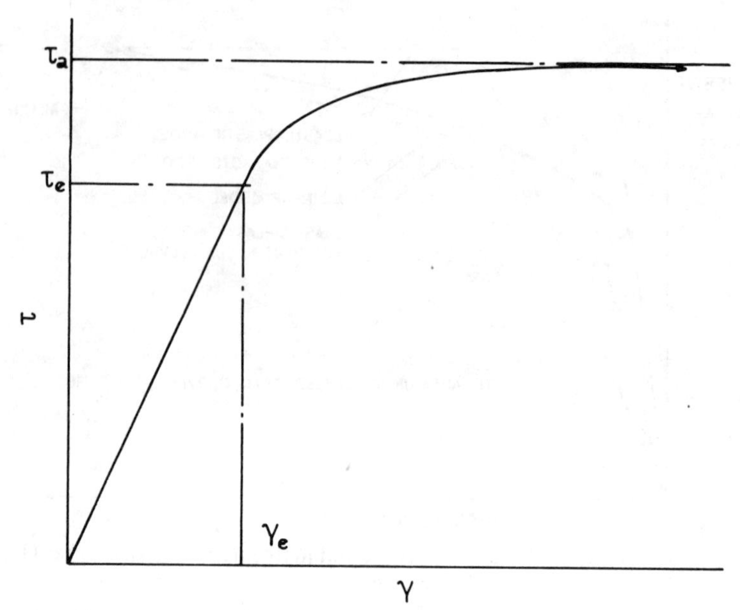

FIG. 9. Theoretical adhesive stress–strain curve (ESDU/Grant model) (from ESDU 79016[28]).

these joints were only 10% weaker than a double lap joint, and much easier to manufacture and inspect.

It should be borne in mind that the PABST/Hart-Smith and the ESDU/Grant design philosophies were developed for a specific application, that of aircraft construction, and may not be generally applicable to other aspects of the engineering usage of adhesives. However, they do represent a major and a successful input to the design of adhesive lap joints.

5.4.3. Finite element methods

When we consider non-linear material properties by a closed-form analysis such as Hart-Smith's, the limitation is the tractability of a realistic mathematical model of the stress–strain curve within an algebraic solution. With the finite element techniques developed for adhesive joints by Adams and his coworkers, the limit becomes that of computing power. The high elastic stress and strain gradients at the ends of the adhesive layer need to be accommodated by three or four eight-node quadrilateral elements across the thickness. However, consideration of non-linear material behaviour requires a much larger computing effort for any given element.

Thus, it becomes necessary to reduce the elements, perhaps to only one across the adhesive thickness, hence reducing the accuracy of the stress distribution. Apart from the computing, it is necessary first to define yield (of the adhesive usually, but it can also be the adherend) and then to adopt a suitable failure criterion.

Let us first consider the yield behaviour of an adhesive. Adams *et al.*[29] analysed double lap joints in which two adhesives of different strengths and strains to failure were used. It is, of course, widely accepted that the yield behaviour of many polymers, including epoxy resins, is dependent on both the hydrostatic and deviatoric stress components. A consequence of this phenomenon is the difference between the yield stress in uniaxial tension and compression. For epoxy resins, Sultan and McGarry[30] obtained ratios of compressive to tensile yield stresses of 1·28 and 1·35: Pick and Wronski[31] obtained a ratio of 1·33. Bowden and Jukes[32] showed that the ratio of compressive to tensile yield stress increased with the amount of plasticiser present in an epoxy. Tests on bulk specimens of AY103 and MY750 have given ratios of 1·27 and 1·14 respectively. This behaviour has been incorporated into the analysis by assuming a paraboloidal yield criterion of the form:

$$(\sigma_1 - \sigma_2)^2 + (\sigma_2 - \sigma_3)^2 + (\sigma_3 - \sigma_1)^2 + 2(|\sigma_c| - \sigma_T)(\sigma_1 + \sigma_2 + \sigma_3) = 2|\sigma_c|\sigma_T \qquad (5.14)$$

where σ_1, σ_2 and σ_3 are a combination of principal stresses causing yield and $|\sigma_c|$ and σ_T are the absolute values of the uniaxial compressive and tensile yield stresses. This type of yield criterion was proposed by Raghava *et al.*[33] and was shown by them to apply to several amorphous polymers over a wide range of stress states. It should be noted that when $|\sigma_c| = \sigma_T$ the paraboloidal yield criterion reduces to the more familiar von Mises cylindrical criterion.

5.5. THE EFFECT OF ADHEREND SHAPE—SCARFED, BEVELLED AND STEPPED ADHERENDS

There are several forms of the lap joint in which the adherends are not parallel-sided, constant-thickness sheets, but have a variety of forms in an attempt to reduce the high stress and strain concentrations which occur at the ends of the lap joint. In these profiled joints, the load line direction must change and, in addition to the tensile stiffness, so the shear and bending stiffnesses of the adherends change.

Whilst profiled adherends can be treated by closed-form techniques, the mathematics is necessarily more complicated than for parallel-sided joints. There have been many notable contributions to this field. Hart-Smith has produced several analyses for tapered and stepped lap joints (e.g. ref. 19), in which the adhesive is considered as elasto–plastic. Lubkin,[34] Wah[35] and Webber[36] have analysed scarf joints. Thamm[37] considered scarf and bevel joints in which the adherends were not tapered fully to an edge since this is very difficult to achieve in practice. He concluded that tapering was of little value unless a fine edge could be obtained. Even an edge reduced to 10% of the original thickness provides little advantage over a parallel lap joint and makes the additional mechanical operation of questionable value.

However, when using the finite element technique, there is no problem in dealing with the complexities of tapered and stepped joints, such as face the closed-form analytical mathematician. Various authors have tackled the problem. Among these, Barker and Hatt[38] used a rather coarse mesh but with a special element to represent the adhesive. Wright[39,40] used triangular elements and an elastic–plastic adhesive.

Essentially, scarfing helps to reduce the stress concentrations in joints but, as Thamm has said, it has to be done fully to a knife edge if it is to be effective. It is impossible to illustrate all the variables possible in a tapered joint; any further detail would be unnecessarily specific.

It has also been proposed that the adherends of a lap joint should be profiled so that the *adhesive thickness* may be varied along the length of the overlap while leaving the adherend thickness essentially constant. Adams *et al.*[17] showed that, by using a quadratic profile, the adhesive shear stress may be made approximately constant along the lap. However, for practical joints, the maximum amount of profiling is less than 0·1 mm, which is difficult to achieve in practice. A simpler, linear profile is possible, but again the problems of precision machining to such fine tolerances outweigh the likely benefits, particularly as further precision in joint alignment is required.

5.6. SOME RESULTS FROM FINITE ELEMENT ANALYSES

5.6.1. Introduction

Adams *et al.*[41] used finite element methods to examine the stress in high-performance composites in symmetrical lap joints with parallel, bevelled, scarfed and stepped adherends. The composite adherends were assumed

TABLE 1
Mechanical Properties of CFRP Adherends

Longitudinal Young's modulus, E_l (GN m^{-2})	135
Lateral Young's modulus, E_t (GN m^{-2})	7
Interlaminar (longitudinal) shear modulus, G_{lt} (GN m^{-2})	4·5
Longitudinal tensile strength (MN m^{-2})	1550
Lateral tensile strength (MN m^{-2})	50
Interlaminar shear strength (MN m^{-2})	110
Longitudinal Poisson's ratio, ν_{lt}	0·3
Lateral Poisson's ratio, ν_{tt}	0·3

TABLE 2
Mechanical Properties of Epoxy Adhesive

Young's modulus (GN m^{-2})	2·8
Poisson's ratio	0·4
Uniaxial tensile yield stress, Y_T (MN m^{-2})	65
Uniaxial compressive yield stress, Y_C (MN m^{-2})	84·5

to be linearly elastic type II carbon fibre-reinforced epoxy composites, with a 0·6% fibre volume fraction. The mechanical properties of this material are given in Table 1.

The adhesive was treated as elastic–plastic with a paraboloidal yield criterion as described earlier in this chapter. Its basic mechanical properties are given in Table 2.

Stress distributions in the adhesive and the adherend were then produced assuming that:

(1) the adhesive was perfectly elastic, or
(2) the maximum adhesive plastic strain was 10%, or
(3) the maximum adhesive plastic strain was 20%.

5.6.2. Linear analyses

The adhesive elastic shear stress distributions for single lap, double lap, double butt strap and double scarf joints between unidirectional type II CFRP adherends are shown in Fig. 10. The adhesive stress concentrations are less than those for joints between aluminium adherends of the same geometry because of the greater longitudinal stiffness of the CFRP adherends. However, as the tensile strength of the CFRP is more than four times the yield strength of the aluminium alloy, the joint efficiencies (defined as the ratio of the joint strength to that of the weakest adherend)

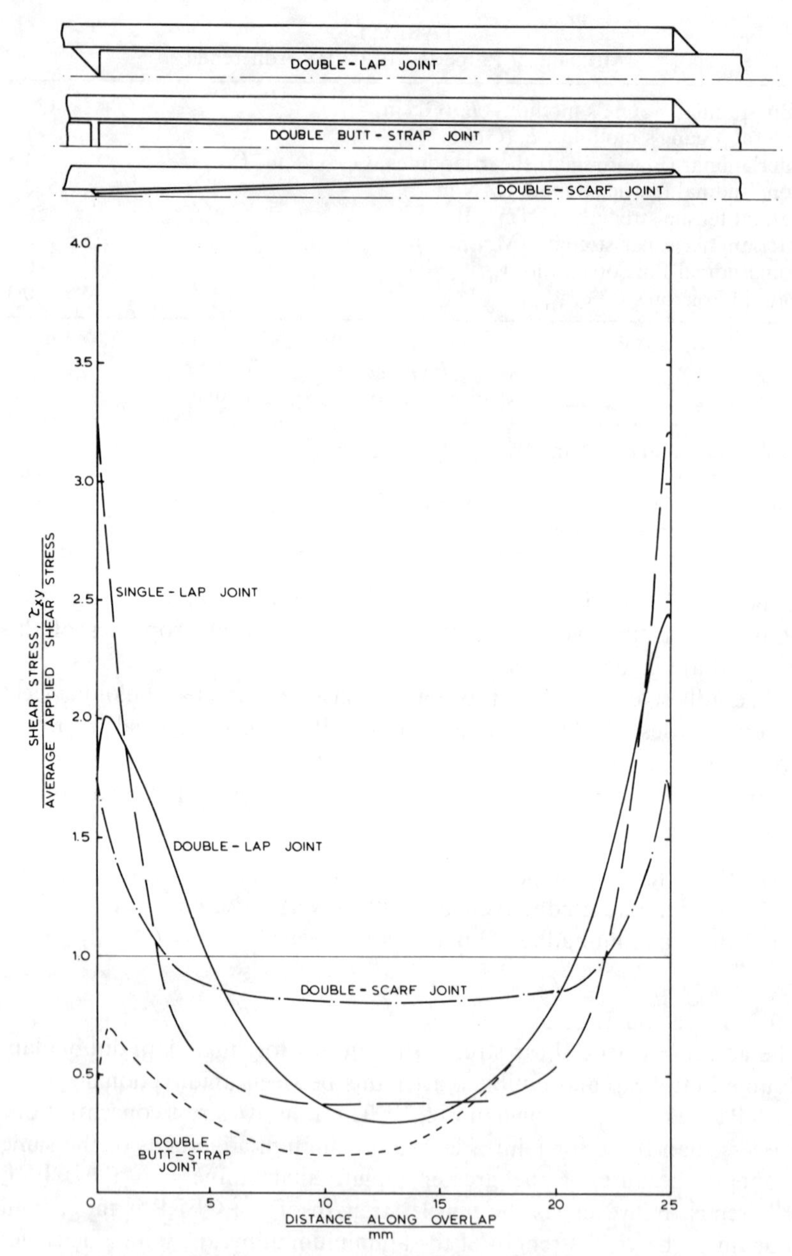

Fig. 10. Adhesive shear stress distributions in CFRP–CFRP joints.

predicted by the linear elastic analyses are much lower than those of the aluminium-to-aluminium joints. The stress concentration at the tension end of both the double lap joint and the double butt strap joint are the same, although in the double butt strap joint the highest stresses are predicted to occur in the adhesive at the butt face. As with the aluminium-to-aluminium joints, the effect of scarfing the adherends is to reduce the stress concentration and to increase the joint efficiency. The stress concentration is reduced to 77% of the parallel lap joint value in the case of the single lap. Comparison with the results for aluminium adherends suggests that there is less benefit to be obtained from scarfing unidirectional composite adherends than there is for isotropic materials such as metals.

As fibre-reinforced plastics have a comparatively low-strength epoxy resin matrix, adherend failure is more likely in composite joints than in metal-to-metal joints. There are three possible modes of failure in the composite:

(i) tensile failure in the fibre direction;
(ii) tensile failure perpendicular to the fibre direction;
(iii) interlaminar shear failure.

The strengths in these three modes were given above for a type II-S (treated) unidirectional carbon fibre composite. Rotem and Hashin[42] suggested that matrix failure depended on the combined effect of the tensile stress perpendicular to the fibre direction and the interlaminar shear stress, and that failure occurs when:

$$\left(\frac{\sigma_{22}}{\sigma_{\mathrm{T}}}\right)^2 + \left(\frac{\sigma_{12}}{\tau}\right)^2 \geq 1 \qquad (5.15)$$

where σ_{22} = tensile stress perpendicular to fibre direction, σ_{T} = lateral tensile strength of composite, σ_{12} = interlaminar shear stress and τ = interlaminar shear strength of composite.

The joint efficiencies calculated from each of these separate failure modes are given in Table 3 for a double lap joint. On the basis of a linear elastic analysis, it appears that adhesive failure is most likely to occur before failure in the adherends.

The tensile stress (σ_x) distributions in the CFRP adherends of a double lap joint are shown in Fig. 11.[43] The highest stress occurs at the surface of the centre adherend and is 1·3 times the stress in the adherend away from the joint, compared with 1·08 times in an aluminium-to-aluminium double lap joint. This is because the composite has a lower shear stiffness

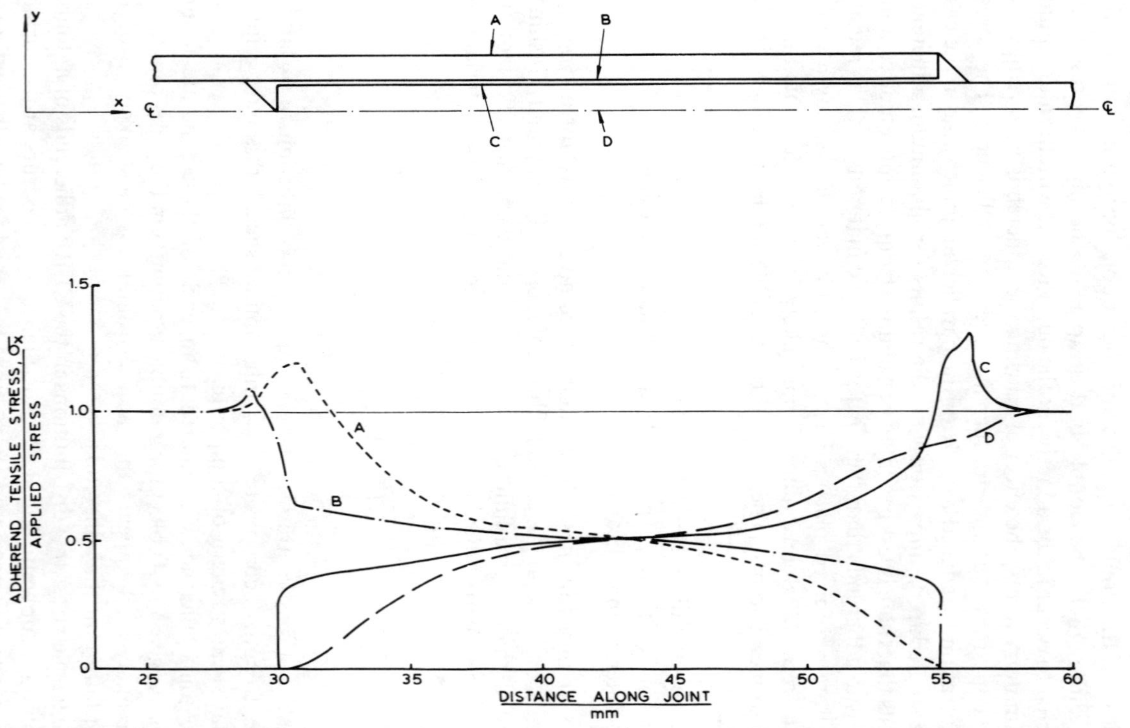

FIG. 11. Adherend tensile stress distributions in CFRP–CFRP double lap joint (from Adams[43]).

TABLE 3
Elastic Joint Efficiencies for Adhesive and Adherend Failure Modes in CFRP–CFRP Double Lap Joint

Mode and locus of failure	*Joint efficiency (%)*
Tensile in adhesive at corner of outer adherends	16
Tensile in fibre direction of composite (centre adherend)	77
Tensile through-thickness in centre adherend at tension end	55
Interlaminar shear in centre adherend	81
Combined tensile and interlaminar shear (matrix failure) in centre adherend	46
Combined tensile and interlaminar shear (matrix failure) at corners of outer adherends	41

than the aluminium. The maximum tensile stress in the outer adherends is in the outer fibres because of bending effects. In a scarf joint, the maximum adherend tensile stress (σ_x) occurs at the same place but is only 1·11 times the average.

5.6.3. Non-linear analyses

If we now allow for non-linear adhesive behaviour, the high adhesive stress concentrations predicted by the linear elastic analysis will be relieved to some extent. Figure 12[44] shows the predicted spread of the yield zone of adhesive at the tension end of a double lap joint as the load is increased. As would be expected plastic flow begins near the adherend corner, and the load corresponds to a joint efficiency of 21%. Each subsequent load increment represents an increase in joint efficiency of 4·4%. When elastic and perfectly plastic behaviour is assumed for the adhesive, a maximum strain criterion for failure seems appropriate. In Fig. 13,[45] the joint efficiency is plotted against the maximum principal strain in the adhesive at each end of a double lap joint. Assuming a failure strain for the adhesive of 5%, the analysis predicts a joint efficiency of 31% for a double lap joint compared with 16% predicted by the linear elastic analysis. Similarly, the non-linear analysis predicts an efficiency of 39% for the double scarf joint compared with 20% predicted by the linear elastic analysis. Although the predicted efficiencies are almost doubled by allowing for non-linear behaviour in the adhesive, failure in the adhesive is still predicted to be more probable than failure in the adherends (Table 3).

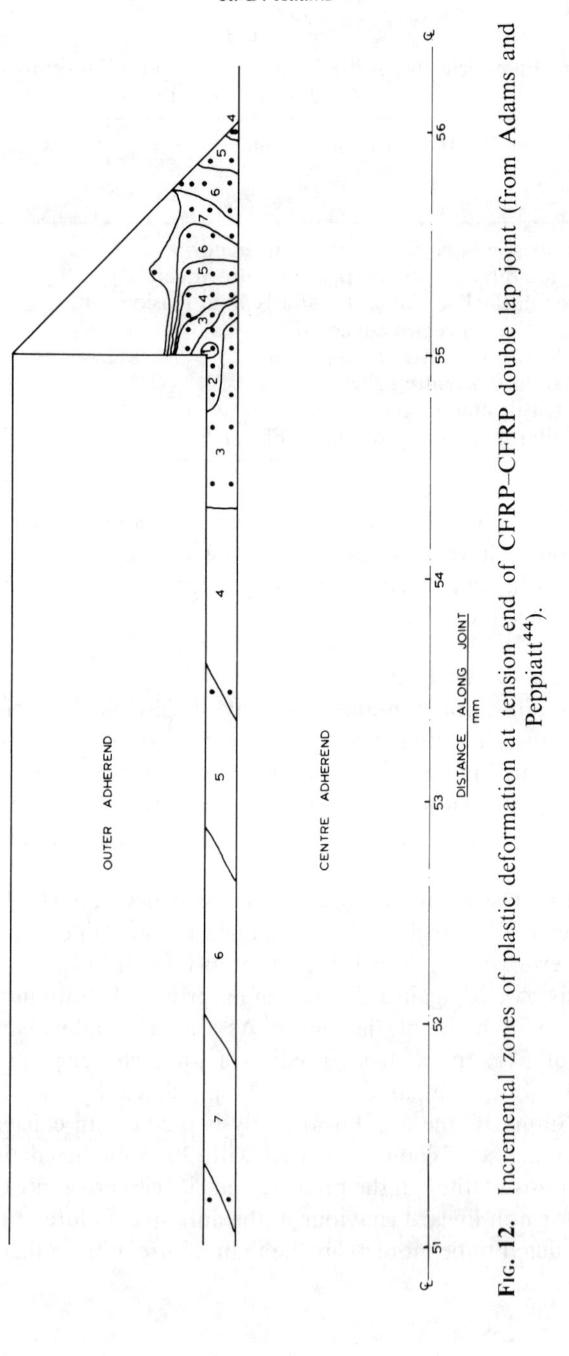

Fig. 12. Incremental zones of plastic deformation at tension end of CFRP–CFRP double lap joint (from Adams and Peppiatt[44]).

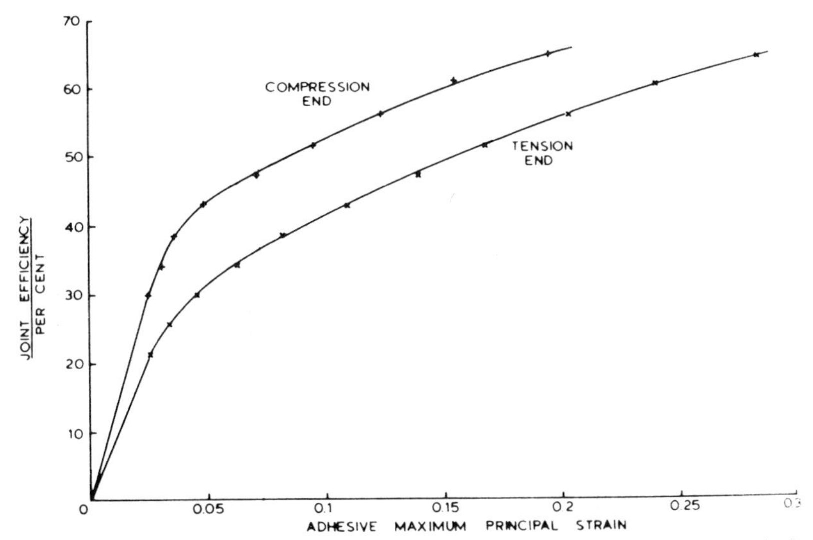

Fig. 13. Joint efficiency plotted against adhesive maximum principal strain in CFRP–CFRP double lap joint (from Adams[45]).

Although the adhesive fails in tension at the end of the joint, most of the load is transferred in the overlap region. It is, therefore, instructive to examine the effect of non-linear behaviour on the adhesive shear stress distribution. The shear stress distributions corresponding to maximum adhesive strains of 10% and 20% in a double lap joint are shown in Fig. 14(a).[46] The shear stress distributions are only modified significantly at high values of maximum adhesive strain, by which time the joint is likely to have failed. The adherend σ_x stress concentrations are reduced, but not significantly, by plastic flow in the adhesive. For a double scarf joint, the elastic shear stress distributions are similarly modified by yielding of the adhesive (see Fig. 14(b)).

The elastic shear stress distribution for a double lap joint between 0/90/90/0 crossply adherends is compared with the shear stress distribution for unidirectional adherends in Fig. 15. The higher shear stresses at each end are caused by the lower tensile stiffness of the crossply adherends. This also produces a higher stress concentration of 10·1 compared with 7·3 for a similar joint between unidirectional type II CFRP adherends.

5.6.4. Composite-to-metal joints

It is not uncommon for composites to be bonded to metals and there are one or two important points to be brought out. Figure 16 shows the

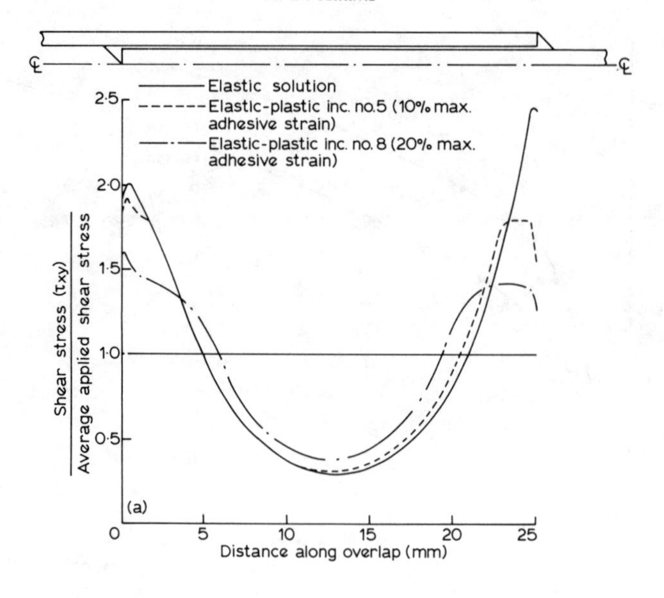

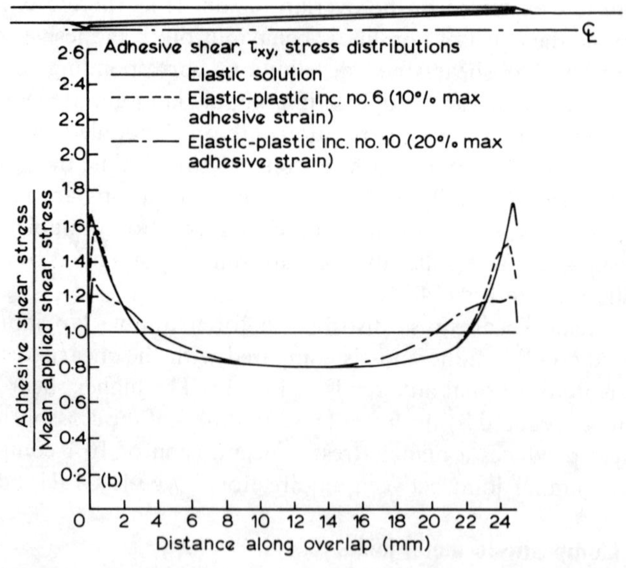

Fig. 14. Adhesive shear stress distributions in CFRP–CFRP joints: (a) double lap joint; (b) scarf joint (from Adams and Peppiatt[46]).

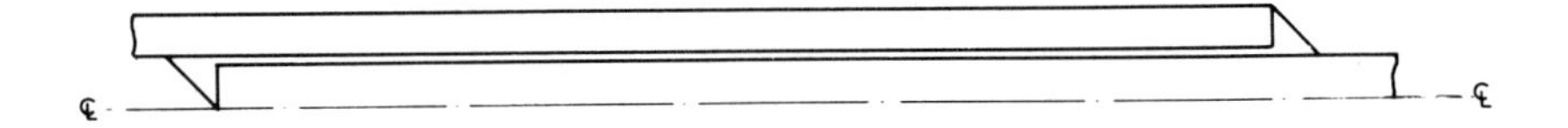

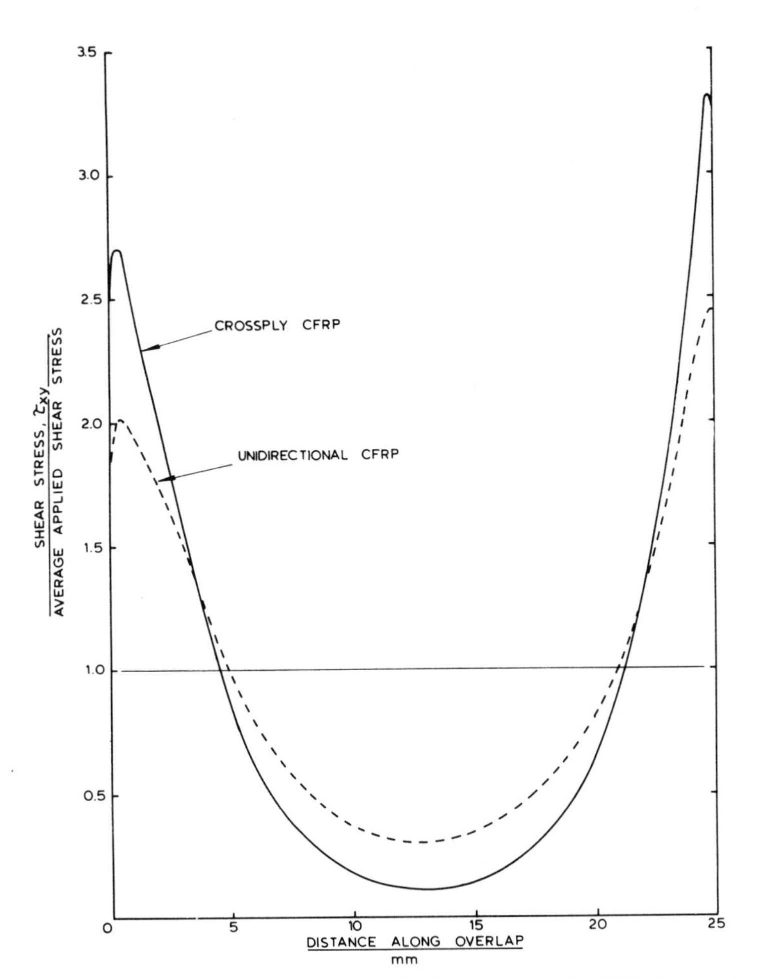

Fig. 15. Adhesive shear stress distribution in crossply CFRP–crossply CFRP double lap joint.

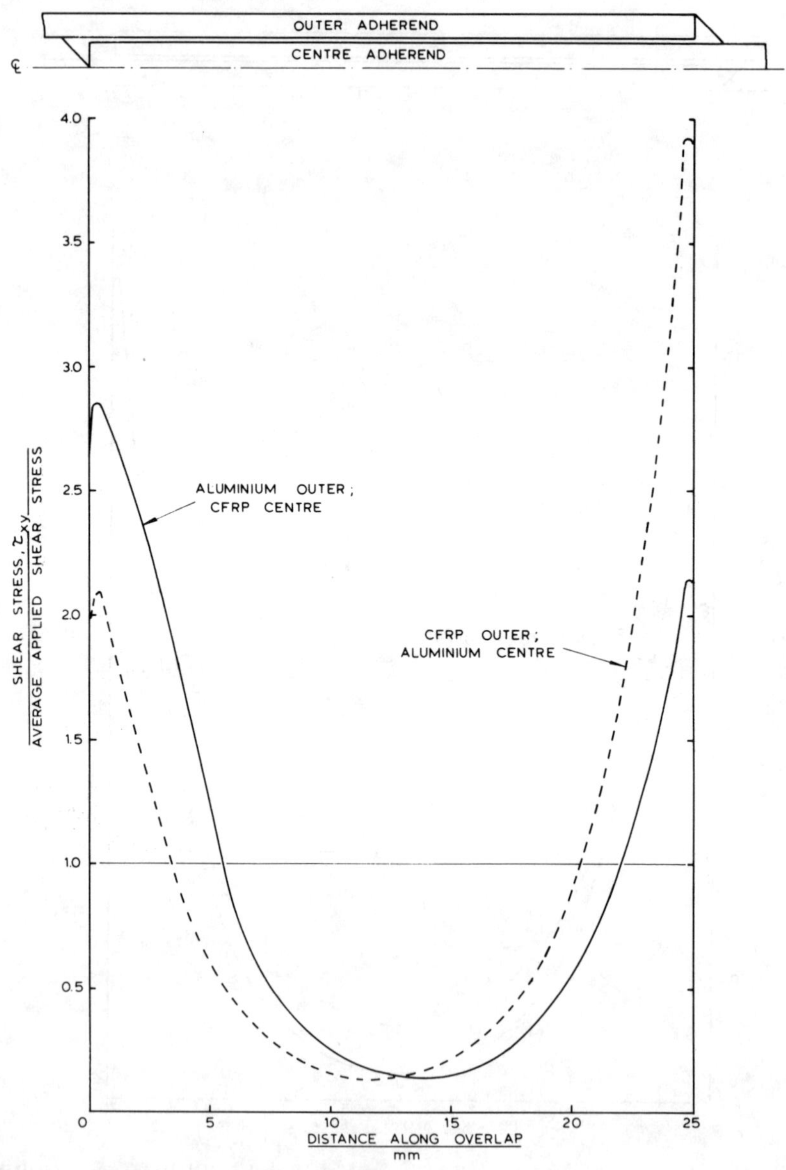

Fig. 16. Adhesive shear stress distributions in aluminium–CFRP double lap joints.

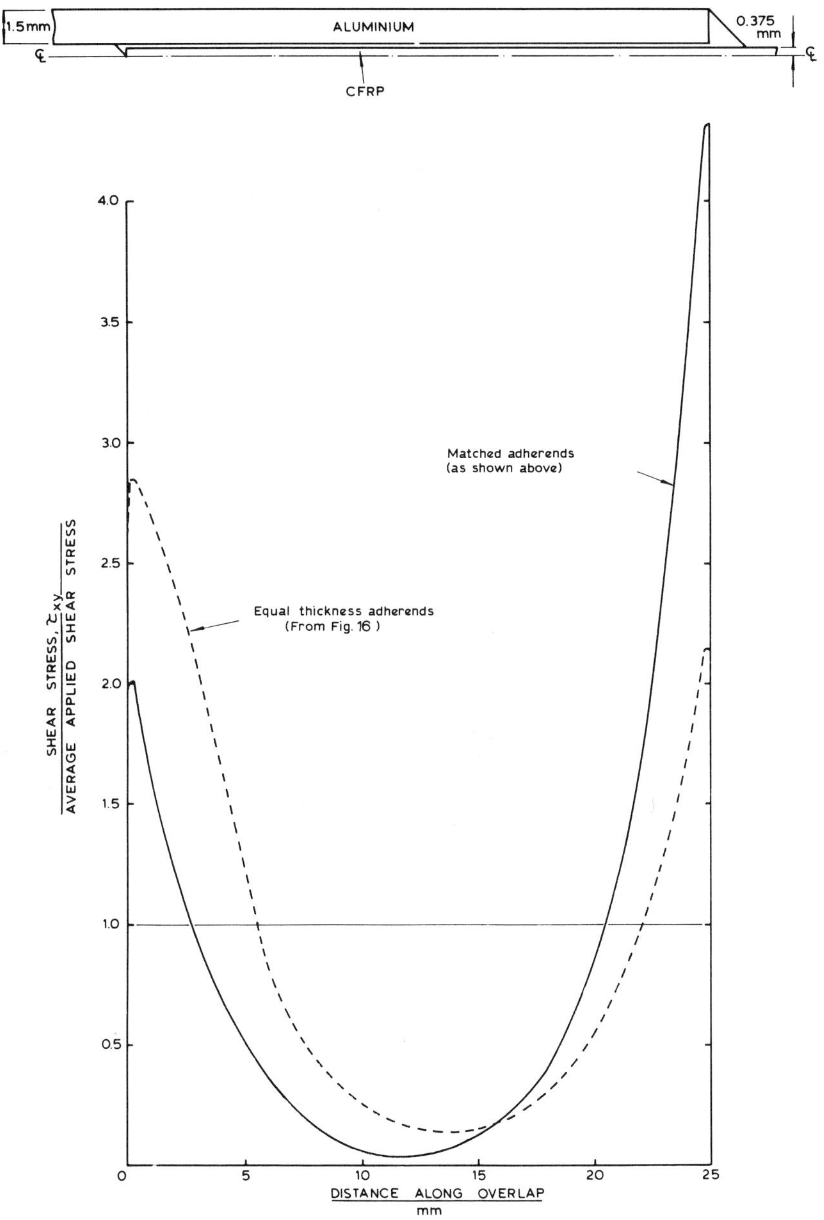

Fig. 17. Adhesive shear stress distribution in aluminium–CFRP double lap joint (matched adherends).

TABLE 4
Elastic Stress Concentrations and Joint Efficiencies for Aluminium–CFRP Joints

Joint type	*Overlap length (mm)*	*Outer adherend*	*Centre adherend*	*Adhesive stress concentration*		*Joint efficiency*[b] *(%)*
				End where outer adherend is loaded	*End where centre adherend is loaded*	
Double lap	25·0	0·9 mm aluminium	1·8 mm CFRP	7·0	6·9	79
Double lap	25·0	0·9 mm CFRP	1·8 mm aluminium	3·2	11·5	48
Double lap	25·0	1·5 mm aluminium	0·75 mm CFRP	5·0	13·4	28
Double butt strap	25·0	0·9 mm aluminium (straps)	1·8 mm CFRP	13·8[a]	6·9	41
Single lap	25·0	0·9 mm aluminium	0·9 mm CFRP	13·8	11·25	—
Double scarf	25·0	0·9 mm aluminium	1·8 mm CFRP	7·6	4·0	74
Single scarf	25·0	0·9 mm aluminium	0·9 mm CFRP	10·2	5·2	—

[a] Between butt faces.
[b] N.B. Joint efficiency is based on strength of aluminium alloy.

adhesive shear stress distributions for double lap joints between aluminium and unidirectional type II CFRP adherends. If the outer adherends are aluminium and the centre adherend is CFRP, the highest shear stress occurs at the compression end of the joint where the aluminium adherends are loaded. This is because the aluminium adherends have a lower tensile stiffness than the composite adherends. However, the adhesive stress concentrations at each end of the joint are similar (Table 4). Therefore, as far as the adhesive is concerned, the joint is well conditioned. Alternatively, if the outer adherends are unidirectional CFRP and the centre adherend is aluminium, then the higher shear stress that occurs at the tension end of a joint with similar adherends is increased still further by the adherend dissimilarity. The stress concentration at the tension end of the joint is now 3·6 times the stress concentration at the compression end, with the result that the joint efficiency, in terms of the strength of the aluminium alloy adherends, is reduced from 79% to 48%.

The joints described above do not attempt to use the high strength of the unidirectional type II CFRP. If the tensile strength of the CFRP is assumed to be 1550 MPa and the yield strength of the aluminium alloy is assumed to be 325 MPa then, if we make full use of the higher strength of the CFRP, the total thickness of the aluminium alloy adherends would need to be approximately four times that of the composite (allowing for a safety factor of 1·2 for the strength of the more brittle CFRP). The adhesive shear stress distribution for a double lap joint with two 1·5 mm thick aluminium outer adherends bonded to a 0·75 mm thick CFRP centre adherend is shown in Fig. 17. The lower tensile stiffness of the thinner centre adherend increases the shear stress and the stress concentration at the tension end of the joint. The joint efficiency, based on the strength of the aluminium adherends, is now only 20% assuming linear elastic failure of the adhesive at a tensile stress of 65 MPa (Table 4).

5.6.5. Scarf and stepped joints

The effect of scarfing the adherends (Fig. 18) is to unbalance the stress concentrations to such an extent that they are much higher at the compression end than at the tension end of the joint. Although the maximum shear stress is about 7% less than in a double lap joint, the adhesive fillets of the scarf joint are much smaller, giving a 9% higher stress concentration (Table 4).

Stepped lap joints are an important method of joining CFRP components. The adhesive shear stress distributions in four-step joints between unidirectional CFRP adherends are shown in Fig. 19. For the small butt

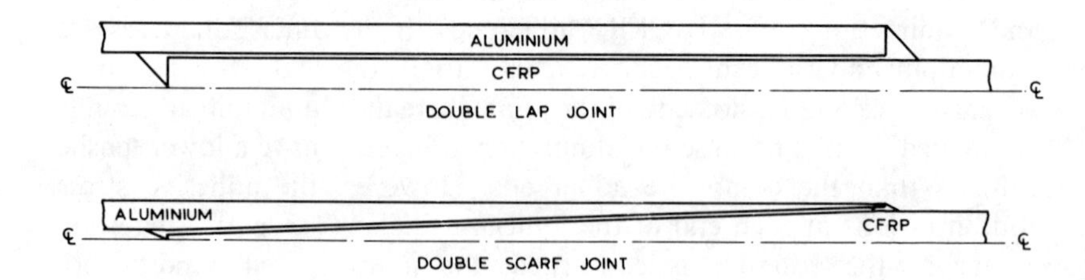

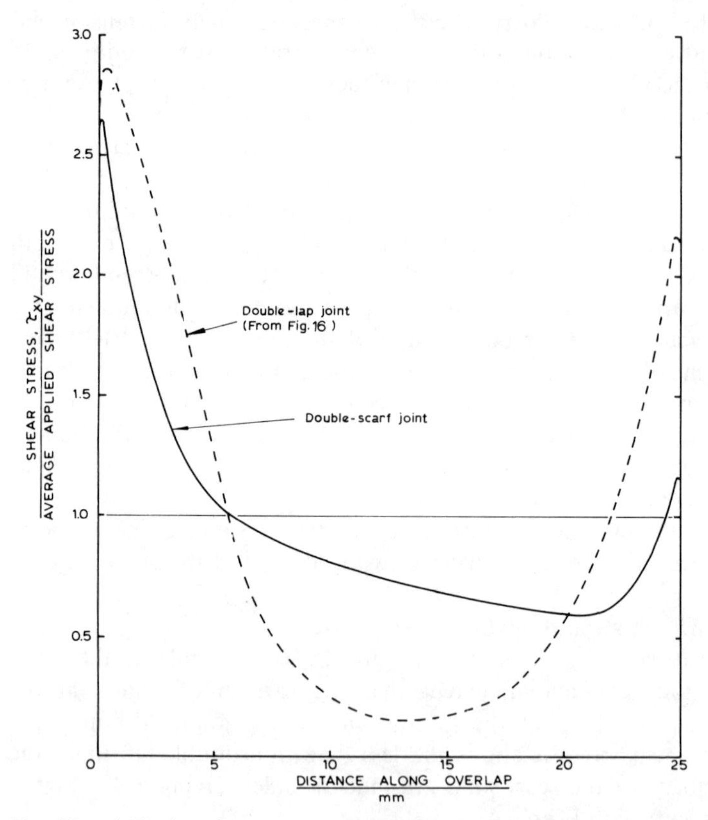

Fig. 18. Adhesive shear stress distribution in aluminium–CFRP double scarf joint.

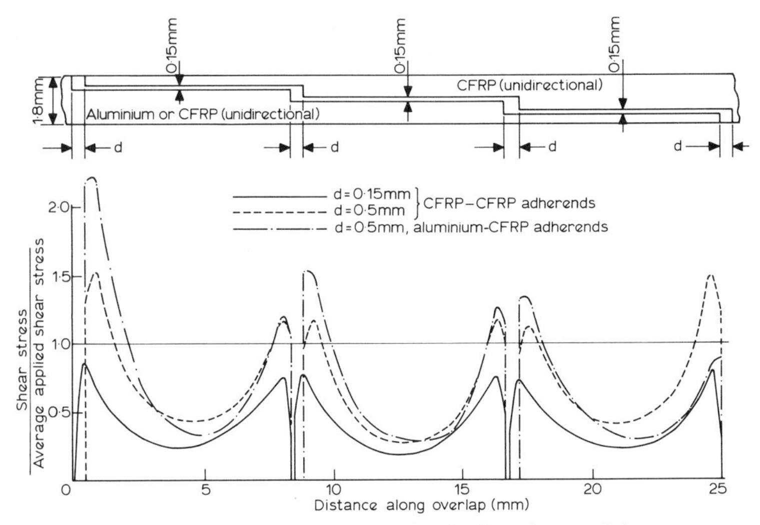

Fig. 19. Adhesive shear stress distributions in step joints.

spacing (0·15 mm), the maximum shear stress at the ends of the steps is less than the average applied shear stress. This is because at least half the applied load is transferred by the adhesive between the butt faces giving stress concentrations in these regions of approximately 10. The proportion of the load transferred by shear, rather than by the butt faces, is increased considerably by increasing the thickness of the glueline between the butt faces to 0·5 mm. Increasing the butt face glueline thickness increases the shear stresses at the ends of the steps but reduces the stress concentrations between the butt faces. The predicted joint efficiency is increased by 6–10% in the case of unidirectional type II CFRP adherends. We also show in Fig. 19 the shear stress distribution in a step joint between aluminium and unidirectional CFRP. The highest shear stress and stress concentrations occur at the end where the lower-stiffness aluminium adherend is loaded.

When elastic–plastic behaviour is assumed for the adhesive, yield is predicted to occur first between the butt faces at each end of the step joint. Figure 20 shows the spread of the yield zones as the load on the joint is increased. At higher loads, the proportion of the load transferred by the butt faces is reduced. If a maximum strain failure criterion is assumed for the adhesive, the load-carrying capacity of the joint is increased by increasing the butt face glueline thickness. For example, assuming a 5% failure strain, the non-linear analysis predicts joint

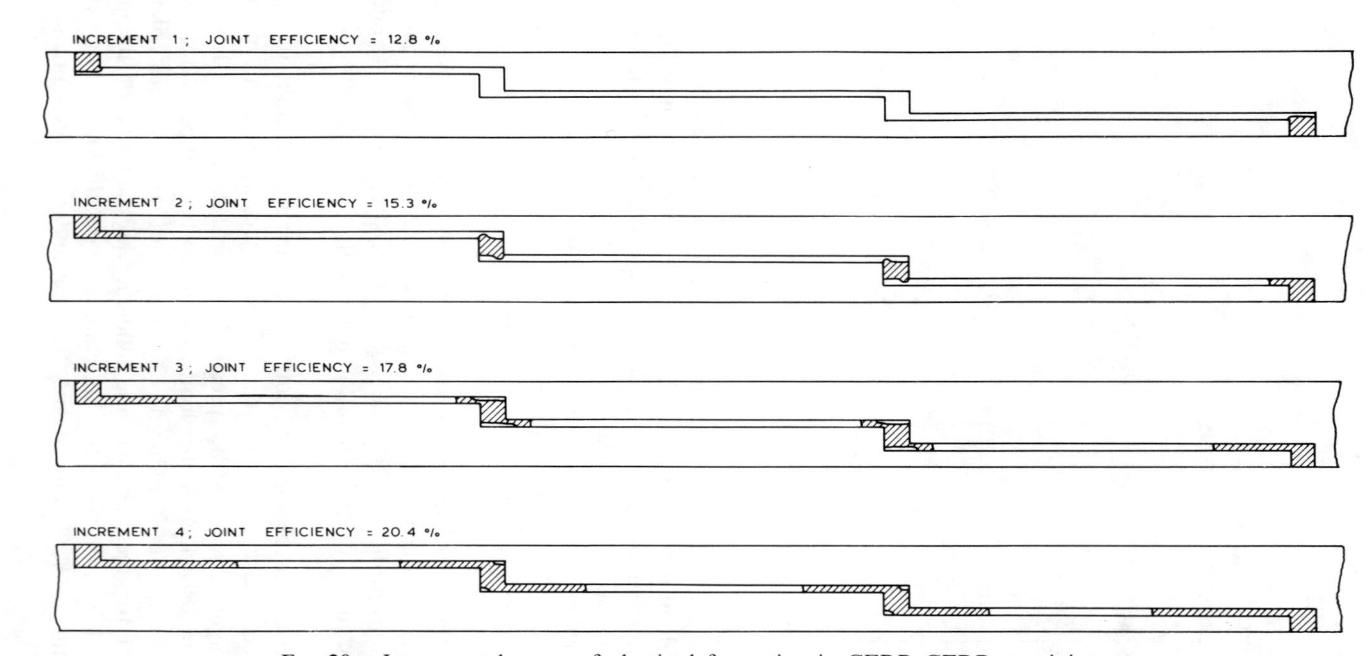

Fig. 20. Incremental zones of plastic deformation in CFRP–CFRP step joint.

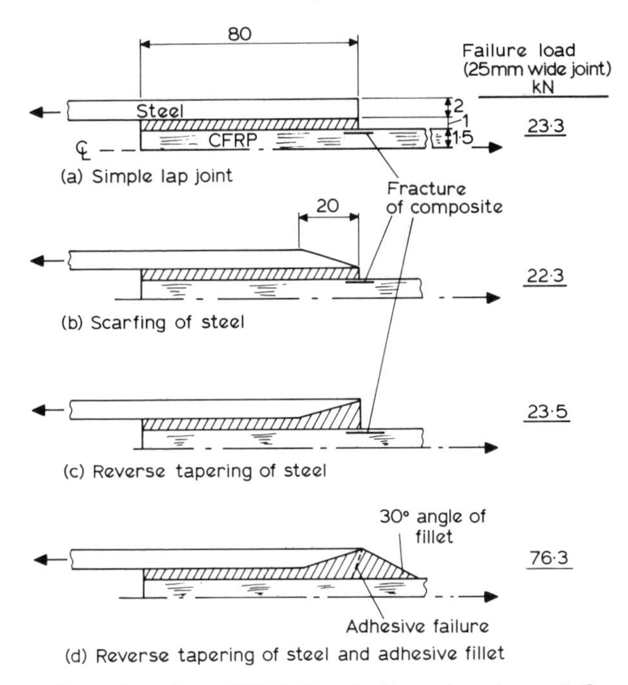

Fig. 21. Bonding of steel to CFRP (length dimensions in mm) (from Adams & Wake[47]). Fracture of composite caused by through-thickness tensile stresses.

efficiencies of 9 and 17% for the two glueline thicknesses, compared with efficiencies of 6 and 10% predicted by the linear elastic analysis. Higher strain adhesives give correspondingly higher joint efficiencies.

An example is given in Adams and Wake[47] of the redesign of an adhesively bonded joint between steel and unidirectional CFRP, using a modern, rubber-toughened epoxy adhesive. A simple lap joint, shown in Fig. 21(a), failed by lateral cracking of the composite. This was due to stress concentrations at the change of section which led to high loads being transferred across the composite to the steel. The steel was therefore finely tapered as shown in Fig. 21(b) to reduce the spatial load transfer rate. The same type of failure was observed at a load only slightly smaller than before. Reversing the taper as in Fig. 21(c) had a similar effect, only slightly increasing the joint strength. Finally, in an attempt to reduce further the peak stresses at the edge of the joint, a fillet of adhesive was formed during manufacture, as illustrated in Fig. 21(d). With an included angle of approximately 30° in the fillet, an increase in joint strength of more than three-fold was achieved. In this case the through-thickness stresses were sufficiently reduced to prevent failure initiating within the composite, and failure occurred by cracking through the adhesive layer as shown in Fig. 21(d).

What this example illustrates is how basic mechanics can indicate the likely mode of failure and provide an indication of how to improve the situation. By using a finite element analysis, it was possible to predict not only the initial mode of failure and the corresponding load, but also how this mode of failure would alter together with a corresponding massive improvement in the joint strength.

5.7. CONCLUSIONS

In order to be able to predict the failure strength of adhesively bonded composites, it is necessary to find which is the weakest link. Sometimes, it will be the adhesive which fails, sometimes the composite. With the composite, the stresses in the joint region must be properly analysed as unexpected failure modes may occur. In the major part of a composite component, failure tends to occur by fibre breakage and/or interlaminar shear. However, where the loads have to be transferred out via the face, lateral tensile and increased interlaminar shear stresses occur. Also, the direct tensile stresses in the composite may be increased just outside the overlap. Composites, therefore, will often fail in or adjacent to a bonded section.

While closed-form mathematical analysis is useful up to a point, finite element techniques provide the only certain means of establishing joint strength at the design stage, provided allowance is made for the non-linear behaviour of the adhesive.

ACKNOWLEDGEMENT

This chapter is based, in part, on material which first appeared in *Structural Adhesive Joints in Engineering* by Adams and Wake, published in 1984.

REFERENCES

1. Demarkles, L. R. Investigation of the use of a rubber analog in the study of stress distribution in riveted and cemented joints, Tech. Note 3413, 1955 Nat. Advisory Cttee Aeronautics, Washington, DC, USA.
2. Matthews, F. L., Kilty, P. F. and Godwin, E. W. *Composites*, 1982, **13**, 29.

3. Volkersen, O. *Luftfahrtforschung*, 1938, **15**, 41.
4. Goland, M. and Reissner, E., *J. Appl. Mech., Trans. ASME*, 1944, **66**, A17.
5. Cornell, R. W., *J. Appl. Mech., Trans ASME*, 1953, **75**, 355.
6. Kutscha, D. and Hofer, K. E., Feasibility of joining advanced composite flight vehicle structures, AD 690 616, 1969, IIT Research Institute, Chicago, Illinois.
7. Sneddon, I. N., The distribution of stress in adhesive joints, *Adhesion* (Ed. D. D. Eley), 1961, Clarendon, Oxford, UK.
8. Renton, W. J. and Vinson, J. R., *J. Adhesion*, 1975, **7**(3), 175.
9. Allman, D. J., *Quarterly J. Mechanics Appl. Maths*, 1977, **30**, 415.
10. Adams, R. D. and Peppiatt, N. A., *J. Strain Anal.*, 1973, **8**, 134.
11. Hahn, K. F., Photo stress investigation of bonded lap joints, Part II: Analysis of experimental data, 1960, McDonnell-Douglas Co. Research Report SM 4000-1, Long Beach, California, USA.
12. Sazhin, A. M., *Russ. Engng J.*, 1964, **44**, 45.
13. Coker, E. C., *Proc. Royal Soc.*, 1912, **86**(A587), 291.
14. Inglis, C. E., *Proc. Royal Soc.*, 1923, **A103**, 598.
15. Mylonas, C., *Proc. Soc. Experimental Stress Analysis*, 1954, **12**, 129.
16. Adams, R. D. and Peppiatt, N. A., *J. Strain Anal.*, 1974, **9**, 185.
17. Adams, R. D., Chambers, S. H., Del Strother, P. J. A. and Peppiatt, N. A., *J. Strain Anal.*, 1973, **8**, 52.
18. Hart-Smith, L. J., Adhesive-bonded single lap joints, CR-112236, 1973, NASA, Langley Research Centre, USA.
19. Hart-Smith, L. J., Adhesive-bonded scarf and stepped-lap joints, 1973, CR-112237, NASA, Langley Research Centre, USA.
20. Hart-Smith, L. J., Adhesive-bonded double-lap joints, CR-112235, 1973, NASA, Langley Research Centre, USA.
21. Hart-Smith, L. J., *J. Eng. Mat. Technol.*, Trans. ASME, 1978, **100**, 16.
22. Hart-Smith, L. J., Adhesive-bonded joints for composites—phenomenological considerations, McDonnell–Douglas Co. Paper 6707, *Technology Confs. Assoc. Conf. on Advanced Composites Technology*, 1978, El Segundo, California, USA.
23. Hart-Smith, L. J., Further developments in the design and analysis of adhesive-bonded structural joints, McDonnell–Douglas Co. Paper 6922, *ASTM Symp. on Joining of Composites*, 1980, Minneapolis, Minnesota, USA.
24. Hart-Smith, L. J., Structural details of adhesive-bonded joints for pressurized aircraft fuselages, Report MOC-J8858, 1980, McDonnell–Douglas Co., Long Beach, California, USA.
25. Hart-Smith, L. J., *Developments in Adhesives—2* (Ed. A. J. Kinloch), 1981, Applied Science Publishers, London, UK, p. 1.
26. Grant, P., Strength and stress analysis of bonded joints, British Aircraft Corp. Ltd Rep. S OR(P) 109, 1976.
27. Grant, P., Analysis of adhesive stresses in bonded joints, *Symposium: Jointing in Fibre Reinforced Plastics*, Imperial College, London, 1978, IPC Science and Technology Press, Guildford, UK, p. 41.
28. Anon., Inelastic shear stresses and strains in the adhesives bonding lap joints loaded in tension or shear, 1979, ESDU 79016 Engineering Sciences Data Unit, London, UK.

29. Adams, R. D., Coppendale, J. and Peppiatt, N. A., *Adhesion 2* (Ed. K. W. Allen), 1978, Applied Science Publishers, London, UK, p. 105.
30. Sultan, J. N. and McGarry, F. J., *Polym. Eng. Sci.*, 1973, **13**, 29.
31. Pick, M. and Wronski, A. A., The effect of triaxial loading on the yielding and fracture properties of epoxy resins, *Conf. Deformation, Yield and Fracture of Polymers*, Cambridge, 1976, Plastics Rubber Inst., London, UK.
32. Bowden, P. B. and Jukes, J. A., *J. Matl Sci.*, 1972, **7**, 52.
33. Raghava, R. S., Cadell, R. M. and Yeh, G. S. Y., *J. Matl Sci.*, 1973, **8**, 225.
34. Lubkin, J. L., *J. Appl. Mech.*, 1957, **79**, 255.
35. Wah, T., *Int. J. Mech. Sci.*, 1976, **18**, 223.
36. Webber, J. P. H., *J. Adhesion*, 1981, **12**, 257.
37. Thamm, F., *J. Adhesion*, 1976, **7**, 301.
38. Barker, R. M. and Hatt, F., *Amer. Inst. Aero. Astro. J.*, 1973, **11**, 1650.
39. Wright, M. D., *Composites*, 1978, **9**, 259.
40. Wright, M. D., *Composites*, 1980, **11**, 46.
41. Adams, R. D., Peppiatt, N. A. and Coppendale, J., Prediction of strength of joints between composite materials, *Symposium: Jointing in Fibre Reinforced Materials*, Imperial College, London, 1978, IPC Science and Technology Press, Guildford, UK, p. 64.
42. Rotem, A. and Hashin, Z., *J. Composite Matls*, 1975, **9**, 191.
43. Adams, R. D., *Adhesion 4* (Ed. K. W. Allen), 1980, Applied Science Publishers, London, p. 87.
44. Adams, R. D. and Peppiatt, N. A., *J. Adhesion*, 1977, **9**(1), 1.
45. Adams, R. D., *Developments in Adhesives—2* (Ed. A. J. Kinloch), 1981, Applied Science Publishers, London, p. 45.
46. Adams, R. D. and Peppiatt, N. A., Fibre reinforced materials, design and engineering applications, Paper 6, 1977, Inst. Civil Engrs, London, UK.
47. Adams, R. D. and Wake, W. C., *Structural Adhesive Joints in Engineering*, 1984, Applied Science Publishers, London, p. 105.

Chapter 6

Design and Empirical Analysis of Bolted or Riveted Joints

L. J. HART-SMITH

Douglas Aircraft Company, McDonnell Douglas Corporation, Long Beach, California, USA

6.1. INTRODUCTION

The purpose of this chapter is to explain the principles behind the design and associated analysis of bolted or riveted joints in fibrous composite structures. Because each manufacturer of composites creates material as the parts are fabricated, no two manufacturers of composite hardware can develop precisely the same structural allowables even from what appear to be nominally the same resin and fibre ingredients. Minor differences in cure cycles, resin content, bagging techniques, lay-up environment, storage conditions, shipping from supplier to manufacturer, and the like, cause statistically different mean strengths and scatter factors. There are, therefore, no universally approved allowables such as in MIL-HBK-5 for metal alloys. Each individual manufacturer of composite aircraft structures, for example, is certified independently and the processing, once approved, is not transferable to other manufacturers' sites without recertification at each new facility. Given this background, it would not be appropriate to try to document precise design-allowable strengths here. In any event, new and improved materials are coming on the market continually.

The approach followed here is instead to explain the philosophy behind the design and associated analysis processes from a general point of view which does not need changing for each different composite material. The variables considered are related to the physical behaviour, to identify which factors must be accounted for and which may be neglected for any particular material, based on what is known of its behaviour. One purpose of such a presentation is the definition of the minimum test programme that can be used to characterise adequately bolted joints in any composite material.

Any material allowables quoted here are usually sufficiently close for preliminary designs and a considerable part of the structures of subsonic transport aircraft can be assessed within the framework outlined here. Other references should be studied for such information as bearing strengths at elevated temperatures. However, once such data have been obtained to establish a new set of coefficients, the analysis methods contained here have been shown to be effective in those situations also.

The material in this chapter is organised in the following sequence. The characteristics of single fastener joints are described first. Next, the effects of fibre pattern and lay-up sequence are discussed. This is followed by an outline of a minimum test plan for bolted joints in new composite materials. This, in turn, is followed by an explanation of an analysis developed to explain the behaviour of bolted composite joints one bolt at a time. Then the analysis is generalised to multi-row bolted joints, which leads to the concept of joint structural efficiency. The concluding sections cover design procedures and miscellaneous considerations.

6.2. SINGLE-HOLE JOINTS

6.2.1. Behaviour of single-hole test coupons

Many of the data on which the design of bolted joints is based have been accumulated from the testing of simple coupons having a single fastener, in double or single shear. The variables considered usually include fastener size, laminate thickness, edge (or end) distance, width, and sometimes clamping torque. Theories developed to explain these observations and to predict the performance of more complex structural joints are customarily checked-out on such data first. Therefore these simple joints provide a logical starting point for this work.

There are various modes of failure for single fastener joints and the most frequently occurring ones are shown in Fig. 1. Of these, only two (bearing and tension-through-the-hole) are considered desirable, in the sense that all the others are premature failures at lower loads. For example, the use of too small a bolt diameter results in excessive bolt bending and an inefficient highly non-uniform bearing stress distribution or, worse still, the bolt head may dig in and pull through the laminate as shown. Shear-out and cleavage failures result from either too small an edge (end) distance, too orthotropic a laminate pattern, or insufficient dispersion of the differently oriented plies. The distinction between bearing and tension failures is established largely by the joint geometry—particularly the width-to-diameter ratio w/d—and, to a lesser degree, by the fibre pattern. Much of the subsequent discussion will be concerned with only these two failure modes.

A further consideration is the difference between loaded and unloaded holes, as shown in Fig. 2. It is customary in design to allow for a higher stress concentration at a loaded hole than at an unloaded hole. More importantly, a distinction must be made between bearing and non-bearing, or by-pass, loads to be able to characterise the internal loads in multi-row bolted joints. The bearing load at each fastener is defined by the shear load transferred by that particular fastener. Any remaining load by-passes that fastener, to be reacted elsewhere. The basis for Fig. 2 will be discussed in a later section.

6.2.2. Effect of joint geometry

The strength of single-hole bolted joints as a function of the joint geometry can be characterised by curves of the type shown in Fig. 3. This information has been extracted from ref. 1 and the explanation of its

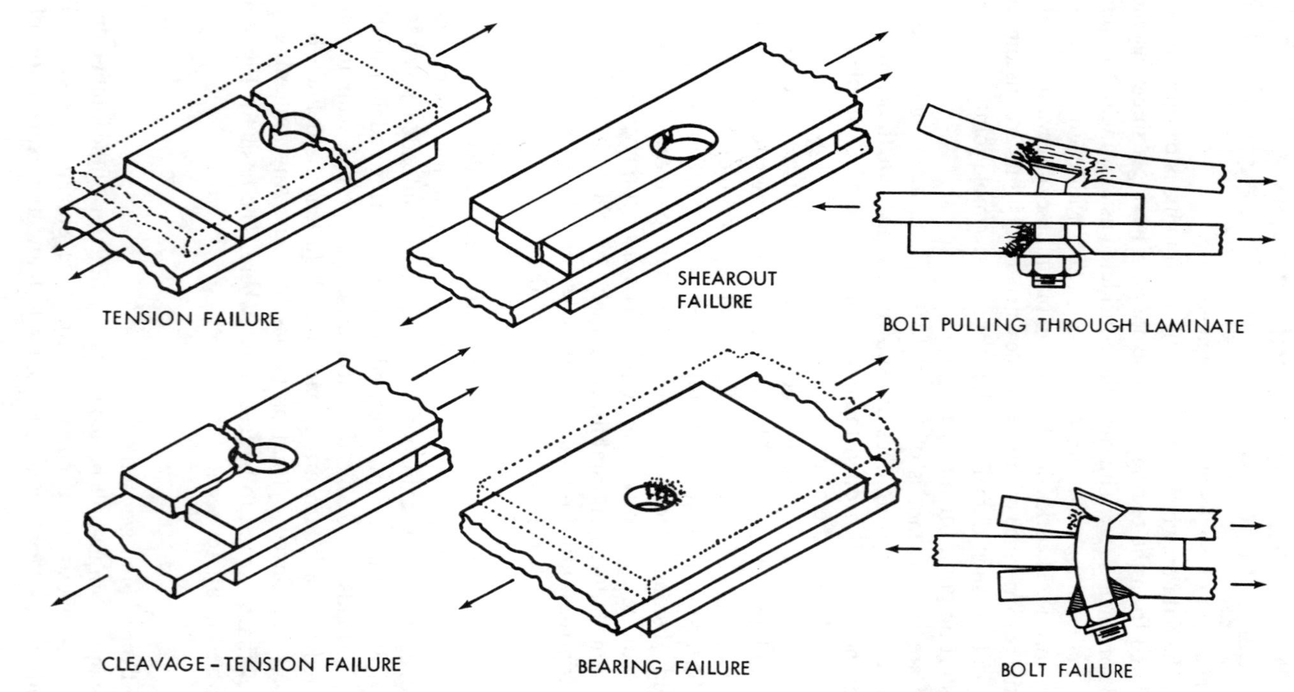

Fig. 1. Modes of failure for bolted joints in advanced composites.

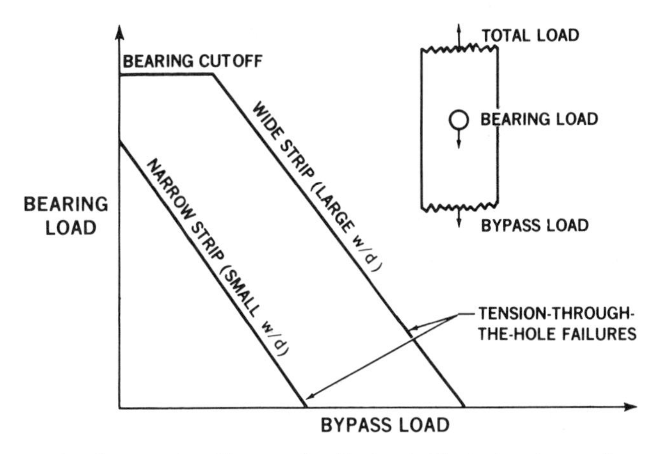

Fig. 2. Bearing/by-pass load interaction for loaded bolts in advanced composites.

derivation is given below. Figure 3 has actually been prepared for quarter-inch (6·35 mm) bolts in double shear in a quasi-isotropic pattern in high tensile strength (rather than high-modulus) carbon/epoxy, but the general form would be the same for almost any fibrous composite. Curves of that type could be derived purely on the basis of test results, with no theory at all. There are bearing failures at the left side of the picture, where the

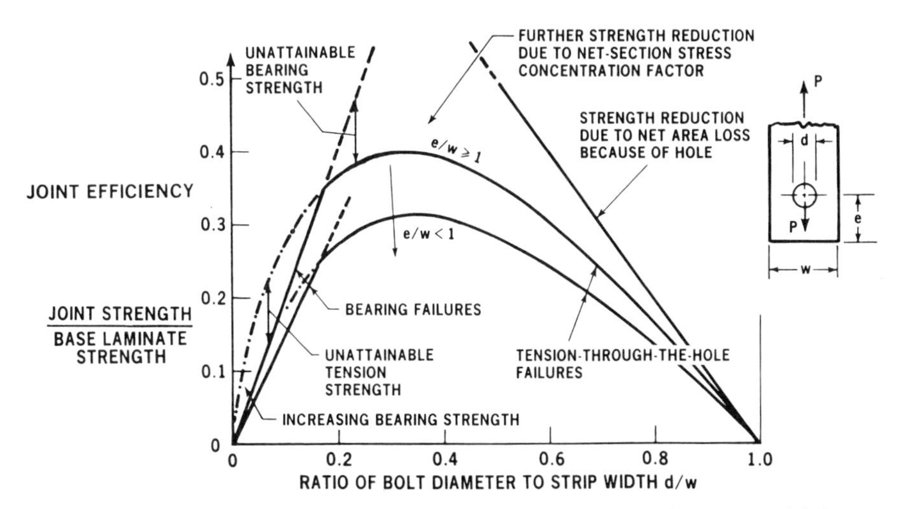

Fig. 3. Bolted joint efficiencies for composite laminates as functions of joint geometry.

strip width (or fastener pitch) is large in comparison with the hole diameter. As the fasteners are moved closer together (or the strip is narrowed), there is a change in failure mode to tension-through-the-hole. Beyond the peak there is a further change in the precise nature of the tensile failure mode but, by then, the fasteners are too close together to be of practical interest. Suffice it to say that the drop-off in strength to the right of Fig. 3 is probably even more abrupt than actually shown. For laminates with a relatively high bearing strength, the regime of tension failures includes a plateau across a range of d/w ratios before the strength drops off steadily due to loss of tension area at larger d/w ratios. Those materials having a relatively low bearing strength would not exhibit that plateau—the strength versus d/w curve would have a sharp peak at some intermediate ratio. Figure 3 shows also the losses of strength due to short edge distance effects—low e/d or e/w ratios.

It is evident from Fig. 3 that the peak strength occurs either at a geometry associated with the tension failure mode or at an abrupt transition between bearing and tension failures, and that, in either case, that peak strength is associated with a definite w/d value, which is usually closer to 3 than the frequent design practice of 4 to 5. Any effort to impose on a bolted composite joint design an arbitrary requirement of failure by bearing only (by using a sufficiently large value of w/d) to ensure a less catastrophic failure mode must necessarily be associated with a significant reduction in joint strength and, hence, in structural efficiency. This applies to both single-row and multi-row bolted joints in fibrous composite structures.

6.2.3. Effect of bolt tightening

The bearing strength of carbon/epoxy laminates has long been known to be improved by torquing the bolts tighter. Furthermore, there is a dramatic difference between the bearing strength with regular torqued fasteners and with pin-loaded holes having no clamp-up whatever, as already noted in Chapter 2. This effect is illustrated in Fig. 4,[2] showing how the bearing strength of carbon fibre-reinforced plastics is doubled by even finger-tight torquing of the bolts. The reason for this increase in strength is that the initially damaged material can still withstand compressive loads under the bolt shank, provided that there is nowhere for the material to be displaced. The initial damage is sometimes detected as a 'blip' on a load–deflection curve which continues linearly to much higher loads, as shown in Fig. 5, or it may often only be heard during the test. As this damage spreads, crushed composite material is displaced laterally

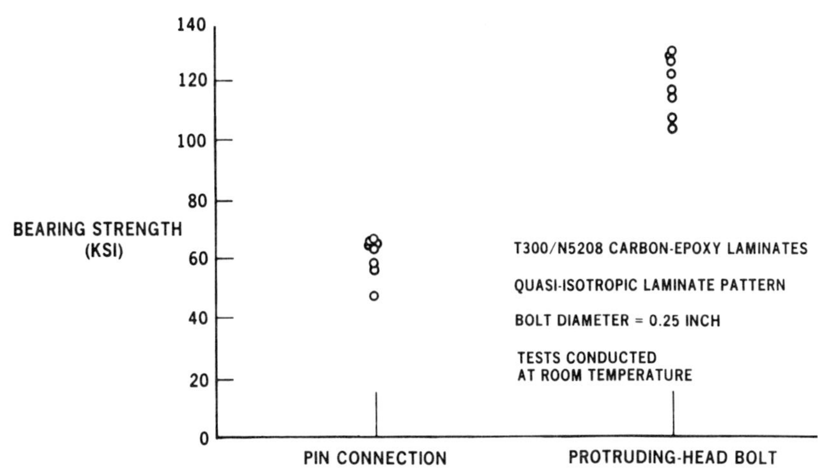

FIG. 4. Comparison between bearing strengths for pin loading and regular (torqued) bolts (100 ksi ≡ 700 MN/m^2).

and induces higher tensile loads in the fastener, compressing the composite material ever more tightly. With a pin-loaded hole with neither fastener head nor nut, on the other hand, the initially damaged material 'brooms out' and does not continue to sustain load. This, in turn, increases the bearing load on the remaining material which then fails at an accelerating rate as the load is increased.

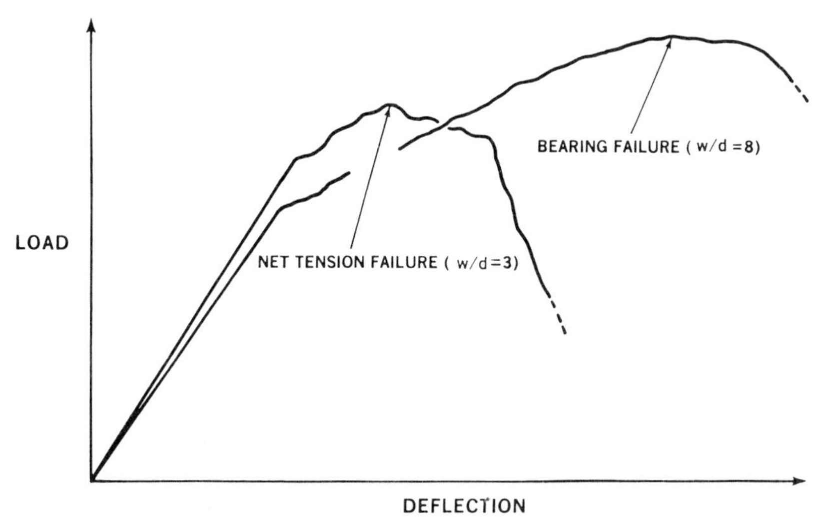

FIG. 5. Load–deflection curves for bolted composite joints.

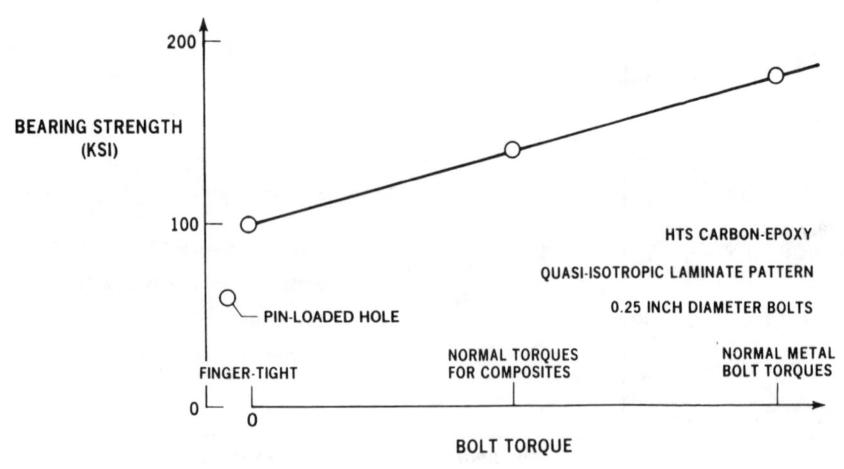

Fig. 6. Effect of bolt torque on bearing strength of fibrous composite laminates (100 ksi ≡ 700 MN/m^2).

The use of flush (countersunk) fasteners permits some clamp-up, but considerably less than under a washer or protruding bolt head, so the benefits of clamp-up are not as evident with countersunk fasteners. In addition, of course, there is a much higher bearing stress associated with the reduced shank area.

Figure 6 shows how still further bearing strength increases can be obtained by even higher torques on the fasteners.[3] Similar phenomena have been documented with glass fibre-reinforced plastic laminates.[4] Reliance on the benefits of torquing beyond finger-tightness (to constrain any damage) has been questioned because some of the additional clamp-up will be relieved by creep over the life of the structure. Investigations into the rate of loss of this clamp-up have been conducted and the decay rate was shown to be fairly slow.[5] The drawback with taking advantage of the added strength associated with the extra bolt clamp-up is that it would take only one under-torqued bolt to impose a significant loss of static strength. The equivalent problem in metal structures—that of one loose bolt in a pattern of interference fit bolts—causes only a loss of fatigue life, not of the static strength.

The effect of increases or decreases of bearing strengths, in relation to tensile strengths, of bolted composite strengths is characterised in Fig. 7, which is a modification of Fig. 3. For carbon/epoxy laminates having a plateau of failure strengths in the tensile failure mode, a higher bearing strength cannot increase the joint strength; the measured increases are

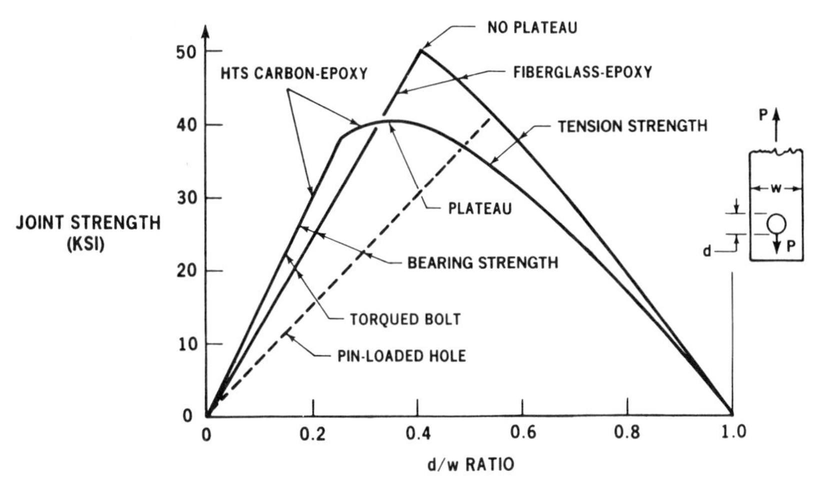

FIG. 7. Relation between bearing and tension failures for carbon/epoxy and fibreglass/epoxy laminates as functions of joint geometry (50 ksi ≡ 350 MN/m^2).

due mainly to additional load transfer by friction and to an effective increase in the laminate strength in tension because of the lateral compression associated with the higher torque and tension in the fastener. However, a substantial loss of bearing strength, as due to failure to tighten the nut down in contact with the laminate, is shown in Fig. 7 to cause a drastic shift in the failure mode as a function of the w/d ratio. This strength reduction is always present, albeit to a lesser extent than shown, whenever flush (countersunk) fasteners are used.

Fibreglass/epoxy laminates tend to have a higher ratio of tension-to-bearing strengths than do carbon/epoxy laminates. Thus there is virtually no plateau in the joint strength versus geometry relation depicted in Fig. 7. The peak strength for bolted fibreglass joints is obtained at the abrupt transition between bearing and tension-through-the-hole failure modes, again at a w/d ratio closer to 3 than to 4. Additional bolt clamp-up is now seen to provide additional strength and an upward shift in the optimum w/d ratio. This effect would be seen in carbon/epoxy composites only for laminates in which the weakest link is the resin, because of excessive orthotropy or inadequate dispersion of the plies.

There is another significant effect associated with clamp-up in a bolted composite joint. Testing[6] has shown consistently that, all other factors being equal, the bearing strength of the laminate in the middle of a double shear sandwich is greater than that of the laminates on the outer faces.

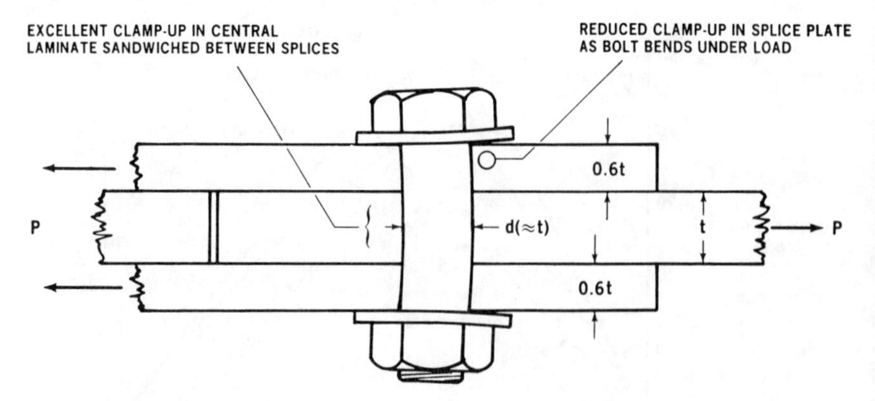

Fig. 8. Need for reinforcement of composite splice plates. The bolt-bending effect shown is actually minimal for $d = t$ but becomes progressively more severe for smaller bolt diameters. This bolt-bending effect is even more pronounced for compressive loads.

This is purely the result of better confinement of any initial damage. It is appropriate, therefore, to design splice plates to be slightly thicker (by about 20%) than a simple balanced design would suggest, as shown in Fig. 8. It is also helpful to use large stiff washers under the fastener head and nut to improve the bearing strength of splice plates.

Indeed it has been shown, in ref. 3 and by testing elsewhere, that the use of such washers can actually alter the failure mode, as shown in Fig. 9. There is often no visible bearing damage at all in the high bearing stress area under the washers, with the damage initiating by delamination at the edge of the washer. Reference 7 contains clear photographs and sections of the same kind of behaviour with glass fibre-reinforced plastics.

The author has observed that most bolted joints designed for fibrous composite structures have a bias towards the use of fasteners of less than optimum diameter. This happens because the shear strength of steel and titanium fasteners is much greater than most composites can develop because of their limited bearing strengths. The consequence of using too small a fastener is usually substantial bolt bending long before laminate failure at a significantly lower load than the laminate could have withstood had a larger and stiffer fastener been installed in the same strip of composite. The fastener diameter can be almost half the strip width (or bolt pitch) before the loss of net tension area becomes more significant than the benefits of a reduction in bearing stress. As a rough rule of thumb, steel or titanium fasteners should have about the same diameter

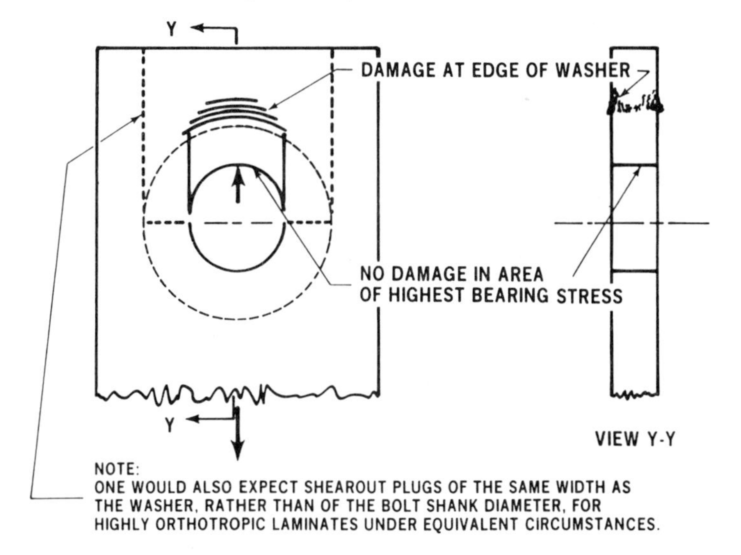

Fig. 9. Composite bearing failures at the edge of a washer rather than at the bolt shank for a highly torqued fastener.

d as the central laminate thickness, *t*, as defined in Fig. 8. The small weight penalty of heavier fasteners is more than offset by the increased operating stress which can then be justified for the entire composite laminate. This is just as true for structural repairs in service as for splices made during initial fabrication.

6.2.4. Single shear joints

The fastener rotation associated with single shear joints decreases the strength of bolted composite joints, with respect to double shear configurations, by two mechanisms. These different bearing strengths must be accounted for in design. Figure 10, taken from ref. 2, gives an indication of the reduction in bearing strength due to bolt rotation and bending without any of the bending of the laminate which would normally be associated with the eccentricity in load path. It is significant that much the same decrease in strength, by 20%, was observed for the splice plates of the double shear joints tested in ref. 6. The mechanism is probably the same in both cases, being the reduction in clamp-up that is shown in Fig. 8. Such laminate bending imposes a severe reduction on the strength of single-fastener single shear test coupons—so much so that designers know not to permit that condition to arise in aircraft structures. Single-row

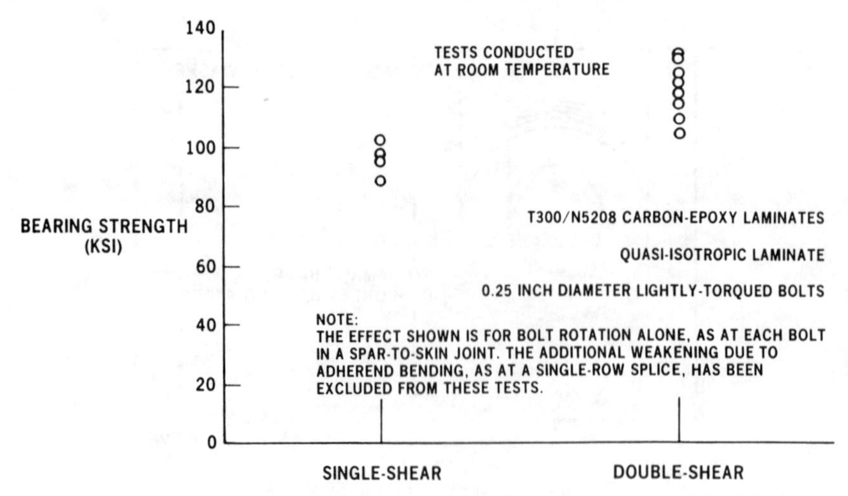

Fig. 10. Difference between bolt bearing strengths under single and double shear ($100\,\text{ksi} \equiv 700\,\text{MN/m}^2$).

single shear joints are used only for in-plane shear load transfer. Any single shear joint transferring tensile lap shear should always have two or more rows of fasteners to reduce the eccentricity in load path, as shown in Fig. 11.

The effects discussed above for carbon/epoxy and fibreglass/epoxy bolted joints are exhibited also by boron/epoxy and even metallic

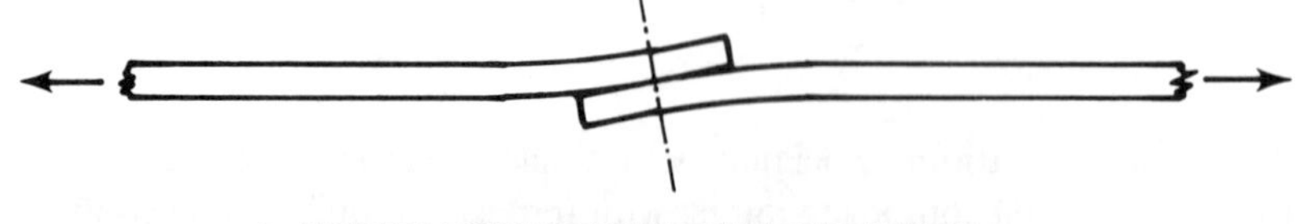

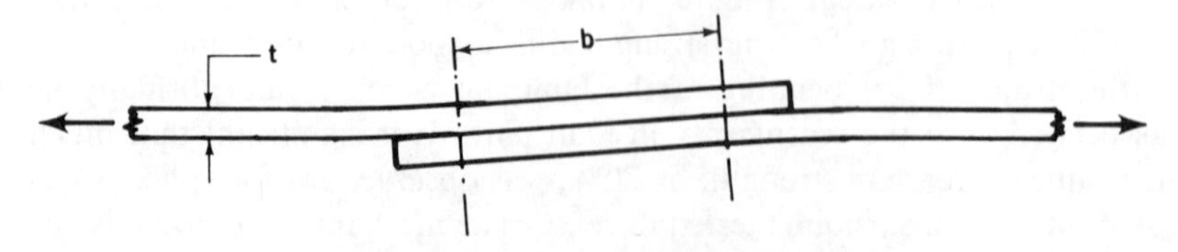

Fig. 11. Alleviation of eccentricity in single shear joints by multiple bolt rows. Note that compressive loading is worse in both cases.

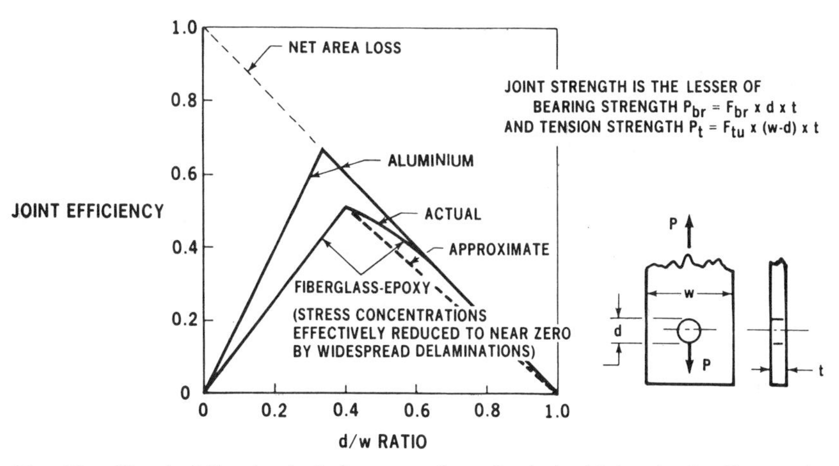

FIG. 12. Simple fully plastic design procedures for bolted joints in ductile metals and fibreglass/epoxy laminates (F_{br} is ultimate bearing strength, F_{tu} is ultimate tensile strength).

structures, with minor shifts in the optimum *w/d* ratios. The analysis and design of bolted or riveted joints in conventional ductile metal alloys are much simpler than those for fibrous composites because of the nearly uniform tensile stresses across the net section through the fastener which is developed prior to failure. The analysis technique is summarised in Fig. 12 and is based on the assumption of virtually complete relief of the stress concentration factors—by yielding in the metal alloys and by massive intra- and inter-laminar failure of the resin (with no fibre damage) in fibreglass/epoxy laminates. A superior analysis for fatigue of metallic joints would account for the non-uniform stresses along the lines explained here. Likewise, the tensile operating stresses in fibreglass laminates should be restricted to control the size of the delaminations around bolt holes in order to retain much of the compressive strength for other load conditions.

The discussion above has referred to tensile shear loading of each fastener. A bolted composite joint behaves differently under compressive shear loading because of the dissimilar stress trajectories, shown in Fig. 13. The bearing strength of most composite laminates is usually found to be higher in compression than in tension. For that reason, and the difficulty of stabilising test specimens under compression, relatively few compression bearing tests are performed. Designs are frequently based on tensile bearing allowables, to be conservative, whenever compressive

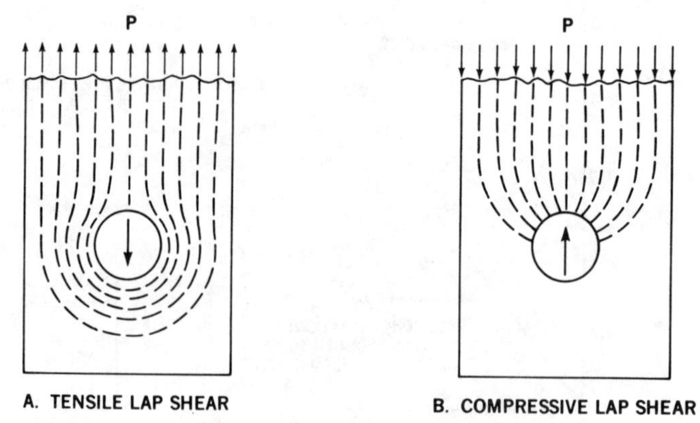

FIG. 13. Stress trajectories around bolts for tensile and compressive lap shear.

bearing data are not available. An exception is made for 'Kevlar' fibre reinforcement, because of its known weakness in compression. The compressive bearing strength is influenced by bolt torque and the presence or absence of washers in much the same manner as for tensile bearing strength.

6.3. EFFECTS OF FIBRE PATTERN ON STRENGTH OF BOLTED COMPOSITE JOINTS

6.3.1. Quasi-isotropic patterns

One of the advantages claimed for fibrous composite structures is that orthotropic properties can be tailored for any specific application. That design philosophy would imply a need for bolted joint strength data for a large variety of laminate patterns. Design of composite structures would then be quite a complex and costly task. However, the motivation for such a philosophy derives mainly from stiffness considerations only; the desire to develop also adequate strength, particularly at bolt holes and cut-outs, tends to restrict the choice of fibre patterns to those that do not deviate very far from a quasi-isotropic pattern. As a general rule there should never be more than three-eighths of the fibres in any one of the basic laminate directions—0°, +45°, −45° and 90°—nor should there ever be less than one-eighth. These fractions refer particularly to highly-loaded unidirectional tape laminates which otherwise tend to split too easily, causing widespread irreparable damage. Similar values were recommended in Chapter 2. Most, if not all, of the major US aerospace companies have similar limits in their design procedures. However, even in 1986, those who have yet to discard their laminate optimisation computer programs still occasionally design structures which violate this principle and, on test, show a distinct unwillingness to accept the load transfer at their joints that had been expected of them. These maximum and minimum fractions are sometimes violated on lightly loaded minimum-gauge structures in which the only alternative is to use a non-standard thinner ply in the laminate. Of course, the choices for all-bonded or integrally stiffened components on one-shot missiles are not constrained at all by any need for repairability or damage tolerance.

The reason for these restrictions on practical laminate patterns for aircraft use is that today's high-temperature-cured brittle resin matrices are not quite adequate to develop the full potential of the high-strength fibres. It is, therefore, necessary to maximise the number of resin interfaces between changes in direction of the layers of fibres and, obviously, specifying a significantly orthotropic lay-up pattern prevents this because parallel plies are inevitably bunched together to at least a minimum extent. Similarly, the use of thicker plies or the unnecessary bunching together of thinner plies has been found by test to impose significant reductions in laminate strengths (which incidentally are not accounted for in most laminate strength theories). It should be noted here that the solution to this problem is definitely not to be found by improving the adhesion between the resin and the fibre—what is needed are tougher, stronger resins, which fail at a much greater strain. Over a decade ago, carbon/epoxy laminates were once made with improved sizing on the fibres. The laminate strengths were reduced dramatically because the most highly loaded fibres then simply broke instead of pulling out of the resin over a short length to alleviate the high fibre strains. This ability of an imperfect bond between the resin and the fibres to average-out local stress concentrations results in more of the fibres being loaded before the final failure. This fibre/resin disbond is vital to the relief of the stress concentrations induced around bolt holes and cut-outs in fibrous composites, as is explained in Fig. 14, taken from ref. 1.

The logical preference for near-isotropic fibre patterns in basic unnotched laminates is even more pronounced at bolt holes. This can be seen in Fig. 15, in which the bearing strengths of HTS carbon/epoxy laminates are plotted for various fibre patterns. Additional data are given in ref. 2, whilst ref. 8 confirms the existence of the presence and height of this plateau for HTS/914 carbon/epoxy laminates. There is an obvious peak strength plateau around the quasi-isotropic pattern in Fig. 15. This

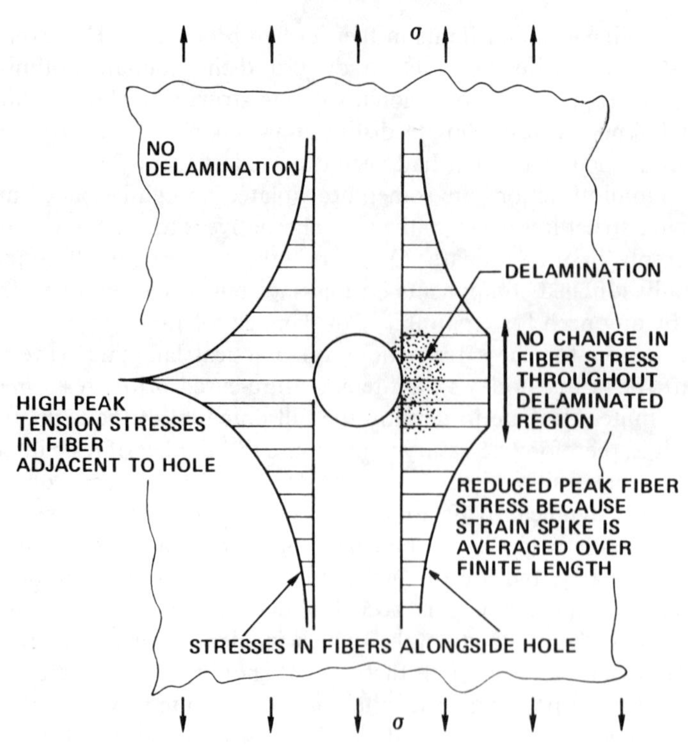

FIG. 14. Stress concentration relief in fibrous composites by delaminations. Taken from ref. 2.

is particularly important since the bearing strength is thus insensitive to small fibre pattern changes in that area, as would occur in laminate thickness transitions. Those patterns with an excess (more than 50%) of 0° plies do not develop the full bearing strengths because of premature failure by shear-out, even at large edge distances (eight to ten times the diameter). The location of the transition in failure mode is known to vary with the fibrous reinforcement and probably also varies with the resin content.

Figure 16 presents the corresponding characterisation of what has rather inappropriately been called the shear-out strength. It can be seen that, despite the short edge distance ($e/d = 2$) and large widths ($w/d = 8$ to 12), all of the failures in the lower part of the diagram were by tension-through-the-hole. The only shear-out failures occurred in the excessively orthotropic patterns at the top of the diagram. It may seem that the

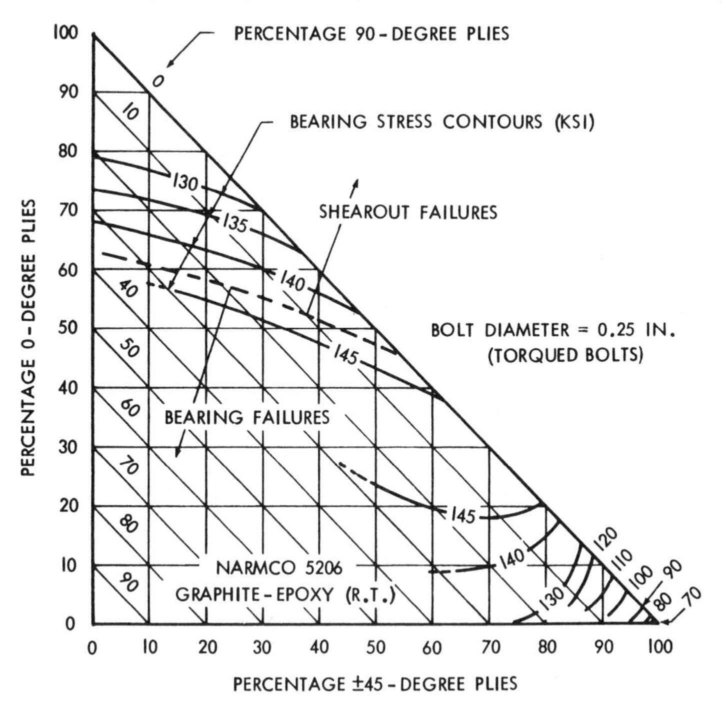

Fig. 15. Bearing stress contours for bolted carbon/epoxy joints (100 ksi ≡ 700 MN/m^2).

absence of shear-out failures for the testing in the lower portion of Fig. 15 would imply that an e/d ratio of 2 is adequate. That is not so because those failures occurred at bearing stresses only about 80% as high as for e/d ratios of 8, which is the value beyond which no further increase can be realised. An edge distance of $3d$ is shown in Figs 37 and 38 of ref. 2 to develop some 90% of the full bearing strength of the laminate. Figures 15 and 16 were constructed from test data on about 15 specific fibre patterns.

The apparent shear-out strengths are not material properties at all and are known to vary with the edge distance. Whereas an e/d ratio of 2 is considered to be the usual minimum for metal alloys, a value of 3 is more appropriate for near-isotropic fibrous composite laminates. (This value of 3 is appropriate for single fasteners or a seam of fasteners subjected to a load perpendicular to the end of the laminate. A lower e/d ratio of 2·5 can be justified for running shear loads parallel to the edge when there

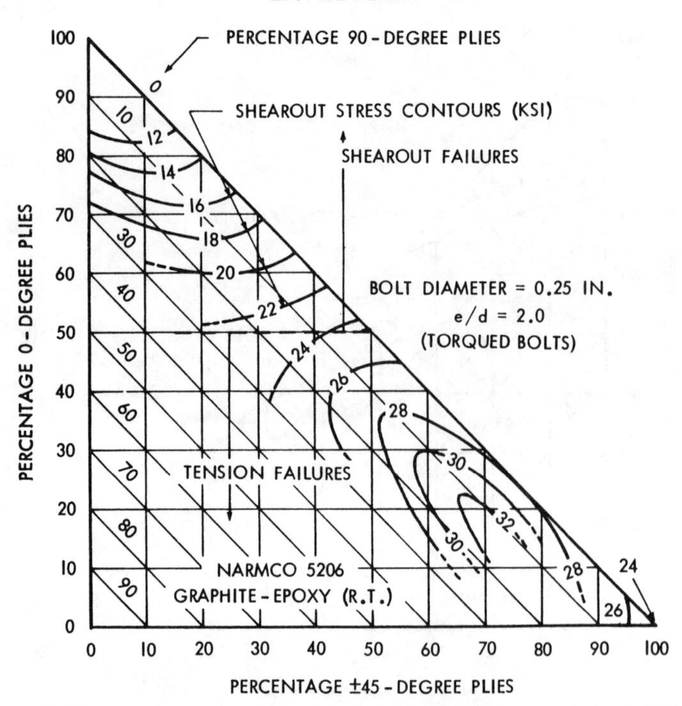

FIG. 16. Shear-out stress contours for bolted carbon/epoxy joints (100 ksi ≡ 700 MN/m^2).

is only a negligible tensile load to consider.) No matter how large the edge distance is, it is not sufficient for highly orthotropic laminates, as can be seen from Figs 17 and 18. The absence of all 90° (or all ±45°) plies in combination with an excess of 0° plies aligned along the load direction results in failures by shear-out of parallel plugs which are often but not always precisely as wide as the bolt hole (see Fig. 19). These failures are premature in the sense that laminates closer to the quasi-isotropic pattern can sustain much higher gross-section stresses without failure, a point also made in Chapter 2. Such shear-out failures have been observed with laminates made from unidirectional tapes in which the reinforcements have been boron, high-modulus carbon and several high tensile strength carbons as well as fibreglass. Woven fabrics should not be so susceptible to this problem because of both the intralaminar weaving and the great difficulty of actually making an excessively orthotropic laminate out of material that inherently contains roughly 50% of cross plies in each layer.

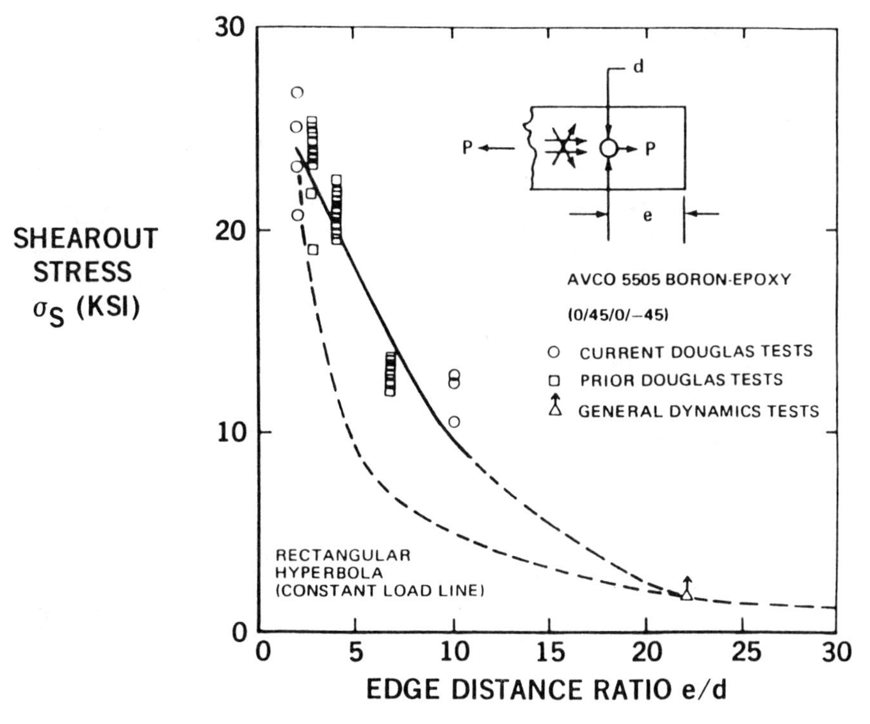

Fig. 17. Shear-out stress as a function of edge distance (10 ksi ≡ 70 MN/m^2).

The coupons in Fig. 18 are reacted at the large central hole and pulled separately on each test bolt hole in turn. The test on the top right was stopped prior to complete failure. It is evident that the initial failure is confined to the immediate vicinity of the bolt. The shear-out failures are only about one bolt diameter long and the edge of the laminate shows no sign of the protruding shear-out plug. Similar but distinct failure modes observed at the McDonnell Aircraft Co., St Louis, did not exhibit a single plug being sheared out. In some of their tests (at a much shorter edge distance), the individual bundles of 0° plies were sheared out after delaminating from the ±45° and 90° plies which remained essentially undamaged.

6.3.2. Stacking sequence effects

After allusion to the strength imbalance between the high-strength reinforcements and brittle, 350°F (177°C) cured, resin matrices, it is

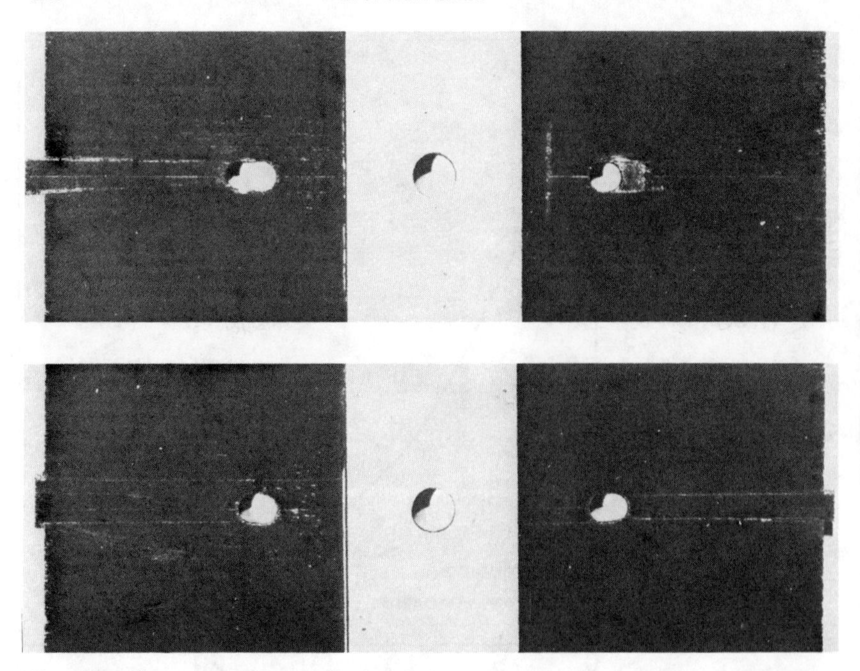

Fig. 18. Shear-out failures for bolted composite joints in highly orthotropic composites.

appropriate to record the highlights from two series of tests which pertain to this subject, even though they did not need any bolt holes to develop the stress concentrations needed to induce premature failure. A series of quasi-isotropic square laminates were made at Douglas Aircraft Company, Long Beach, starting with basically 0·005 in (0·13 mm) unidirectional tape (T300/N5208). All laminates contained the same total number of plies and were, therefore, of the same thickness. However, one specimen was made with a change in fibre direction after each ply (a 'homogeneous' lay-up)—to maximise the number of effective resin interfaces—whilst the others were laid up two plies at a time, four plies at a time, eight plies at a time, and finally, sixteen plies at a time, corresponding to a 0·080-in (2·0 mm) tape. The laminate with sixteen-ply stacks disintegrated before it ever got out of the autoclave. After the remaining square laminates had been trimmed carefully by a diamond slitting wheel, the edges of the laminate made from bundles of eight parallel plies were so badly split by the contraction of the resin during the cool-down after cure as to have substantial resin damage, which was visible to the naked eye. Even the

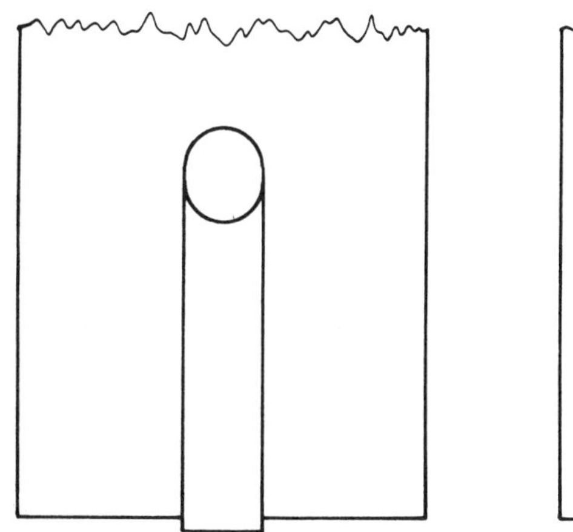
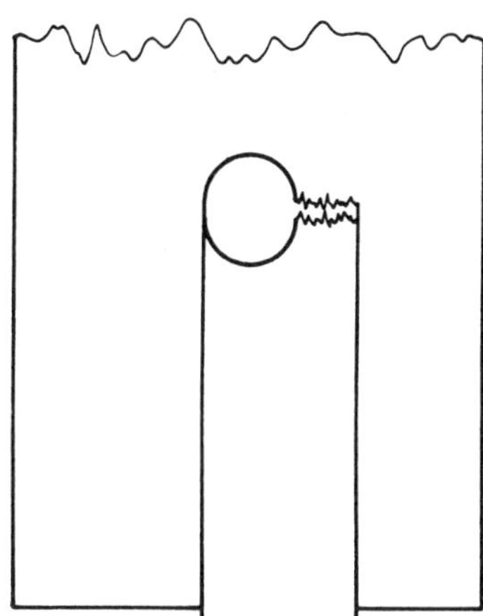

- SHEAROUT FAILURES FOR LARGE EDGE DISTANCES
- NEAT PARALLEL PLUG SHEARED OUT AT LOW LOAD
- FAILURE LOAD INSENSITIVE TO EDGE DISTANCE
- PREVALENT FAILURE MODE FOR FIBER PATTERNS CONTAINING TOO MANY 0° PLIES

Fig. 19. Shear-out failures for bolted joints in composites.

laminate made from four-ply bundles had edge cracks that were detected under a low-magnification eyepiece. Only the specimens laid up one or two parallel plies at a time were considered free from such internal delaminations.

The second series of tests was performed at NASA, Langley,[9] and consisted of tension–tension fatigue testing of laminates containing three 0·005 in (0·13 mm) layers of 90° fibres as a bunch sandwiched between surface layers each having four plies in the $\pm 30°$ direction. This laminate was deliberately selected to aggravate the problem of resin weakness at a change in fibre direction. Despite notch-free sides and the absence of any stress raiser such as a hole, by the end of the test the laminate had been reduced to three distinct wafers and the 90° core had been repeatedly split transversely between the fibres. However, not a single fibre had been broken, proving even by exception the remarkable fatigue strength of fibrous composite laminates.

The significance of these delamination failures is that, in bolted composite joints, those resin interfaces are needed to share the load as the fibres change direction around the hole. In doing so, there is already considerable intralaminar and interlaminar resin failure even with thoroughly interspersed thin-ply lay-ups, as indicated in Fig. 14. Examples of such delaminations are shown in ref. 4 and have been reported in ASTM proceedings in the USA also. It is standard practice at McDonnell Aircraft Co., St Louis, to define the onset of such delaminations as the limit-load capacity of bolted composite structures. The consequent load redistribution has always been found to be sufficient to generate at least a further 50% more load capacity prior to catastrophic failure, which is consistent with the test data in Fig. 4. The ultimate failures of the pin-loaded holes correspond with the onset of damage (i.e. limit load) around the torqued bolts. The additional increment of load carried by the torqued bolts is the result of through-thickness clamp-up, which retards the spread of initial delaminations.

As larger and thicker composite structures are studied and being built, there is considerable pressure on engineering designers to reduce fabrication costs by thickening the basic plies—after all, how can composites compete with a single slab of aluminium alloy for a wing skin that is 10 ft (3 m) wide, 100 ft (30 m) long and over 0·5 in (1·2 cm) thick, particularly if the composite skin alone (let alone the stiffeners) entails the laying-up of over 100 layers of composite tape or about 50 layers of woven fabric? As long as the resin remains the weak link in the composite, it is probable that these pressures will not achieve an appreciable reduction in the number of plies to be laid up but will, instead, merely double the thickness and weight of large bolted composite structures! In addition to this, the design philosophy of the aircraft regulatory agencies throughout the world has been to allow no permanent structural damage at less than limit load for conventional ductile metal alloys. How would they cope with thick-plied composite structures that contained millions of small but detectable internal delaminations before such an aircraft was even assembled? Worse yet, any trend towards thickened plies would have the effect of increasing the minimum gauges (or cost, by using what would by then be non-standard thin plies) on those lightly stressed components such as control surfaces and fairings which are entering service today. It seems to the author that a far more fruitful approach for large subsonic transport aircraft would be to move away from the brittle 350°F (177°C) cured resins which are full of micro-cracks towards ductile 250°F (120°C) cured resins of much greater toughness and to accept *those* associated manufacturing savings instead and leave the ply thicknesses alone. It is interesting to note in this regard that the *only* all-composite aircraft actually already certified by the FAA—the Windecker Eagle—is made from a *room-temperature*-cured, wet lay-up, resin and has not suffered measurable degradation in over ten years of use.

6.3.3. Repair

Before leaving the related subjects of laminate fibre pattern and lay-up sequence, a few words on the provision for repair in service are warranted. At one extreme are the optimised highly orthotropic laminates, with local modifications giving closer to quasi-isotropic patterns, and local build-ups in thickness in the areas of fastener seams. By including as well the use of glass softening strips in basically carbon/epoxy laminates, Douglas has so reduced the stress concentrations at bolt holes as to demonstrate, by test, structural efficiencies in excess of 90% of the unnotched laminate strength (see ref. 10). Unfortunately, thick laminates ($\frac{1}{4}$ in (6·3 mm) or more) of such fibre patterns are to all intents and purposes irreparable and, therefore, quite unsuitable for most aircraft applications. The other extreme is exemplified by the carbon/epoxy wing on the AV-8B Harrier. This wing skin has been made from the quasi-isotropic pattern (25% 0°, 50% ±45°, 25% 90°) throughout, precisely to permit simple bolted repairs, even under battlefield conditions, without the need to seek help from the manufacturer for each and every location of damage.

6.4. EXPERIMENTAL TEST PROGRAMME

6.4.1. General approach

Various experimental investigations into bolted joints in fibrous composite laminates have repeatedly shown that the most efficient mechanical joints are associated with near quasi-isotropic fibre patterns. One might expect then that the analysis of such joints would involve little more than the determination of the appropriate geometric stress concentration, particularly since those unnotched laminates exhibit virtually no non-linear behaviour prior to failure. Such an approach has been shown to be far too conservative. Fibrous composites exhibit considerable non-linear behaviour in the immediate vicinity of stress concentrations and free edges, as discussed in ref. 1. The reason is that the composites have distinct fibre and matrix phases and, whilst each behaves essentially linearly until failure, local failure of the interface between the two can cause significant modification of the internal stress distributions, as explained in Fig. 14. Further 'non-linear' behaviour prior to eventual failure is associated with local splitting of the resin between the fibres in a single layer and with local delaminations between the plies at a change in fibre orientation. Those effects could actually be accounted for with a linear analysis by modelling the composites not as homogeneous orthotropic materials but as individual constituents and interfaces. The complexity of such an idealisation will continue to prevent its widespread use. Nevertheless, the usual analysis of fibrous composites as single-phase materials should not be misinterpreted as implying that they actually behave that way just prior to failure.

Indeed, the interpretation of bolted joint data in refs 1 and 2 makes it clear that analyses based on either perfectly elastic or fully plastic behaviour would be seriously erroneous. The actual results were found to lie roughly halfway between those widely different extremes, as shown in Fig. 20. The correct behaviour could not be predicted on the basis of a minor perturbation of either the linear elastic or fully plastic analysis. The key to relating the test results with the analyses in refs 1 and 2 was an experimentally determined correlation factor.

Another analytical approach for bolted composite joints is to be found in ref. 11. This analysis of the stress field around the bolt holes is linear, but the prediction of failure is made not at the surface of the fastener but at a small offset, of the order of 0·020 in (0·5 mm) away from the hole.

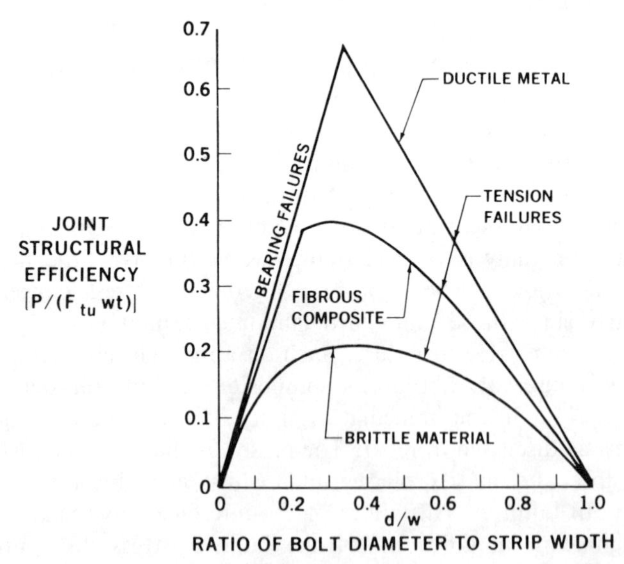

FIG. 20. Relation between strengths of bolted joints in ductile, fibrous composite and brittle materials.

The philosophy behind this approach is that any non-linear behaviour prior to limit load will be confined to the immediate vicinity of the hole (even though damage may spread more than a diameter away from the hole just prior to failure). A perfectly linear elastic analysis should apply beyond that zone. The magnitude of that offset must be established empirically on the basis of specific test data.

This need for a major correlation factor, rather than just a minor perturbation, with respect to some basic analysis means that the reference analysis can be quite simple and still achieve the same end results as a more complex analysis combined with a slightly smaller empirical modification. It seems logical, therefore, that the reference analysis should be that for linear elastic isotropic materials, since so many such analyses are already available. This is the thinking behind the interpretation of the test data reported in ref. 2 in a form which permitted the test results to be generalised to other configurations for which specific test results were not available.

The efficacy of this approach has been demonstrated in ref. 6, in which large multi-row bolted composite joints were designed on the basis of single-row test data. Before-the-fact predictions of joint strength were often within a few per cent of the test results (except for some joints which needed reanalysing because of premature delamination of the splice plates at spot faces around the bolt holes), and the best of the joints so designed attained gross section failure strains of 0·005 at load levels of about 50 000 lb/in (87·56 kN/m).

6.4.2. Planning the test programme

The first step in planning a test programme to characterise the behaviour of bolted joints in a particular composite laminate is the selection of the fibre pattern and lay-up sequence. Wherever possible this should be from within the area identified in Fig. 21, throughout which the joint response has consistently been found to be well behaved. The designer is sometimes forced into the other area by such considerations as minimum gauge requirements and the area identified should not be regarded as surrounded by abrupt strength losses. However, that other area combines significant strength losses with known anomalies in behaviour and generally requires both more extensive testing and more detailed analyses. As an example, consider the simple problem of an unloaded hole in a large composite panel under uniaxial load. Existing test data[1,2] showed that for the quasi-isotropic fibre pattern, the theoretical isotropic stress concentration factor of 3·0 had been reduced to 1·5 for carbon/epoxy

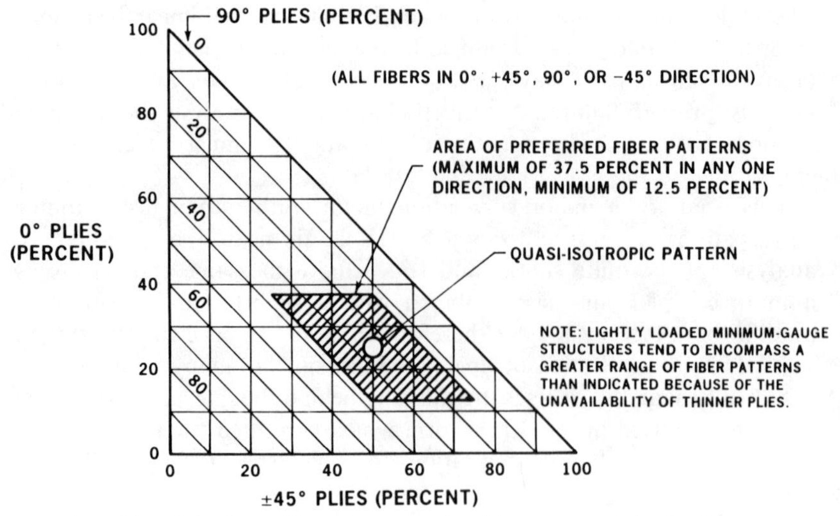

Fig. 21. Selection of lay-up pattern for fibrous composite laminates.

laminates. For two different laminates having three-eighths of the fibres in the 0° (load) direction, the corresponding stress concentration factor had been determined by test to be 1·75 at failure of the composite. (Note that this increased stress concentration factor is associated with an even higher unnotched laminate strength, leading to a net strength gain of about 15% with respect to the quasi-isotropic laminate.) On the basis of these known results, a stress concentration factor of 2·0 had been predicted for a tape laminate containing 50% 0° fibres and 50% at 90°—a laminate that does not seem to be too far outside the shaded area in Fig. 21. The measured strength corresponded to a net section stress concentration factor very close to 1·0. The reason was long ($3d$) cracks between the 0° fibres originating at the tangents to the hole, which effectively unloaded the laminate throughout a strip as wide as the hole. Not a single fibre had been broken by the tests and only extremely small inter-ply delaminations were detected. Similar behaviour has been reported in the literature for other orthotropic laminates on a fairly regular basis.

Having identified the fibre pattern, or patterns, to be tested for a particular fibrous material, the next step is to identify the minimum set of joint geometries to be tested. A useful starting point is the single composite laminate, sandwiched between two metal splice plates, and loaded in double shear to develop the highest possible strengths—which can subsequently be modified by knock-down factors based on other tests found in the literature. Supplementary tests should be run to establish the reduced strengths for flush fasteners, for example, and to verify the strengths of specific highly loaded multi-row joints after they have been designed based on data from single-hole tests. Depending on the needs of each design application, and the presence or lack of some pertinent data base, the tests may have to be run for a single fastener diameter or for a range of sizes. The area of softening caused by the delamination and splitting of the resin is related to the ply thickness and fibre diameter and, therefore, decreases proportionally as the fastener diameter is increased.

For carbon/epoxy laminates, the ultimate tensile bearing strengths are achieved for a laminate having a width (w) and edge distance (e) both equal to eight bolt diameters (d). The transition between bearing and tension-through-the-hole failures has been found to occur at a strip width somewhere in the range of four to six bolt diameters, depending on lay-up (see Chapter 2). Narrower test strips than $8d$ will consequently develop lower bearing stresses at failure (in tension). The laminate thicknesses should be roughly equal to the bolt diameter except that proportionally smaller thicknesses will be needed for small bolt sizes to prevent excessive bending or failure of the bolts. If the design is subject to geometric constraints that preclude an adequate edge distance or, in particular, if the laminate pattern lies outside the preferred area in Fig. 21, one should repeat the bearing test at a short edge distance of $2d$ or so. If the fibre reinforcement is 'Kevlar', which is weak in compression, or if the load spectra warrant it, it may be appropriate to repeat the above tensile bearing test in compression, along the lines described in ref. 2. Glass fibres have much higher strains-to-failure than the so-called advanced composites, so the specimen width and edge distance could perhaps be reduced to $6d$. However, the transition range does seem to be larger for some forms of GFRP (see for example Figs 14–16, Chapter 2) so care should be exercised here.

The next tests cover tension-through-the-hole failures for both loaded and unloaded holes. For the loaded bolt holes, the strips of carbon fibre-reinforced plastic should be about $3d$ wide, corresponding with the highest possible single-hole joint strengths. For glass fibres, a width of $2\frac{1}{2}d$, or perhaps as little as $2d$, might be more appropriate. These particular tests should be as free as possible from any influence of bearing failure. Again, the laminate should be sandwiched between metal clevises and the bolt should be only lightly torqued. It has not yet become usual to run these tests under compressive loading as well as tensile loading because the stress concentration factor is reduced by the load transferred through the fastener in bearing. In the absence of specific test data, the author advocates the use of a stress concentration allowance for compression that is half as severe as for tension. Experience to date suggests that, just as for ductile metal alloys, the compressive design cases for bolted joints are usually less severe than for tensile shear loading. Unloaded hole tests should be run at a strip width of $4d$, taking advantage of light clamp-up by a bolt in the hole. The precise fit of the fastener in any of the single-hole tests described above is known not to be critical because of the substantial non-catastrophic damage sustained by the laminates prior to eventual failure. These unloaded-hole data are needed in the design of multi-row bolted joints, in which the bolt rows are typically $5d$ apart in carbon/epoxy or $4d$ apart in fibreglass/epoxy. If all of the tensile tests above have failed with a visibly similar failure mode, preferably clearly by tension-through-the-hole, the need for further testing will be minimised. The results should be compared against known results in the literature to identify any anomalies or need for retesting.

On the basis of the test programme outlined above, one might need as little as three test geometries for each bolt diameter and material combination. The reasons for recommending the specific joint geometries above will become apparent in the following section, in which the reduction of these data is discussed.

6.5. CORRELATION BETWEEN TEST AND ANALYSIS

The hypothesis on which the author's analysis of bolted composite joints[1,2] is based is that the stress concentration relief due to resin damage prior to fibre breakage is proportional to the intensity of the stress concentration factor itself. In other words, the stress concentration factor k_{tc} on the net section of the composite laminate is proportional to (and less severe than) the elastic isotropic stress concentration factor k_{te} for an ideal joint having the same geometry. This is explained in Fig. 22, which compares hypothetical but typical stress concentration factors observed at failure in quasi-isotropic carbon/epoxy laminates with the corresponding theoretical predictions for perfectly elastic isotropic materials. Actual plots of results such as these are given in refs 1, 2 and 6. It is immediately evident that there is substantial relief from the theoretical predictions; this is expressed more explicitly by the higher strengths shown in Fig. 20. The reasons for picking the specific values of the w/d ratios are also evident. On the right side of the picture, the results for the bearing failures (large w/d ratios) should lie consistently above the line drawn through the tension-through-the-hole failures by a steadily increasing increment.

Actually, the consistency of the loaded- and unloaded-hole data is not always as good as shown in Fig. 22. However, there are as many sets of results for unloaded holes distinctly above the line through the loaded-hole data as there are below this line. Also, some of those results substantially below that line are known to have a markedly different failure mode.

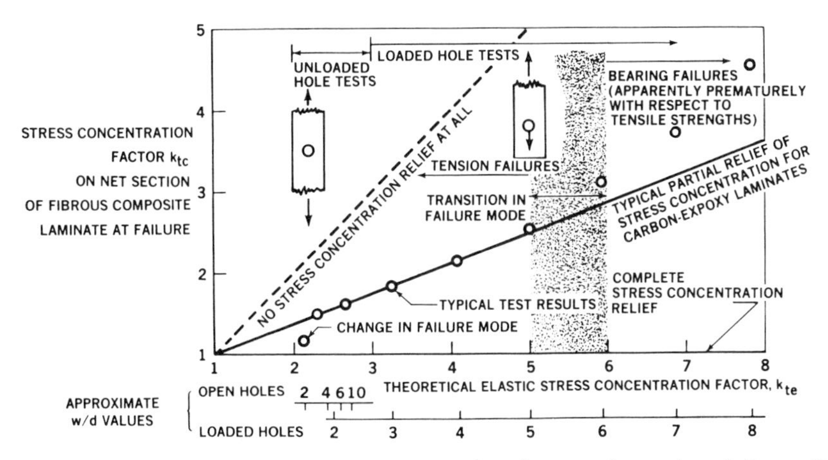

Fig. 22. Relation between stress concentration factors observed at failure of fibrous composite laminates and predicted for perfectly elastic isotropic materials.

Relationships of the type shown in Fig. 22 have been prepared for different carbon/epoxy materials, for different fibre patterns, and for different fastener diameters. The results are subject to many influences—particularly to bolt clamp-up—and the scatter is too great to justify the fitting of anything other than a straight line through the data. Nevertheless, this approach has made it fairly simple to predict reliably the strength of bolted joints which have not been tested, on the basis of a very limited number of tests. The influence of composite material orthotropy on the stress concentration factor might have been accounted for by modifying the abscissa k_{te} by an appropriate factor. However, provided that the failure mode and location are the same, the same result is achieved by absorbing that additional factor into the slope of the line in Fig. 22.

Thus,

$$k_{tc} - 1 = C\,(k_{te} - 1)$$

where the coefficient C provides the necessary correlation between test and theory. As an example, C has a value of about $\frac{1}{4}$ for quasi-isotropic carbon/epoxy laminates made from 0·005 in (0·13 mm) thick tape in brittle (350°F, or 177°C, cured) resins. A thorough quantitative treatment of this aspect of the problem is given in refs 1 and 2. The expressions for the stress concentration factors k_{te} are quite simple. For example, for an unloaded hole of diameter d in a strip of width w,

$$k_{te} = 2 + \left(1 - \frac{d}{w}\right)^3$$

Likewise, for a loaded hole in a similar laminate, with the centre of the hole a distance e from the edge of the laminate,

$$k_{te} = \left(\frac{w}{d} + 1\right) - 1{\cdot}5\,\frac{(w/d - 1)}{(w/d + 1)}\,\Theta$$

in which

$$\begin{aligned} \Theta &= 1{\cdot}5 - 0{\cdot}5/(e/w) \quad &\text{for} \quad e/w \leq 1 \\ \Theta &= 1 \quad &\text{for} \quad e/w \geq 1 \end{aligned}$$

The stress concentration factor at failure is computed as

$$k_{tc} = \frac{F_{tu}(w - d)t}{P}$$

in which P is the failing load and F_{tu} the unnotched ultimate tensile strength of the laminate.

With these formulae, together with the bearing strength cut-off, it now becomes possible to predict the behaviour of each fastener in a multi-row joint, even without testing such more complex joints. Tests of that type which have been run already[1,2] have shown a linear interaction between the bearing and by-pass loads (as explained in Fig. 2) from as long ago as 1970, when what appears to have been the first such tests were run at General Dynamic Convair in San Diego, USA. That linear interaction is the basis of the unkinked straight line in Fig. 2 for narrow strips. For wider strips the kinked straight line can be formed by combining the bearing strength cut-off with a theoretical extension of the formulae above to the unmeasurable tensile load that the laminate could have endured

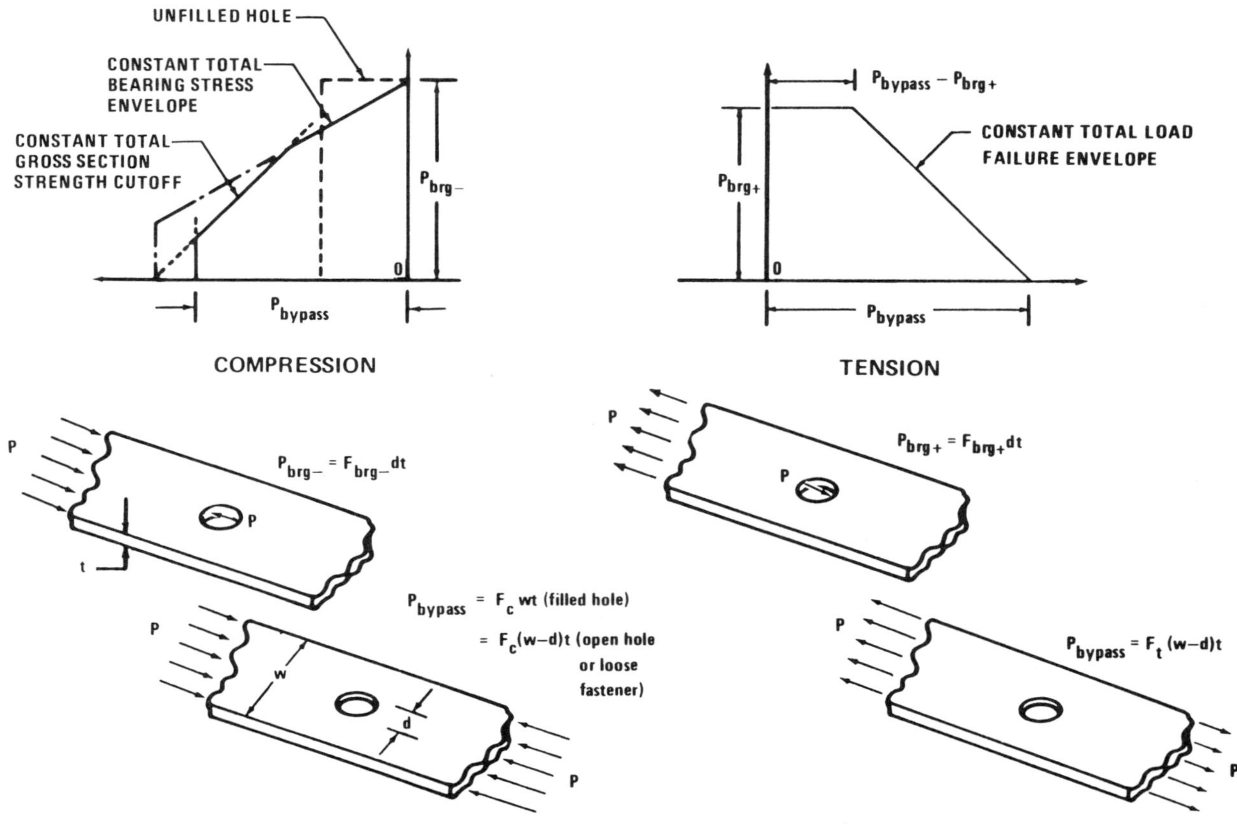

FIG. 23. Outer envelope of bearing–by-pass load interactions, for ductile metal alloys.

had it not failed first in bolt bearing. The BJSFM analysis method[11] developed at McAIR also predicts a straight line for each segment of the bearing–by-pass interaction, with any kink associated with a change in failure mode.

Similar bearing–by-pass interactions can be developed for compressive loading also, as shown in Fig. 23, again on the basis of very few tests. References 6 and 12 contain a more detailed discussion of this aspect. The corresponding treatment of running shear loads does not appear to have been performed yet.

6.6. EFFICIENCY CHARTS FOR DESIGNING BOLTED COMPOSITE JOINTS

The combination of the formulae above and the correlation factor, together with the bearing strength cut-offs, make it possible to predict the joint strengths for any geometry of single- or multi-row bolted joints subjected to tensile or compressive double shear loads. When such joint strengths are divided by the unnotched gross-section laminate strengths, one would predict the corresponding joint efficiency. Such assessments have shown that the most carefully designed fibrous composite joints will achieve an efficiency of no more than 40–50%, no matter how many fasteners or how much design finesse or analysis effort is expended. To place that number in perspective, the usual design range for ductile metal alloys is an efficiency of 70–80%. The lower numbers in Fig. 12 refer to single-row joints.

Now, if that joint efficiency were to be multiplied by the ultimate fibre strain-to-failure, the strength of bolted joints in fibrous composites could be characterised in a manner that is very easily related to the conventional design process for composites. This has been done in Fig. 24 (adapted from ref. 1), which encompasses the entire gamut from unloaded holes at the top to the transfer of 100% of the load on a single fastener at the bottom. The interval between covers all possible forms of multi-row bolted joints. Figure 24 was actually prepared for 0·25 in (6·35 mm) bolts in quasi-isotropic HTS carbon/epoxy laminates, but is quite representative of the characteristics of other fibre patterns and materials. (Note, however, that fibreglass laminates would not exhibit the tensile-failure plateau on the top of the single-hole curves, but would be more triangular in form.)

Many important conclusions can be drawn from Fig. 24. As a reference, the ultimate unnotched fibre strain can be taken to be 0·009,

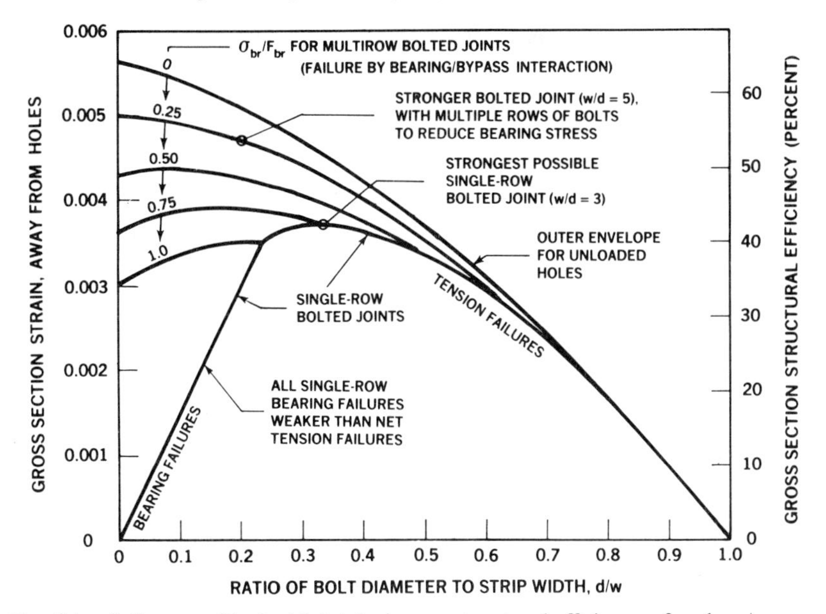

FIG. 24. Influence of bolted joint design on structural efficiency of carbon/epoxy composite structures.

so that the strongest single-hole bolted joint, at $d/w = 0{\cdot}3$, has a joint efficiency of 40% and a gross-section failure strain of about 0·0035. The associated failure mode is tensile and is inherently stronger than any single-bolt bearing failure, which would require smaller values of d/w ($<0{\cdot}2$) to realise. That optimum single-bolt joint is hard to improve upon without considerable complexity being added to the joint. For example, in refs 1 and 2, there is an account of testing joints which were identical to the single-bolt tests in all other regards except that they contained two bolts in tandem. The test results showed consistently only a 10% increase in strength, which is quite consistent with the predictions of Fig. 24 in which, for the same $3d$ width, the bearing stress had been halved by sharing the load between two bolts. This seemingly surprising result is caused because the stress concentration factor on the by-pass load is very nearly as high as that for the bearing load.

That 40% best efficiency for single-bolt composite joints is not competitive with major aluminium alloy primary structures on a weight basis—and the equivalent value for an orthotropic laminate with three-eighths of the fibres in the direction of the principal load is barely competitive. One can avoid this problem on small aircraft by eliminating the major splices, as on the one-piece wings of the AV-8B Harrier and LearFan aircraft. Alternatively one can employ an efficient but costly titanium stepped-lap adhesive bonded splice at the wing root, as on the wing of the F/A-18 aircraft. Local reinforcement could permit simple one-row bolted splices at the time of manufacture (actually, two rows or one staggered row would be needed to stabilise compressive loads), but at the cost of repairability of the basic thinner laminate. However, Fig. 24 identifies the only possible way of overcoming this limitation—by moving up and to the left from the location of the best single-bolt joint. That move necessarily involves multiple rows of fasteners, an increase in the bolt separation (a smaller value of d/w, down to 1:5 or 1:6), but more importantly, a drastic reduction in the bearing stress to no more than 25% of what the laminate could withstand. Moving the lines of bolts further apart seems at first sight to be incompatible with a major reduction in the bearing stress which would normally be accomplished by using fasteners of a larger diameter, which is equivalent to moving the bolts closer together. Fortunately, this challenge is not as daunting as it first seems because the bearing stress needs to be reduced at only the most critically loaded bolt stations. The other stations are rendered less critical by the progressive transfer of load out of the basic laminate into the splice plates.

The possibility of thereby achieving joint efficiencies slightly in excess of 50% has been verified by the test results reported in ref. 6. Whilst difficult in terms of joint complexity, that represents a substantial 25% weight saving for the entire laminates (with respect to the best single-row joints) for the trivial penalty of a slightly heavier splice. The significance of that seemingly minor accomplishment is that, for the first time, there has been an actual demonstration that major composite structures so large as to need mechanical splices can, in fact, be made lighter than equivalent well-designed aluminium alloy structures while not giving up the possibility of practical bolted repairs. The attainment of such improved efficiencies for four-row bolted joints relied upon the accurate load-sharing computations permitted by the non-linear computer program A4EJ (see ref. 12) in conjunction with bilinear fastener stiffnesses deduced from test results and a formula proposed in ref. 6. It would be premature to say that the design of major mechanical splices in composite primary aircraft structures is straightforward yet.

Nevertheless, some definite trends have been identified by analytical studies of such joints. Figure 25 shows the bolt load distributions for four different four-row bolted joints which were analysed to compare their

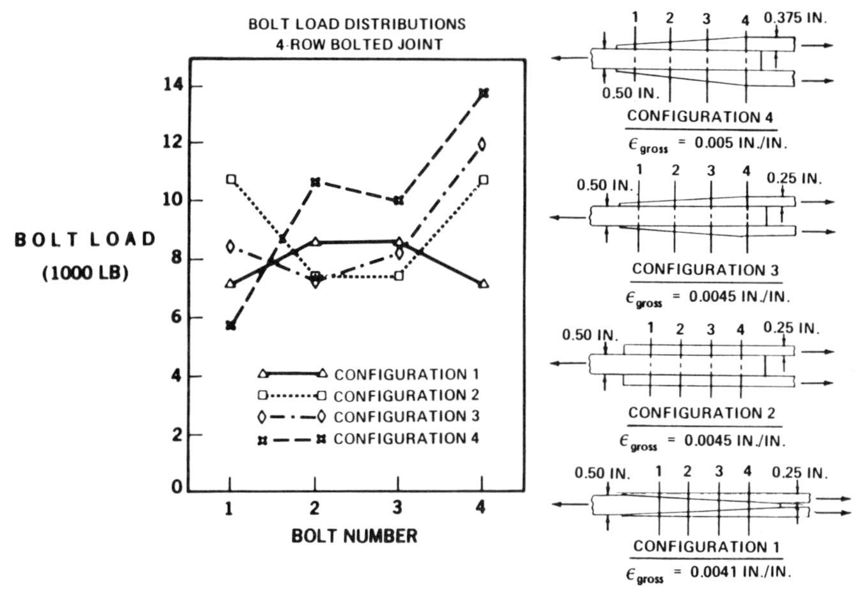

FIG. 25. Effect of joint configuration on bolt load distribution (1000 lb = 4·448 KN, 1 in ≡ 25·4 mm).

strengths. Somewhat surprisingly, the weakest joint was the scarf joint, with both the basic laminate and splice plates tapered (configuration 1). This weakness was due to the loss of thickness at the first fastener station, with respect to the basic laminate outside the joint. The joint with uniformly thick members was predicted to have the second highest strength. That joint having tapered but unreinforced splice plates sandwiched around a uniform skin was found to be the second weakest. With respect to the joint having untapered splice plates, all that the tapering accomplished was to reduce the load transfer through the outermost bolts while not changing the criticality of the most highly loaded bolts at the middle of the splice plates, adjacent to where the skins butt together. The strongest joint analysed combined tapering of the splice plates with their reinforcement, while leaving the basic laminate at a uniform thickness. The objective of such a design is to maximise the load transfer on the last bolt in the skin, where there is no by-pass load, while minimising the load transfer on the first bolt, where the by-pass load is greatest. The variables with which to accomplish this are the amount of reinforcement, the minimum taper thickness and selective variation of fastener diameter. It should not be surprising that this joint configuration was shown to be superior. It has long been known that the techniques for achieving high static joint strengths in fibrous composites are precisely the same as those employed to develop long fatigue lives in metal aircraft structures—and that superior composite joint has all of the same features as the wing skin splice at the side of the fuselage on the DC-10.

The design of highly loaded joints such as described in Fig. 25 is a specialised task requiring data not yet generally available. However, aircraft structures contain many more lightly loaded fasteners and quick simple procedures have been developed throughout the industry for such cases as the running shear loads transferred by the bolts joining wing skins to spars, for example. Several companies have accomplished such tasks by means of a chart of the form of Fig. 26, but with minor variations in the values shown (see, in particular, ref. 13). If this diagram is interpreted in the context of Fig. 24, it can be seen that the allowable design region is safe to use without deeper insight into the effects of the joint variables and that, with a few simple restrictions on the d/w ratios, the 0·004 strain cut-off could well be raised to 0·005. Conversely, doing so would imply a need to ensure that there were no abnormalities in the fastener bearing stresses—for instance due to material thickness changes or some very loose bolt transferring its load into the two adjacent bolts—so it should be said that the procedures in Fig. 26 are well founded, even though higher strains will be needed to make highly loaded primary

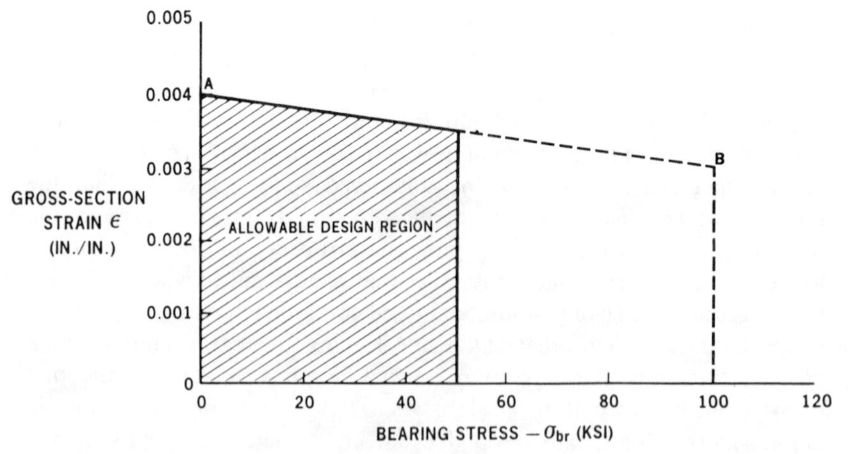

Fig. 26. Design technique for bolted carbon/epoxy structures (100 ksi ≡ 700 MN/m^2).

composite aircraft structures competitive with tomorrow's improved metal airframes.

The design of all but highly loaded primary structural bolted or riveted joints in fibrous composite laminates is thus seen to be straightforward in most cases. The treatment of the exceptions is discussed in the following section. After the choice of material has been made, the first question to be resolved is the selection of laminate lay-up pattern. A good general-purpose pattern is the quasi-isotropic pattern (25% 0°, 50% ±45°, 25% 90°). If there is a dominant load direction, identified with the 0° direction, a more appropriate laminate would contain about three-eighths of all the fibres in that 0° direction. The associated choice between 50% ±45° with $12\frac{1}{2}$% 90° or $37\frac{1}{2}$% ±45° with 25% 90° would then be determined by the relative severity of perpendicular loads (in the 90° direction) or of torsional requirements (in the ±45° direction). The next question to be resolved is whether or not there needs to be provision for repair of the structure. If so, the selection of the best laminate pattern in the area of the joints also defines the laminate pattern and thickness in the spaces in between. Experience has indicated that the opposite approach of first selecting some idealised pattern for the basic laminate usually compromises the strength of the joints and hence the integrity and repairability of the structure.

However, if there is no need for either damage tolerance or provision for repair, only those plies required in the basic laminate need be continuous—the others in the joint area then become local build-ups. In such one-shot applications, the use of glass softening strips can provide even greater structural efficiencies for basically carbon fibre-reinforced laminates. Glass softening strips involve the substitution of local strips of 0° carbon fibres by strips of glass fibres about four diameters wide. No carbon fibres are cut since those in the ±45° and 90° directions remain continuous. Tests on such hybrid laminates have confirmed the virtual elimination of stress concentrations at bolt holes. The improved strengths of bolted joints in such materials are characterised by the test results in ref. 2, which also explains how to deduce the stress concentration alleviation factor on the basis of extremely limited test data.

The thickness of the basic laminate in the presence of lightly loaded fasteners is then determined by the local load intensity in conjunction with the laminate stiffness and strain level of 0·004 from Fig. 26. For bearing stresses as high as 50 ksi (350 MPa), the laminate strain would be restricted to 0·0035. If that meant thickening of the laminate in such an area, a trade-off would be made between the weight of such thickening and the weight of larger fasteners to reduce the bearing stresses. As a general rule, the fastener diameter should be about the same as the laminate thickness, being greater for laminates of up to $\frac{1}{4}$ in (6·36 mm) and less for laminates thicker than about $\frac{3}{4}$ in (19·05 mm). The shear strength of steel or titanium fasteners should usually not be even close to critical, although the stiffness of the fasteners and their bending yield strength can be of particular importance to the load sharing in multi-row joints.[6] The apparent excessive shear strength of fasteners in well-designed bolted composite joints has at times led to the redesign of the joint to use smaller and lighter fasteners (to remove that strength margin), at the expense of a considerable portion of the overall joint strength. That tendency needs to be resisted since the associated reduction in laminate strength is clearly evident in Fig. 26.

Finally, the strength of composite laminates containing bolt holes can be improved by clamping the laminate well between large washers and protruding-head fasteners wherever possible. Likewise, some conservatism should be exercised whenever such clamp-up is impossible. Reliance on the extra strength associated with severe, rather than lighter, bolt torques should be avoided except in tightly controlled special applications.

The straightforward design procedure above becomes more complicated in cases which, for various reasons, violate the rules for maximum possible efficiency in bolted composite joints. Such cases will usually require additional testing and some instances are discussed in Section 6.7.

6.7. MISCELLANEOUS CONSIDERATIONS

Deviations from the straightforward approaches above can sometimes occur due to truly unavoidable constraints. A number of these and their consequences are discussed in turn. Specific tests are often necessary in such cases, to generate data for design allowables.

6.7.1. Countersunk fasteners

The most frequent limitation on joint strength is due to the need for flush fasteners to minimise aerodynamic drag. The strength of such bolted joints in composites is reduced both by the non-uniform bearing stress associated with the usually concurrent rotation of the fastener under single shear and by the reduction in additional strength after the initial failure owing to the greatly reduced clamp-up. Whilst no completely general explanations have been found (so specific tests are usually necessary), it was shown in ref. 14 that a reasonable strength prediction could be made for near-isotropic laminates by ignoring the bolt head completely and considering that the shank was able to develop a bearing stress of about 100–120 ksi (700 to 800 MPa), just as for double shear joints. Obviously, this approximation would break down in the extreme case of knife-edge countersinks. The countersunk head should penetrate no more than one-third to one-half of the laminate thickness, depending on the operating bearing stress on the shank.

Another problem arises for structures that are so lightly loaded that it is not necessary to use as many as eight of the minimum gauge tape plies to sustain the loads. In the same category are structures in which there is no significant shear load on the fasteners because the loads acting on the structures are mainly normal, rather than in-plane. Sometimes, tension-head (rather than shear-head) fasteners are needed to prevent the bolt head from being pulled through the laminate. That condition can arise under both normal pull-off and pure shear loadings. Such larger bolt heads would require local thickening of the laminate. Such cases require specific testing of any bolted joints, if the weight of the structure is critical, because the failure modes are often not as well behaved as for those of the preferred fibre patterns in Fig. 21. Consequently, bolted joints in such composite structures fail at lower stresses.

If a fibrous composite structure has been designed as a replacement for an existing metal structure, there is often an undesirably low edge distance caused by the mating with an adjacent structure. This condition is particularly bad if it is associated with a fibre pattern outside those recommended in Fig. 21. Reference 2 gives some idea of the reduction in joint strength due to this condition. As a guide, the bearing strength for a $2d$ edge distance is only about 80% of that for a large ($8d$) edge distance. The short edge distance causes the stress trajectories in Fig. 13 to bunch up much closer to the fastener.

6.7.2. Fatigue

It has been found from considerable experience that well-designed bolted composite joints are not sensitive to cumulative damage under fatigue loads. Another way of expressing this is that those factors which reduce the fatigue strength of joints in ductile metal structures (and have little effect on their ultimate strengths) impose a substantial reduction on the static ultimate strength of fibrous composite structures. The same techniques to relieve stress concentrations are needed in both situations. Actually, gentle tensile–tensile fatigue of composites tends to *increase* the

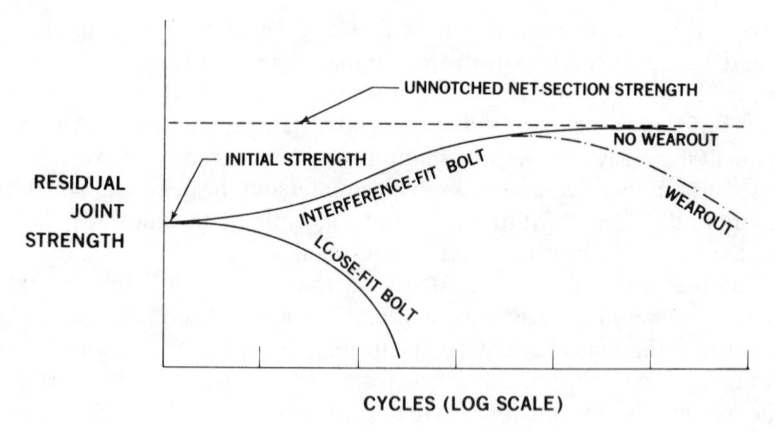

FIG. 27. Fatigue of notched composites or bolted composite joints (wear-in may or may not be followed by wear-out).

residual strengths of composite structures, as shown in Fig. 27, because of progressive relief of the stress concentrations. Tests at NASA Langley and elsewhere have shown that the strength can increase asymptotically towards the unnotched net-section strength. That is to say that there is a 'hole-out' factor with no stress concentration, or notch-insensitive situation, just as for ductile metals. Obviously, one cannot rely on preconditioning to increase the joint strength but at least there is usually no concern about loss of strength under such conditions throughout the life of the structure. Unfortunately, the same cannot be said of the life of a composite laminate containing loose bolts pounding back and forth in their holes under reversed (tension–compression) fatigue loading. It is most important that the fasteners in composite laminates should not be loose and that the laminates should not be damaged by the installation of fasteners with an excessive interference.

6.8. CONCLUDING REMARKS

It has been suggested[4] that the present author is in favour of a 'continued reliance on art, skill and empiricism' in the design of bolted composite joints. In the period of more than five years that has passed since the writing of the original and more explicit final paragraph in ref. 1—which the author of ref. 4 summarised so succinctly—considerable progress has been made in this subject. Yet today the design of highly efficient multi-row bolted joints is still very much an art, even though the scientific computer programs to determine the load sharing and to assess the combination of the bearing and by-pass loads provide invaluable tools in that task. Furthermore, the considerable supply of test data for precise elastic analysis of bolted composite joints has not prevented the reporting of tension-through-the-hole strength as a material (net-section) allowable depending only on the material and fibre pattern (as would be the case for a ductile metal), as if it were independent of the joint geometry (the w/d ratio in particular) and the bearing/by-pass load interaction. Worse, that same error has been perpetuated in a widely distributed composites design manual. There is still a need for considerable skill in interpreting and integrating the bulk of largely unco-ordinated testing of bolted composite joints, with many investigators interested in only specific aspects of the problem. Finally, no accurate predictions of the strengths of bolted composite joints have yet been made without reliance on either a substantial empirical correction factor or the benefit of directly applicable test data. And, given the lack of a laminate theory accounting for separate fibre and resin constituents and possible failures at the interface, this situation is likely to persist. In other words, science alone still cannot get the job done without a considerable dose of art, skill and empiricism!

The purpose of this chapter has been to provide an integrated framework within which the behaviour of bolted composite joints can better be understood. The analysis and design of bolted joints in conventional ductile metal alloys benefits from gross yielding of the metals to such an extent as to completely mask any imperfections in the rather simple analyses used for that task. The design and analysis of bolted joints in fibrous composite laminates will always remain a far more complex problem because of the extremely low strain-to-failure of such materials. Nevertheless, the 'non-linear' behaviour that is created by the non-catastrophic local failures between the resin and fibre phases of composite materials provides a considerable increase in strength beyond that which could be achieved by equivalent elastic homogeneous materials.

Although it does not provide a large source of specific composite material data, this chapter has identified relatively simple procedures to cover the design and analysis of bolted composite joints throughout the regime in which they are simultaneously the strongest and the best behaved.

ACKNOWLEDGEMENT

Whilst this chapter alludes to information from several sources, the primary source of the material presented here has been three US

Government-sponsored investigations in which the author participated. The first is the NASA Langley Contract NAS1-13172, *Bolted Joints In Graphite Epoxy Composites*, which was reported in NASA CR-144899[2] in January 1977 and forms the basis of Douglas Paper 6748A, November 1978. The second is the USAF Flight Dynamics Laboratory Contract F33615-79-C-3212, *Design Methodology for Bonded–Bolted Composite Joints*, to be found in AFWAL-TR-81-3154, dated February 1982, and of which the computer codes in Volume II are not available for unlimited distribution (see ref. 15). The third research programme was the NASA Langley Contract NAS1-16857, *Critical Joints in Large Composite Aircraft Structure*, documented *inter alia* in Douglas Paper 7266, January 1983.[6]

The author would also like to take this opportunity to thank his many friends throughout the international aerospace community who have kindly provided him with insight into their own researches and problem areas and thereby contributed to the understanding of the subject conveyed in this chapter.

REFERENCES

1. Hart-Smith, L. J., Mechanically-fastened joints for advanced composites—Phenomenological considerations and simple analyses, Douglas Aircraft Company Paper DP6748A, November 1978. Also contained in *Fibrous Composites in Structural Design* (Ed. E. M. Lenoe, D. W. Oplinger and J. J. Burke), Plenum Press, New York, 1980.
2. Hart-Smith, L. J., Bolted joints in graphite–epoxy composites, Douglas Aircraft Company, NASA Langley Report CR-144899, January 1977, NASA, Washington DC, USA.
3. Crews, J. H., Jr, Bolt bearing fatigue of a graphite/epoxy laminate, *Joining of Composite Materials*, ASTM STP 749 (Ed. K. T. Kedward), 1981, American Society for Testing and Materials, pp. 131–44.
4. Godwin, E. W., Matthews, F. L. and Kilty, P. F., Strength of multi-bolt joints in GRP, *Composites*, **13**(3), July 1982, 268–72.
5. Shivakumar, K. N. and Crews, J. H., Jr, Bolt clampup relaxation in a graphite/epoxy laminate, TM 83268, January 1982, NASA, Washington DC, USA.
6. Nelson, W. D., Bunin, B. L. and Hart-Smith, L. J., *Critical Joints in Large Composite Aircraft Structure*, Douglas Aircraft Company Paper DP 7266, January 1983.
7. Matthews, F. L., Wong, C. M. and Chyrssafitis, L., Stress distribution around a single bolt in fibre-reinforced plastic, *Composites*, **13**(3), July 1982, 316–22.
8. Collings, T. A., On the bearing strengths of CFRP laminates, *Composites*, **13**(3), July 1982, 241–52.

9. O'Brien, T. K., *Characterization of Delamination Onset and Growth in a Composite Laminate*, NASA Langley TM 81940, January 1981, NASA, Washington DC, USA.
10. Nelson, W. D., *et al.*, *Composite Wing Conceptual Design*, Douglas Aircraft Company USAF Contract Report AFML-TR-73-57, March 1973.
11. Garbo, S. P. and Ogonowski, J. M., *Effect of Variances and Manufacturing Tolerances on the Design Strength and Life of Mechanically Fastened Composite Joints*, McDonnell Aircraft Company, USAF Contract Report AFWAL-TR-81-3041, V. 1–3, April 1981.
12. Hart-Smith, L. J., *Design Methodology for Bonded–Bolted Composite Joints*, Douglas Aircraft Company, USAF Contract Report AFWAL-TR-81-3154, February 1982.
13. Garbo, S. P., *Effects of Bearing/Bypass Load Interaction on Laminate Strength*, McDonnell Aircraft Company, USAF Contract Report AFWAL-TR-81-3144, September 1981.
14. Thompson, C. E. and Hart-Smith, L. J., *Composite Material Structures—Joints*, Douglas Aircraft Company, IRAD Report MDC-J0638, July 1971.
15. Hart-Smith, L. J., *Bonded–Bolted Composite Joints*, Douglas Aircraft Company, Paper DP 7398, May 1984; see also: *Journal of Aircraft*, **22**(11), Nov. 1985, 993–1000.

Chapter 7

Design of Adhesively Bonded Joints

L. J. Hart-Smith

Douglas Aircraft Company, McDonnell Douglas Corporation, Long Beach, California, USA

7.1. INTRODUCTION

The use of adhesive bonding in aircraft structures dates back to the dawn of aviation history. Wooden structures contained glued scarf joints in the primary members; wing spars were often constructed from laminated strips of wood to improve the mechanical properties; and plywood skins and shear webs contained much glue in themselves and transferred their loads to the primary structural members entirely through bonded joints. In the infancy of the development of airframes, mechanical fasteners were used at areas of relatively high load intensity (for which the glued joints were not considered strong enough) and to provide manufacturing breaks to facilitate storage and repair. One is tempted to suggest that the understanding of our great-grandfathers about the respective uses of adhesive bonding and of mechanical fastening in aircraft structures seems to have been lost with the passage of time.

At all stages of development since then, the science of predicting the strength of adhesive bonded joints has always been adequate. There are virtually no known instances of failures associated with improperly proportioned joints. Unfortunately, there are far too many well-known service problems involving the interaction between moisture and the glue (or resin). Such problems have occurred with wooden structures with the early casein glues. They recurred later with some US adhesively bonded metal structures, particularly those with a perforated honeycomb core which had not been treated to enhance its corrosion resistance. Even today, moisture absorbed within composite laminates has an adverse effect on secondary bonding and on the second-stage curing of stiffeners against a pre-cured skin, for example. It should be noted that all of these problems are such as to be beyond analysis. They are not load-sensitive and the only solution is to eliminate the problems—not to establish some knock-down factor and tolerate the existence of such conditions.

The introduction during the 1960s of the advanced fibrous composites—boron/epoxy and carbon (or graphite)/epoxy—led to a resurgence of interest in the design and analysis of bonded joints. This has been particularly so for highly loaded joints between composites and stepped titanium plates because of a desire to avoid the reduction in structural efficiency associated with bolt holes in such brittle materials. However, there has been a subsequent recognition that such all-bonded thick fibrous composite structures are all but impossible to repair back to the original structural capabilities. Therefore, the extra-efficient composite structures are likely in the future to be confined to one-shot applications, such as missiles, while reusable vehicles like manned aircraft are likely to make more use of mechanical fasteners at areas of high load intensity. Nevertheless, the material presented in this chapter covers both possibilities.

The discussion below starts with a description of the various joint geometries, each of which is appropriate for a different load intensity. There is a discussion of the pitfalls in trying to correlate the measurements of bonded test coupons designed to force a failure in the adhesive with the performance of structural bonded joints in which the objective is to ensure that the adhesive itself never fails. The elimination of induced peel stresses by tapering the edges of the adherends is explained. Also, the distinction between repairable and non-repairable bonded structures is explained. Some adhesive bonding is used primarily not to transfer load but to increase structural efficiency by the use of laminated structures. Further applications of adhesive bonding, as between skins and stiffeners, transmit very little load provided that the structure remains intact but become very important with respect to containing any local damage without widespread disbonding. One pleasing characteristic of adhesively bonded joints is that both testing and detailed analyses have confirmed that they have considerably greater tolerance to local flaws than is generally recognised.

7.2. ADHESIVELY BONDED JOINT CONFIGURATIONS

7.2.1. General

Adhesively bonded joints are used in many different configurations, of which the most common are shown in Fig. 1. Their relative uses are

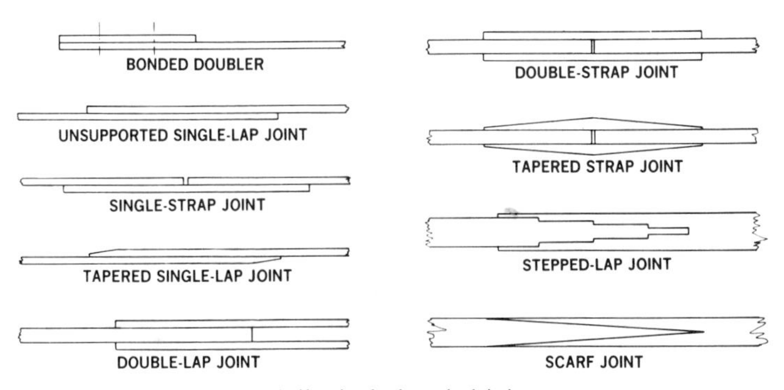

Fig. 1. Adhesively bonded joint types.

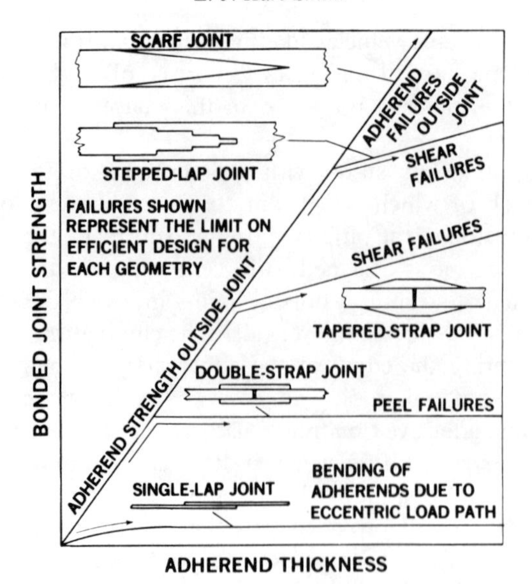

Fig. 2. Relative uses of different bonded joint types.

placed in perspective by Fig. 2, from which it is clear that there is no purpose in using an unnecessarily complex joint for the lower load intensities. Conversely, it is obviously hopeless to expect that the configurations that are simpler and cheaper to build could ever sustain high load levels, no matter how much quality workmanship was employed in their manufacture.

The simplest and most common joint is the bonded doubler and it is often not even considered to be a structural joint at all. Yet, as is shown in Fig. 3, the load transfer through the adhesive in load-sharing bonded doublers is just as intense as in the case of a full-transfer bonded joint. Perhaps the equivalence of these situations has escaped attention because bonded doublers are usually confined to thin-gauge structure whereas bonded joints can also be applied to much thicker and more heavily loaded structures. Bonded doublers are sometimes used to provide local thickening for countersunk fasteners in thin skins. They are also used to provide resistance to acoustic fatigue of thin metal structures. They are also used, particularly at British Aerospace (previously de Havilland) in the UK and at Fokker in The Netherlands, to increase the structural efficiency of stiffened structures subject to shear or compressive loads. In that context, stiffener flanges bonded to the skins act much like a doubler as far as the adhesive is concerned.

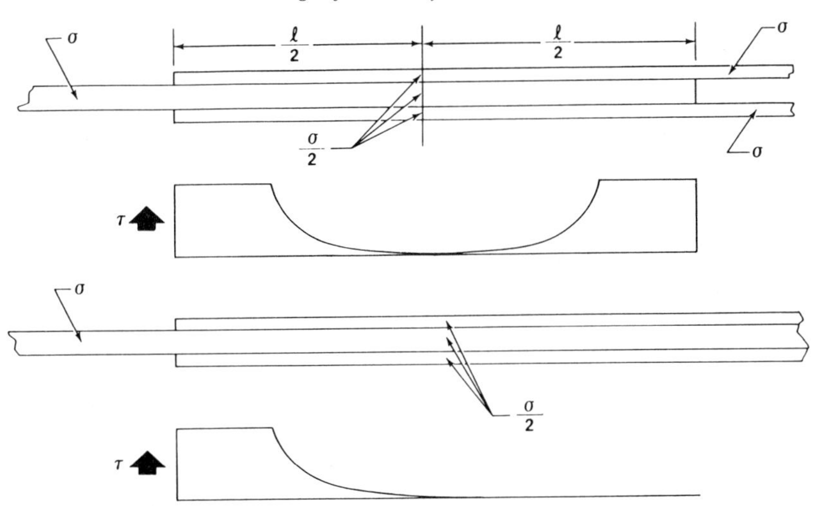

Fig. 3. Doublers versus joints. There are the same adhesive stresses in each case, and the same maximum adhesive shear strain for the same adherends and metal stresses.

7.2.2. Single lap joints

The unsupported single lap joint can never be as strong as the members being bonded together. Nevertheless, with an adequate overlap-to-thickness ratio (of at least 50 to 1 and preferably as much as 100 to 1) the gentle transverse deflections under tensile load can relieve the eccentricity in load path and develop structural efficiencies high enough to compete with the loss of net-section of mechanical splices. Unsupported single lap joints are also efficient for in-plane shear transfer but should never be used for compressive loads—the initial eccentricity becomes progressively worse as the load is increased and the joint should be stabilised in such cases.

The single strap joint in Fig. 1 barely qualifies as a joint: a better description would be a severe built-in stress concentration. The abrupt eccentricity where the skins butt together is necessarily associated with a high bending moment in the middle of the splice plate. In addition, the same area is subject to high adhesive peel stresses and an associated tendency to delamination of fibrous composite adherends. The analysis of peel stresses in that and other joints is given in ref. 1.

Thicker single lap joints need the ends of the adherends tapered, as in Fig. 1, primarily to prevent a premature failure by induced peel stresses before the full potential shear strength of the adhesive can be developed.

A secondary benefit is that, provided both ends are tapered, there is an actual increase in that shear strength because of a reduction in the severity of the peak in the shear stress distribution.

7.2.3. Failure modes

It is appropriate at this point to introduce the concept of the three distinct failure modes by which adhesively bonded joints can fail. The strongest joints do not fail in the adhesive at all; they fail outside the joint area at a load level equal to 100% of the strength of the adherend. The next highest strength occurs when the load that can be developed is limited by the shear strength of the adhesive. The weakest failure mode is associated with failure of the adhesive under peel loads, or with the even more premature failure by delamination of fibrous composite adherends. Now, the strength of the joint is proportional to the laminate thickness (to the first power) if it fails outside the joint under direct tension, compression, or in-plane shear, in the absence of bending. The shear strength of the adhesively bonded joint, on the other hand, is proportional to only the square root of the laminate thickness. The peel strength drops off even faster, being proportional to the quarter-power of the laminate thickness. These effects are explained in Fig. 4. Consequently, seemingly similar

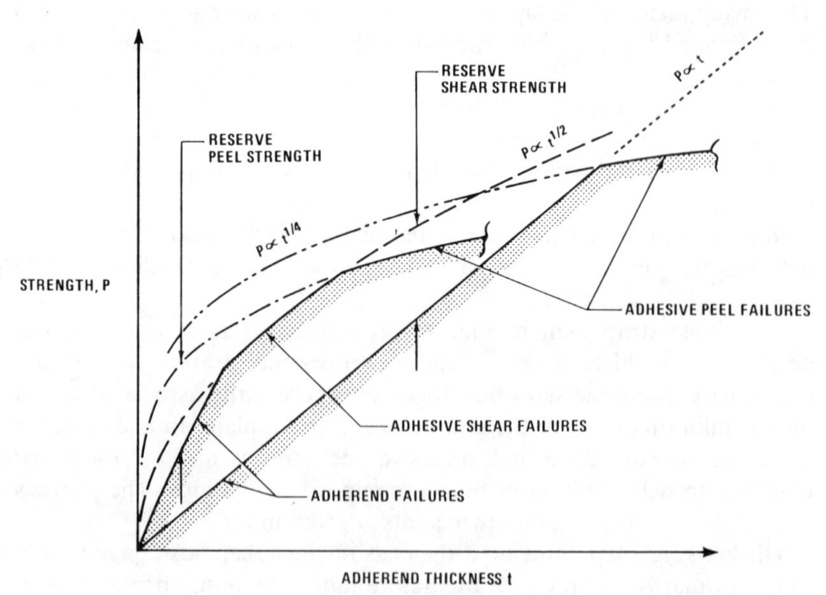

Fig. 4. Relative severity of adhesive shear and peel stresses.

adhesively bonded joints will fail by different modes at quite different structural efficiencies, primarily as a function of the thickness of the joint. The reason for this is that, despite the ability to vary all other dimensions of the joint, adhesive layers are most efficient only throughout the thickness range of 0·005–0·010 in (0·125–0·25 mm). Thicker bond lines could theoretically develop higher strengths, but it is virtually impossible to make them without an intolerable level of flaws or porosity. Simply tapering the outer adherends locally, down to a tip thickness of $0{\cdot}020 \pm 0{\cdot}010$ in ($0{\cdot}5 \pm 0{\cdot}25$ mm) with a taper of one-in-ten, is a very effective means of suppressing premature failures of adhesively bonded joints that virtually prevents failures by induced peel stresses.

7.2.4. Thicker laminates

Laminates of somewhat greater thickness than those for which single lap joints are adequate require two bond surfaces to transfer the strength of the adherends. The transition in thickness is not exact, being a function of the adhesive as well as of the thermal environment, but is typically of the order of 0·06–0·07 in (1·5–1·75 mm) for aluminium alloy adherends or for near quasi-isotropic carbon/epoxy laminates. Lower thicknesses would apply for stronger steels or unidirectional composite members. The optimum overlap-to-thickness ratio for double lap or double strap joints (see Fig. 1) is of the order of 30:1 for ductile adhesives used with the structure of subsonic transport aircraft. Despite the lack of any obvious gross eccentricity in load path, such joints are also subject to peel strength cut-offs, as explained in Fig. 5.

However, just tapering the outside of the splice plates (while leaving the basic members at a uniform thickness) will not create any additional shear strength in the joint. (Neither will it usually cause any decrease in the shear strength, but it could if the overlap were too short.) The tapered strap joint shown in Fig. 1 employs deliberate thickening of the splice plate at its middle to achieve simultaneously the alleviation of peel stresses and an increase in shear strength of the adhesive joint. The uniform double strap joint is limited to the joining of about $\frac{1}{8}$–$\frac{3}{16}$ in (3–4·5 mm) of aluminium alloy, or equivalent. The use of the tapered splice straps might push that up to as high as $\frac{1}{4}$ in (6·35 mm) without a significant increase in the difficulty of manufacturing.

Only the stepped lap joint or scarf joints should be considered for joining of members more than $\frac{1}{4}$ in (6·35 mm) thick by adhesive bonding. The suggestion in Fig. 2 that the theoretical scarf joint can always develop the entire strength of the laminate is usually violated by the necessity of

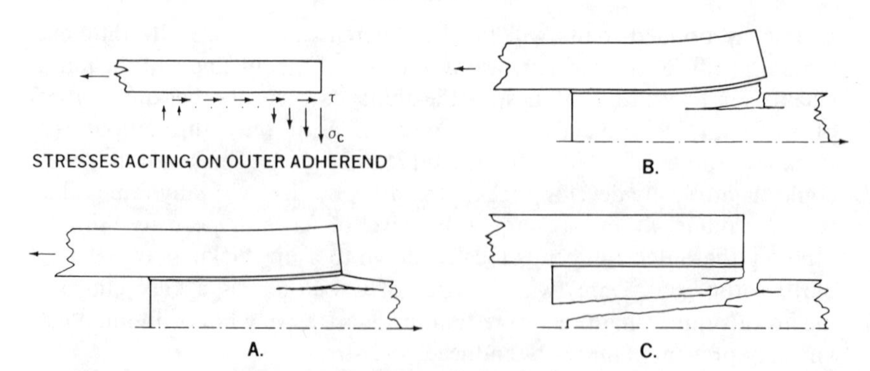

FIG. 5. Peel stress failure of thick composite joints. A, B and C indicate the failure sequence.

a finite thickness at the tips of the taper. One researcher has even calculated that a tip thickness of only 0·001 in (0·025 mm) is equivalent to a stress concentration of 25% in the elastic adhesive shear stress distribution. For this reason, the analysis of scarfed joints is often better performed by a stepped lap joint programme rather than one for scarf joints. This point is discussed in ref. 2, which explains also when a stepped lap joint should not be treated as an approximation to a scarf joint. The key difference is that the shear strength of a scarf joint continues to increase as the scarf angle is decreased, whilst the strength of a stepped lap joint with a fixed number of steps does not continue to increase indefinitely as the overlap is increased. The prime factor in determining the strength of a stepped lap joint is the number of steps, provided that there is not some gross error in the dimensioning of some single critical step.

7.3. ADHESIVE STRESS–STRAIN CURVES IN SHEAR

7.3.1. Non-linearity and test coupons

The basis of the analysis of the shear strength of adhesively bonded joints is the non-linear stress–strain curve of the adhesive layer in shear. This refers to the actual characteristics, as measured on the napkin-ring or thick-adherend test specimens, which are described in Figs 18 and 19 of ref. 3. Such characteristics are illustrated in Fig. 6 for both ductile adhesives (as used on subsonic transport aircraft) and brittle adhesives (as needed for the much higher service temperatures developed by supersonic fighters

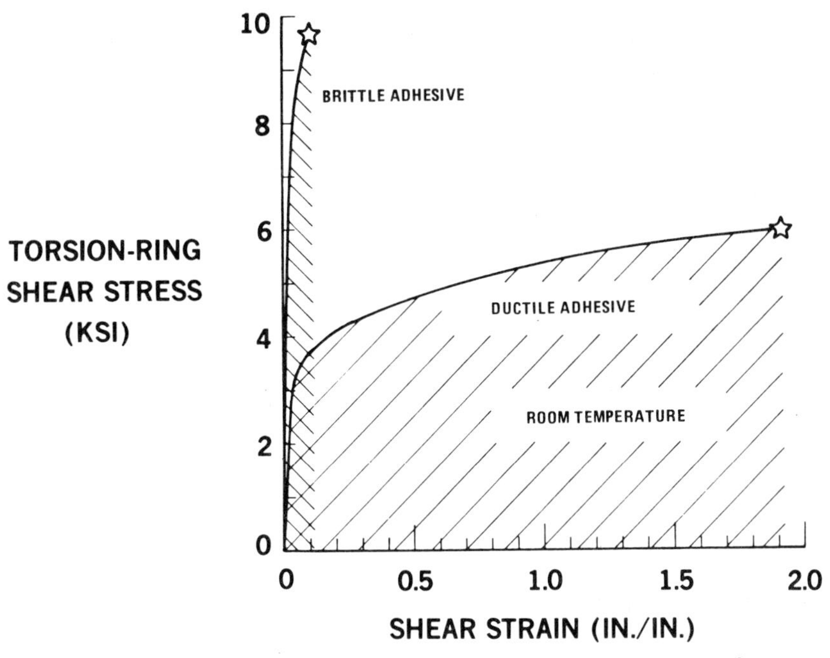

FIG. 6. Adhesive stress–strain curves in shear (10 ksi ≡ 70 MN/m^2)

or in the vicinity of engines). It is shown in ref. 4 that the shear strength of adhesively bonded structural joints can be expressed uniquely by the strain energy-to-failure per unit bond area of the adhesive layer, rather than by any of the individual properties such as peak shear stress. Therefore, even for the brittle adhesive in Fig. 6, the majority of the load transfer is accomplished by the non-linear behaviour; in the case of the ductile adhesive the contribution of the linearly elastic behaviour may be as little as 10%. Consequently, any perfectly elastic analysis would require a very large correction factor in order to correlate with test results.

The adhesive stress–strain curves are quite sensitive to the environment, as can be seen from Fig. 7. Note, however, that the area under each of those curves is much more nearly the same than are any of the individual properties such as peak shear stress or strain-to-failure. Consequently, the strengths of structurally proportioned joints would be insensitive to the environment, even though measurements on the standard test coupons would suggest otherwise. The pitfalls of trying to relate the performance of structural bonded joints to results from test coupons are so substantial

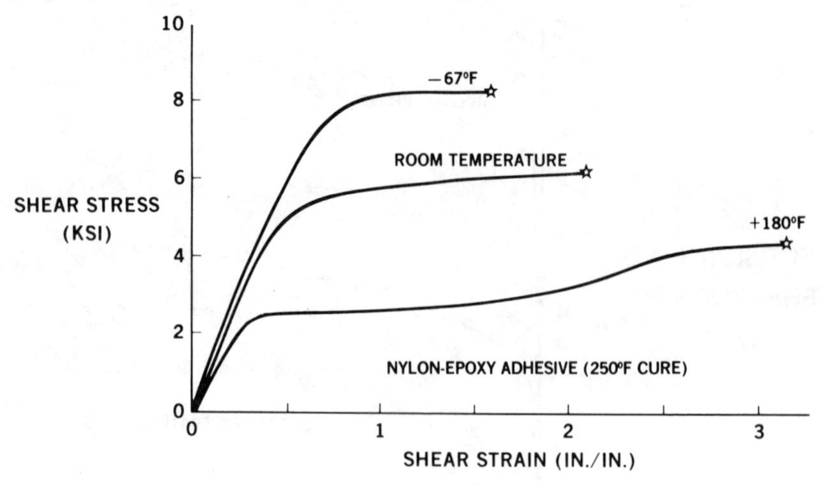

FIG. 7. Effect of temperature on adhesive stress–strain curves in shear ($10\,\text{ksi} \equiv 70\,\text{MN/m}^2$).

that ref. 3 was written just on that issue. Indeed, there is almost a complete lack of any one-to-one correlation, even though certain test data are obviously needed as the basis for design. Reference 3 presents a strong case that a totally different approach is needed to characterise adhesives in such a way that a rational selection can be made for any given application. Some of the reasons for this will become evident from the discussions in this chapter.

7.3.2. Model behaviour

Various linear and non-linear adhesive models are shown in Fig. 8. The elastic–plastic model shown passes through the same failure stress and strain and has the same strain energy-to-failure. It permits accurate predictions of the ultimate strength of such bonded joints, but does not give an accurate representation of the internal stresses associated with much lower loads. The perfectly elastic model is to be preferred at very low load levels. Improved accuracy can be obtained by use of a modified model for intermediate loads, as shown. The peak stress and strain in the adhesive are again matched, and the elastic portion of that model is selected by equating areas again. Theoretically, a new model should be prepared for each load level, but that is obviously impractical. Considerable simplifications would ensue if the single bilinear characteristic shown could be used throughout. Unfortunately, the complexity of the associated

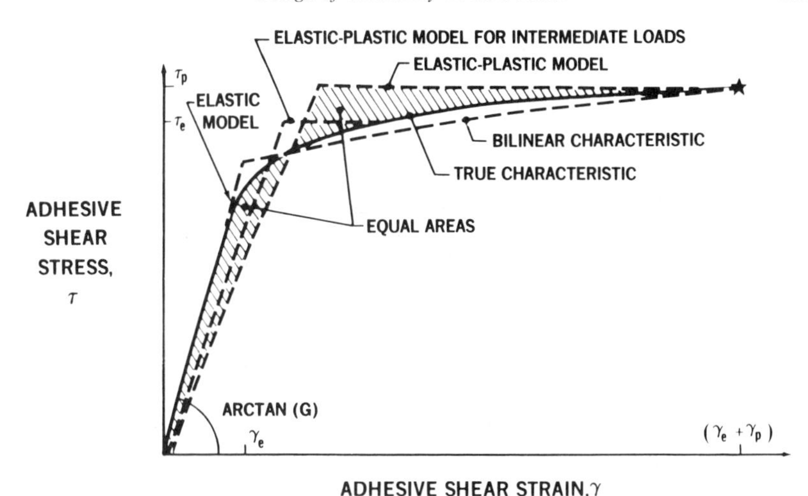

Fig. 8. Representations of adhesive non-linear shear behaviour.

mathematics is increased greatly. Nevertheless, such an explicit closed-form solution for uniform double lap joints has established that *any* two-straight-line representation having the same failure stress and strain as well as the same strain energy-to-failure would predict precisely the same joint strength. A refinement such as the bilinear model would therefore not enhance the accuracy of ultimate strength predictions but would simplify the computation of the internal stresses associated with specified lower loads.

7.3.3. Practical considerations

Adhesive characteristics can also be affected by porosity, which occurs as a result of the bond layer being slightly thicker than nominal; gross bond flaws result if the gap between the adherends becomes so large that the adhesive is not retained by either capillary action or external dams. Reference 6 contains examples of stress–strain curves for porous bonds in both ductile and brittle adhesives. Interestingly, it is shown there that naturally occurring porosity does not usually result in the failure of those portions of the bonds which would be suspect on the basis of ultrasonic inspection. Porous bonds, being thicker than nominal, are also softer. Therefore they transfer some of the load they should have transmitted to adjacent thinner bonds which thus become more highly loaded than they would be in a uniform adhesive layer. Actually, porous bonds are rarely created in areas of naturally high load transfer as at the ends of a bonded overlap; rather, they tend to occur away from those edges, in areas in which there would be little load to transfer, even if the bond were not porous there.

The considerable straining of the ductile adhesives beyond the knee in their stress–strain curves is typified by Fig. 6. However, it is not considered good design practice to use all that strength for normal design conditions, even if the minimum strain is adequately restricted. Whilst the exact limits have yet to be determined scientifically, a fairly standard design philosophy has evolved in which the knee of the stress–strain curve is not exceeded by frequently occurring loads such as the differential cabin pressure loads at cruise altitudes or 1-g normal level-flight loads. There is, therefore, apparently some non-linear behaviour at less than limit loads, but the upper limit on the minimum strain prevents the occurrence of any permanent damage. That considerable non-linear behaviour is available for load redistribution due to local manufacturing imperfections of various forms. The strength of such bonded joints in practical structures would therefore be greatly diminished if this reserve of strength were not available.

Restricting the peak strain in the adhesive to the knee in the curve for normal conditions has the effect of limiting the adherend thicknesses for which the simpler joint configurations are satisfactory; the use of carefully designed stepped lap joints, even with this restriction, still permits the transfer of substantial loads. Such joints, between boron/epoxy or carbon/epoxy laminates and titanium step plates, keep the skins attached to the tails and wings of several modern fighter aircraft.

7.4. THE NON-UNIFORMITY OF LOAD TRANSFER IN ADHESIVELY BONDED JOINTS

The most important single fact to learn about the design and operation of structural adhesively bonded joints is that the load transfer is not, and must not be, uniform. To those intent on optimising joint proportions and performance mathematically, this must seem to imply a need for an unnecessarily long bonded overlap. Nevertheless, the need is very real and is explained in Fig. 9. In the short overlap test coupon in Fig. 9(b), the adhesive is seen to be uniformly stressed and nearly uniformly strained. Indeed, that is the usual intention when such a specimen is used—to know the exact state of stress and strain of the adhesive when it fails. No such claims could be made for the adhesive in the much longer overlap

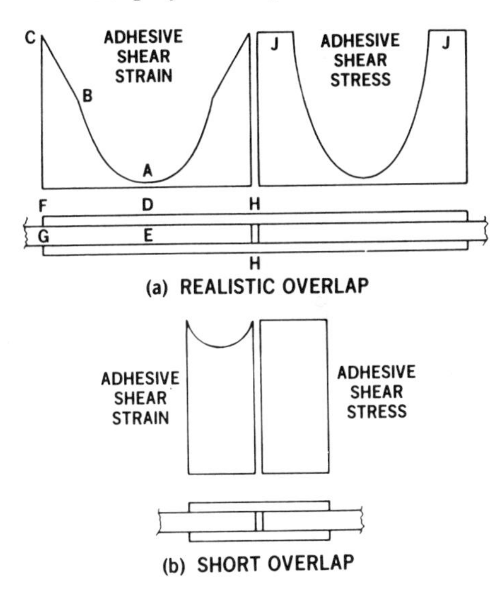

Fig. 9. Non-uniform stresses and strains in bonded joints.

of Fig. 9(a). There is very considerable variation in both the adhesive stress and strain throughout the bonded overlap.

The importance of this variation is that, provided the minimum adhesive shear strain at point A is sufficiently *low*, the bonded joint will be able to survive more than one or two load cycles or more than a few minutes of high sustained load. The short overlap joint is incapable of doing that because the minimum adhesive shear strain, at the middle of the overlap, is so high that such a joint would fail rapidly because of creep rupture. That is not to imply that there would be no adhesive creep, under sustained load, between the adherends at points F and G in the long overlap joint. Obviously, the adhesive shear strain at point C is so high that creep must be inevitable there. How then does such a structural bonded joint not fail in the same manner as a short overlap coupon? The reason is that, whilst the creep occurs, it does not accumulate.

Consider what happens when the load is removed after 1 h, 8 h or even 14 h (to represent the longest flight durations on commercial transport aircraft). It is fair to assume that there has been no creep between points D and E across the middle of the bonded overlap, and that neither adherend has changed length due to creep between points D and F or E and G. Otherwise there would be no long-life bonded structural joints. Yet what appears to have been creep has been measured experimentally

CENTRAL SHEET THICKNESS t_i (IN.)	0.040	0.050	0.063	0.071	0.080	0.090	0.100	0.125
SPLICE SHEET THICKNESS t_o (IN.)	0.025	0.032	0.040	0.040	0.050	0.050	0.063	0.071
RECOMMENDED OVERLAP[1] ℓ (IN.)	1.21	1.42	1.68	1.84	2.01	2.20	2.39	2.84
STRENGTH OF 2024-T3 ALUMINUM (LB/IN.)	2600	3250	4095	4615	5200	5850	6500	8125
POTENTIAL ULTIMATE BOND STRENGTH (LB/IN.)[2,3]	7699	8562	9628	10,504	10,888	11,865	12,151	13,910

[1] BASED ON 160°F DRY OR 140°F/100-PERCENT RH PROPERTIES NEEDING LONGEST OVERLAP.

VALUES APPLY FOR TENSILE OR COMPRESSIVE IN-PLANE LOADING. FOR IN-PLANE SHEAR LOADING, SLIGHTLY DIFFERENT LENGTHS APPLY.

[2] BASED ON −50°F PROPERTIES GIVING LOWEST JOINT STRENGTH AND ASSUMING TAPER OF OUTER SPLICE STRAPS THICKER THAN 0.050 IN. STRENGTH VALUES CORRECTED FOR ADHEREND STIFFNESS IMBALANCE.

[3] FOR NOMINAL ADHESIVE THICKNESS $\eta = 0.005$ IN. FOR OTHER THICKNESSES, MODIFY STRENGTHS IN RATIO $\sqrt{\eta/0.005}$.

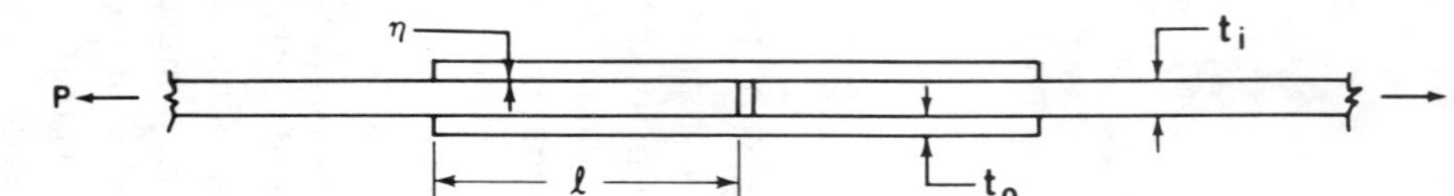

Fig. 10. Design overlaps used for PABST skin splices. (1 in = 25·4 mm; 10 000 lb/in = 1·75 MN/m.)

between points F and G as long as the load was maintained. What has happened is that the viscoelastic behaviour of the adhesive in the vicinity of point C causes an elastic redistribution of the stresses within the intervals FD, DE and EG. So, when the load is removed, elastic residual stresses are induced in the adherends and, as these try to relieve themselves, they push the adhesive at point C back towards its original unstrained state. There is no mechanism to accomplish that in the short overlap joints because of the creep which would have occurred under load between the equivalent of points D and E.

It should be clear now that in the design of adhesively bonded joints it is even more important to limit the *minimum* adhesive shear strain than it is to limit the maximum! The deep elastic trough between the plastic load transfer zones J in Fig. 9 should not be regarded as an inefficiency to be eliminated by 'improved' design. Much research remains to be done to establish a scientific basis for setting that minimum value of adhesive shear strain or stress. As a matter of record, the design procedures employed on the Primary Adhesively Bonded Structure Technology[7] (PABST) programme were based on restricting the minimum shear stress to no more than 10% of the maximum at the ultimate load level. That corresponded with an even smaller fraction for the normal operating stresses. Those double strap joints which are characterised in Fig. 10 performed without failure under both full-scale testing and artificially severe coupon testing for four years in a hot/wet environment under slow-cycle testing. The need for slow-cycle testing (half an hour or more per cycle) arises because even the short overlap test coupon appears to have an infinite fatigue life if tested at 30 Hz; the load is then being removed so rapidly that the adhesive has no opportunity to creep.

The transition between the extremes of behaviour in Fig. 9 is explained in Fig. 11 as a function of the bonded overlap for double lap joints. The joint strength is initially proportional to the short overlap, as at A. Then, as the overlap is increased still further, the strength increases only very slightly as the elastic trough is developing, at B. Finally, no matter how much longer the overlap is made, as at C, the joint strength remains constant and so does the maximum adhesive stress and strain. The minimum adhesive stress decreases asymptotically towards zero, and the load transfer zones keep the same width, but move steadily further apart.

The design of such joints is thus reduced to the simple task of identifying the overlap for which that minimum stress has become low enough. Any further increases in overlap cannot increase the joint strength. This process is explained in Fig. 12 and the results of such calculations are given in

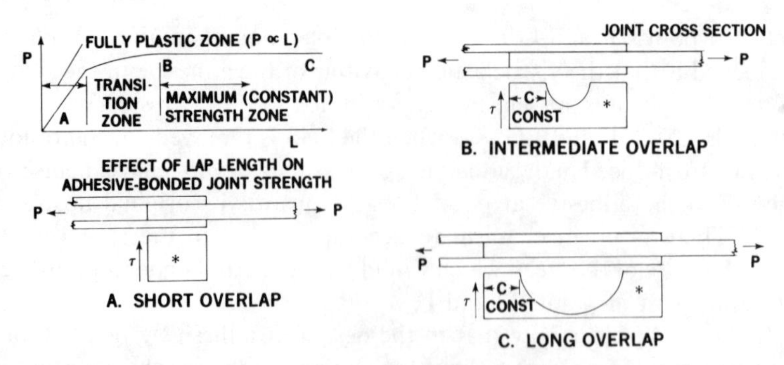

Fig. 11. Influence of lap length on bond stress distribution. *, Adhesive bond stress distribution.

Fig. 10, which was developed for the adhesive bonding of aluminium alloy adherends during the PABST programme. The results can be summarised by the simple approximation that each bonded overlap is roughly 30 times as long as the thickness of the inner adherend. Much the same ratio would apply also to typical cross-plied carbon/epoxy laminates. The slightly shorter overlaps, for essentially the same strength, that were given as optimum designs in Fig. 16 of ref. 4, Fig. 11 of ref. 8 and Fig. 3 of ref. 5 should now be considered superseded. The $3/\lambda$ in Fig.

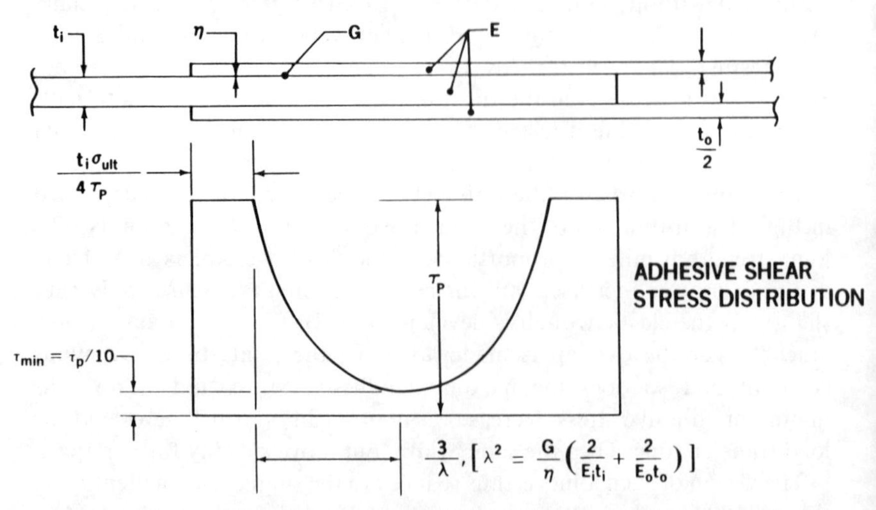

Fig. 12. Design of double lap bonded joints. Checks should be made that the plastic zones are long enough for the ultimate load, that the elastic trough is wide enough to prevent creep at the middle, and that the strength is adequate.

12 here replaces an increment of $1/\lambda$ used in those references above before the role of the elastic trough had been fully understood.

7.5. DOUBLE LAP AND DOUBLE STRAP JOINTS

7.5.1. Introduction

The pioneering analysis of adhesively bonded joints is that by Volkersen.[9] That linear analysis and its non-linear development by the present author[5] rely on continuum mechanics analyses and are expressed in terms of one mathematical variable, with no account of the minor variations in stress through the thickness. That level of analysis had led to an extremely thorough understanding of the stress state within such bonded joints. References 10 and 11 provide comprehensive illustrated explanations of the behaviour of such joints. They cover the origin of the non-uniformity in the load transfer across balanced double lap joints, the strength losses associated with adherend stiffness imbalances or thermal mismatches, and the grossly dissimilar behaviour of short overlap test coupons and long overlap structural joints. The consequences of refining such analyses to account for variation in the adhesive stresses across the thickness of the layer are discussed in Fig. 4 of ref. 11, where there is an explanation which justifies neglecting such effects in actual joints because of the naturally occurring fillets in the adhesive. There is also a significant shear-lag effect that varies the adherend stresses across the splice plate thickness over the point where the skins butt together in double strap joints. This effect has been investigated in depth in relation to the use of adhesively bonded boron/epoxy patches to retard the growth of cracks in metallic structures (see ref. 12). The effect was also sufficiently pronounced during the fatigue testing of aluminium alloy adherends during the PABST programme to result in a strong bias towards a failure in the middle of the splice plate rather than in the central sheet at the other end of the bond, where the adherend stresses were nominally equal. This led to an increase by half a gauge in the thickness of the splice plates, with respect to a nominal thickness half that of the basic member. The same kind of reinforcement is recommended for the bonding of fibrous composites also.

7.5.2. Peel and shear stresses

The key characteristics of the state of adhesive stress and strain in an adhesively bonded double lap joint are the peak adhesive shear strain, the minimum adhesive shear strain and the peak-induced adhesive peel stress. The latter will probably delaminate the composite adherend, as shown in Fig. 5, before the adhesive could peel apart. The detail modifications in Fig. 13 to eliminate the peel stress problem, where necessary, are quite simple yet effective. The tip thickness of the splice plate should not exceed 0·020 in (0·50 mm) for composites. The local thickening of the adhesive layer makes the glue more flexible there, reducing the peel stresses by half or more and also reducing the peak adhesive shear strain without any loss in the shear strength of the joint. The upper limit on layer thickness shown is because heat-cured adhesives would run out of that area under capillary action unless there were special dams. The degree of such thickening is not at all critical, as is explained in ref. 11. The adherend tapering on the outside is not critical either, and this is shown in Fig. 14. The critical condition is then moved to the other end of the joint and remains constant because there are no modifications there.

Once any possible premature failures by induced peel stresses have been eliminated, it is possible to consider the adhesive shear stresses alone. Figure 15 shows the maximum and minimum adhesive shear stresses as functions of the bonded overlap. (Figures 2 to 4 of ref. 3 show the additional influence of adherend thickness.) It is immediately noticeable that there is an abrupt transition in the adhesive behaviour between those overlaps at points A and B. The minimum adhesive shear strain is almost as high as the maximum for short overlap joints and those configurations should be used only for certain test specimens and never for structural bonds. To the right of those transitions on the graph, the peak adhesive shear strain for the thin and moderately thick adherends is restricted to values insufficient to permit failure by the limited strength of the adherends,

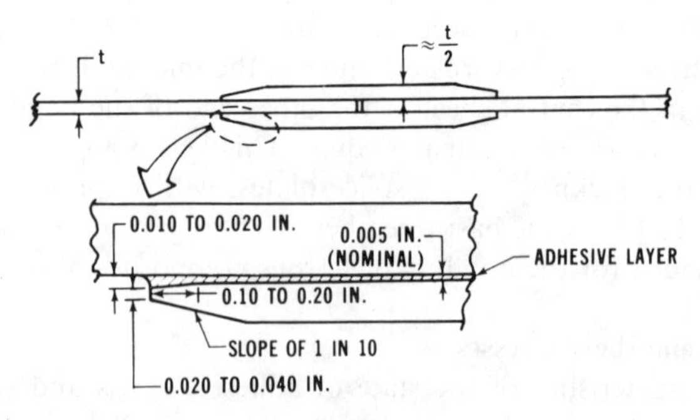

FIG. 13. Tapering of edges of splice plates to relieve adhesive peel stresses (0·01 in ≡ 0·25 mm).

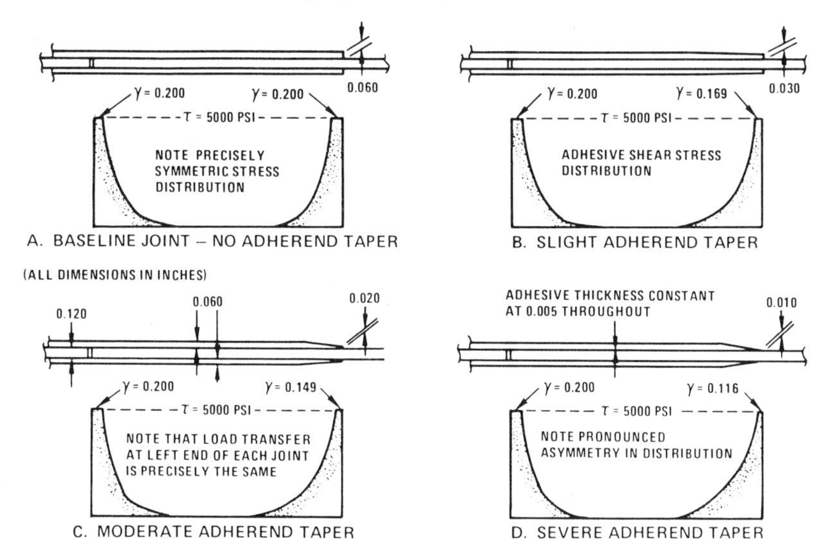

Fig. 14. Insensitivity of adhesively bonded joint strength to modifications at one end of the joint only. The adhesive strain at the right-hand end of the joint decreases with more taper. All dimensions are in inches (0·01 in ≡ 0·25 mm; 1000 psi ≡ 7 MN/m^2).

which would inevitably fail first. The design exercise is thus reduced to the establishment of that overlap, to the right of such a transition, on the graph (Fig. 15) for which the minimum adhesive shear strain is sufficiently low. Any further increase in overlap could not add to either the strength or the life of such a joint. Those calculations form the basis for the recommendations in Fig. 10. (The computations on which Figs 10 and 12 are based were actually performed for aluminium alloy bonded construction but, since the adherend modulus E has a value which is representative of the range of practical fibre patterns for HTS graphite/epoxy shown in Fig. 21 of ref. 13, the overlaps are representative of those for fibrous composites also.) Figure 10 can be summarised by the approximation, as stated before, that the overlap should be about 30 times the central adherend thickness. The effect of varying the adhesive ductility or the strength of the fibrous reinforcement has much more of an effect on the maximum thickness of adherends that should be bonded than on this overlap-to-thickness ratio.

The reason why there is an upper thickness cut-off in Fig. 10 can be understood from curves C in Figs 2 to 4 of ref. 3. There is no abrupt transition in adhesive behaviour, no matter how long the overlap is made, and consequently, an imperfection in the bond could cause a catastrophic unzipping, even if analysis suggested that there was adequate strength for a nominally perfect bond. The calculations on which Fig. 10 is based were performed for that environment, which is usually the hot/wet one, which needs the longest overlap. The actual calculations are particularly simple, as shown in Fig. 12. The adhesive properties are represented by the peak shear stress τ_p, the elastic shear modulus G and the thickness η.

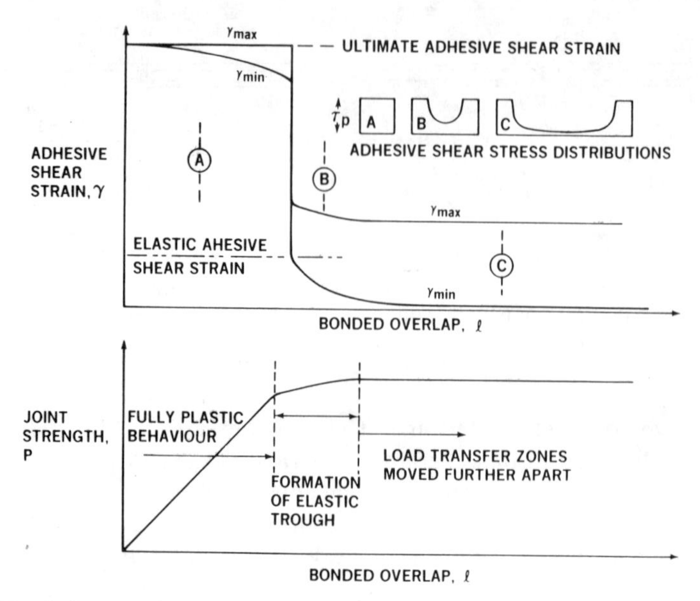

Fig. 15. Influence of overlap on maximum and minimum adhesive shear strains in bonded joints.

The suggested check for adequate bond strength to fail the adherends outside the joint implies an upper limit on the adherend thickness for which this simple joint configuration should be used. Although Fig. 12 implies an equal transfer of load through the two plastic zones at the ends of the joint, the adherend tapering shown in Fig. 14 for thicker adherends alters the widths of those plastic zones in such a manner that the combination of the two zones is essentially constant because the total load is still limited by the adherend strength. Therefore the technique in Fig. 12 still applies, even though the actual adhesive stress and strain distributions would be modified as shown in Fig. 14.

7.6. SINGLE LAP JOINTS

7.6.1. Basic considerations

The design of single lap joints in fibrous composite structures is actually easier than for double lap joints, even though the mathematical analysis of the former is more difficult. Goland and Reissner were the first to account for the out-of-plane bending associated with the eccentricity in the load path in what is now considered one of the all-time classic bonded joint analyses.[14] Their determination of the bending moment in the adherend at the ends of the bonded overlap was improved by the present author in ref. 15, which includes adhesive plasticity as well as the earlier perfectly elastic behaviour.

Figure 16 illustrates the key to the difference between the behaviour of unsupported single lap joints and double lap bonded joints. The single lap joints incur significant bending moments in their adherends in conjunction with the average (uniform) stress component due to the applied loads. The presence of such bending stresses requires such long overlaps to permit the adherends to bend gently rather than abruptly, that the possible criticality of the adhesive in shear becomes secondary. Actually, the same eccentricity in load path also induces more severe peel stresses in the single lap joints than occur in double lap joints. Again, beyond the tapering of the adherends shown in Fig. 13, the only mechanisms with which to reduce the peel stresses are an increase in the overlap, or transverse support at the ends of the overlap to react against

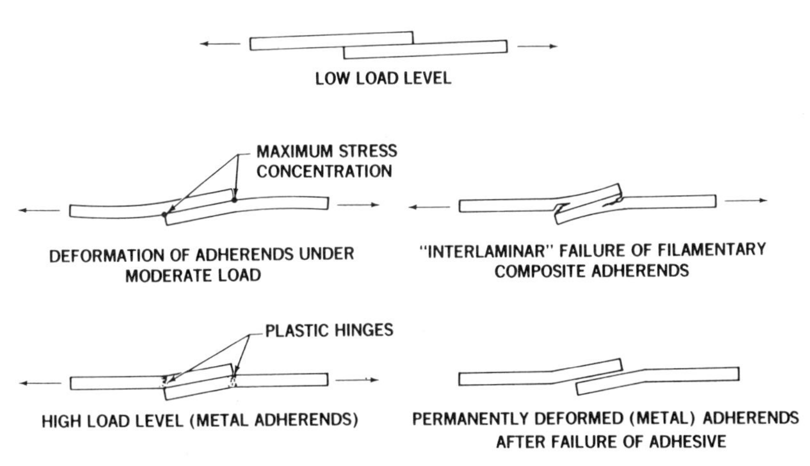

FIG. 16. Failure of single lap bonded joints (brittle and yielding adherends).

the eccentricity in load path. These supports would be absolutely necessary to stabilise such joints under compressive shear loading.

7.6.2. Influence of adhesive

Figures 17 and 18 show how the structural efficiency of adhesively bonded single lap joints varies as a function of joint geometry for ductile and brittle adhesives, respectively. These composite adherends are assumed to be uniformly thick and have no peel stress relief from any tapering. Consequently, the thicker adherends show considerable decrease in strength. Not a single failure of the adhesive in shear is predicted in Fig. 17 for the ductile adhesive—which is why the design of such joints

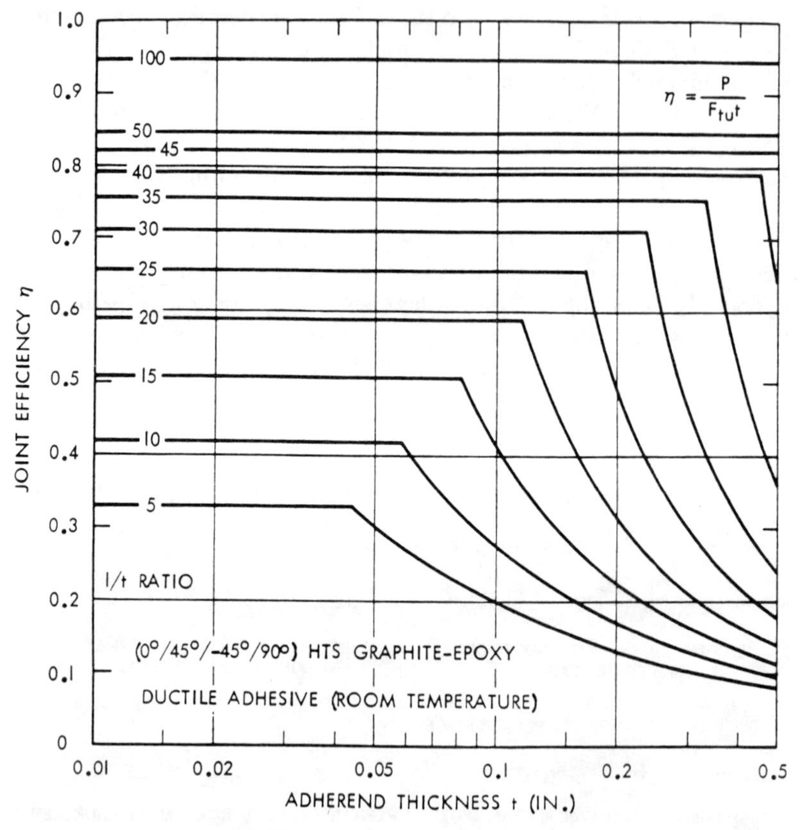

FIG. 17. Joint efficiency chart for single lap composite joints (P = applied load; F_{tu} = ultimate tensile strength of adherends; 0·1 in ≡ 2·5 mm).

is so simple. The brittle adhesive in Fig. 18 is weaker, as well as more brittle, and there is a small area of shear failures predicted in the upper right-hand corner of the graph. The effect of the use of a brittle adhesive is to decrease the thickness of adherends that can be bonded well as single lap joints. For the thin adherends, to the left of those figures, the efficiency is limited by adherend bending only and is strongly dependent of the l/t ratio. A value of 80, corresponding with a structural efficiency of 90%, was adopted for the PABST programme. Note how, for the typical l/t ratio of 8 used in the standard lap shear test coupon with a $\frac{1}{2}$ in (12·7 mm) overlap and aluminium alloy adherends 0·063 in (3·2 mm) thick, the structural efficiency is limited to less than 40% by that abrupt eccentricity.

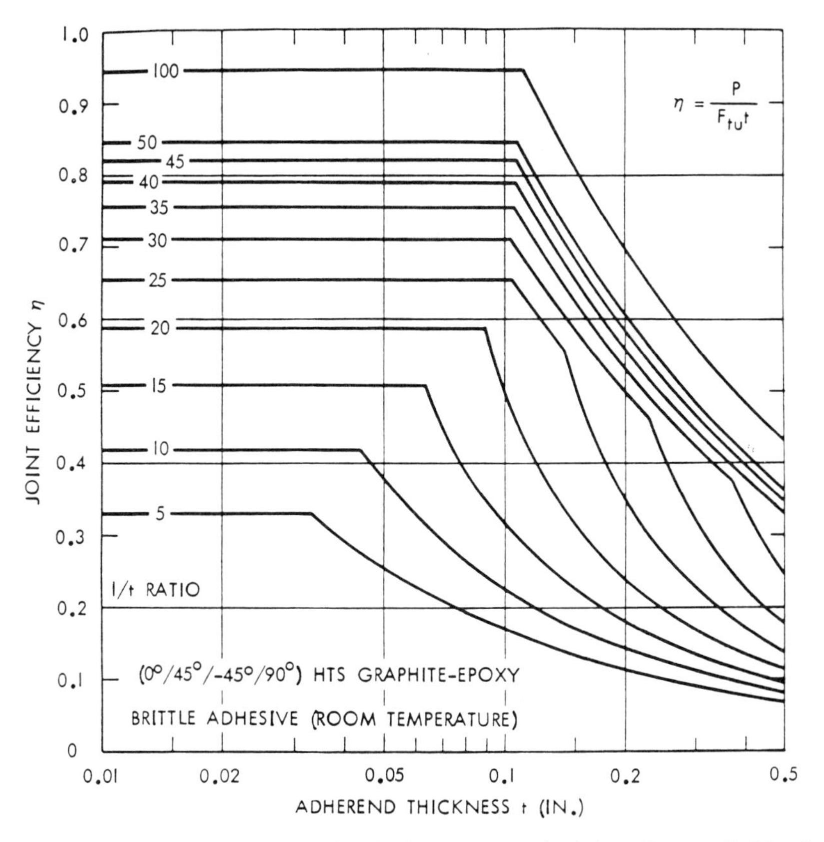

Fig. 18. Joint efficiency chart for single lap composite joints (P = applied load; F_{tu} = ultimate tensile strength of adherends; 0·1 in ≡ 2·5 mm).

It is of special significance that neither Fig. 17 nor Fig. 18 would predict that the failure of such a test coupon would be in any way influenced by the shear properties of the adhesive. Yet that is the very specimen which is most widely (mis)used in the belief that it does characterise the shear strength of adhesives and permit meaningful comparisons. It is the basis of many MIL specifications also. The reasons for its continued use for such purposes have not been clear for at least a decade, although it should be acknowledged that it still serves as a practical quality control coupon. Reference 3 explains that the apparent shear strength indicated for adhesives by that specimen is influenced strongly by the adherend thickness and material, and invariably the failure is by peel rather than by shear. Actually, the use of such data to aid in adhesive selection has become extremely difficult and misleading for newer stronger adhesives. The author has suggested in ref. 3 that a more meaningful characterisation of adhesives for selection would be a rating of the type that adhesive A could fail a given adherend material outside the joint up to a thickness of 0·05 in (1·25 mm), say, while the stronger adhesive B might fail the same material up to a thickness of 0·10 in (2·5 mm). In neither case would there be an adhesive failure at all because the purpose of designing bonded joints is that they should not fail. Such an approach to characterising adhesives would be unambiguous and remove all of the difficulties of trading-off the higher peak shear stress-at-failure of brittle adhesives against the much greater strains-to-failure of ductile adhesives.

The design of structural single lap joints is then seen to be reduced to the specification of an adequate overlap (l/t in the range 50–100) and to the determination of an upper limit on thickness (of the order of 0·06 in or 1·5 mm) even after the peel stresses have been alleviated by tapering the ends of the adherends. These recommendations refer in particular to applications to subsonic airframe structures. The same techniques would apply for the different adhesives used in the extreme environment of outer space or of the blast-off of a rocket, for example, but the thickness cut-off would probably be more severe and the adherend tapering may become more critical.

7.7. STEPPED LAP BONDED JOINTS

7.7.1. Background

At the time when the so-called advanced composites (boron/epoxy and carbon/epoxy) were first being introduced into aircraft structures over a decade ago, little was known about how to design bolted joints in such materials. In the case of boron/epoxy it might fairly be added that drilling holes was so difficult as to discourage that approach anyway! Also, the use of adhesive bonding offered the chance to obtain higher structural efficiencies, and to use less of what was then an extremely costly material. That was the environment which led to the derivation of powerful analytical computer programs with which to design composite-to-titanium stepped lap bonded joints in the early 1970s. The first such analysis seems to be the elastic derivation by Corvelli,[16] which was followed soon after by the present author's non-linear (elastic–plastic adhesive) program A4EG.[17] The latter has been developed much further [18,19] and now provides for non-uniform adhesive thicknesses and mechanical properties as well as accounting for local flaws and porosity. Many worked examples derived by A4EI, the latest version, are contained in refs 20, 19 and 2. Joints of this kind have found production applications in the tails of the F-14, F-15, F-16 and F-18, as well as in the wing of the F-18. There are also numerous applications of such joints in research and development programmes. The current state-of-the-art is about 1 in thick (2·54 cm) laminates, transferring load intensities of about 30 000 lb/in width (5·25 MN/m). However, that is not a theoretical upper limit; production applications demanding stronger joints of this type have not arisen yet.

7.7.2. Design variables

The strength of these bonded joints is known to be particularly sensitive to a number of factors. Perhaps the most powerful one is the need for adherend stiffness balance from one end of the joint to the other, as explained in Fig. 19. The joint on the right has almost twice the strength of

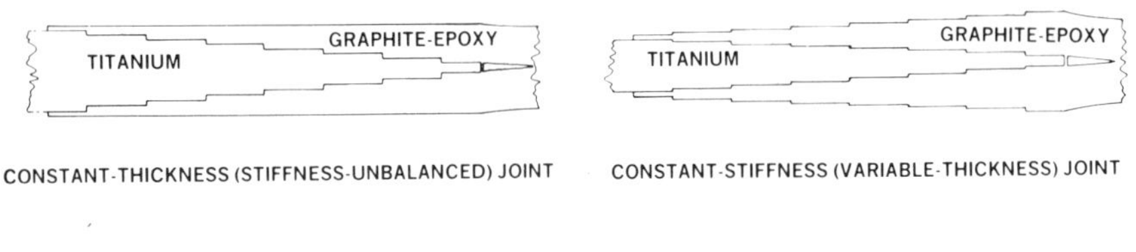

Fig. 19. Stepped lap bonded joints (strength improvement by matching stiffnesses).

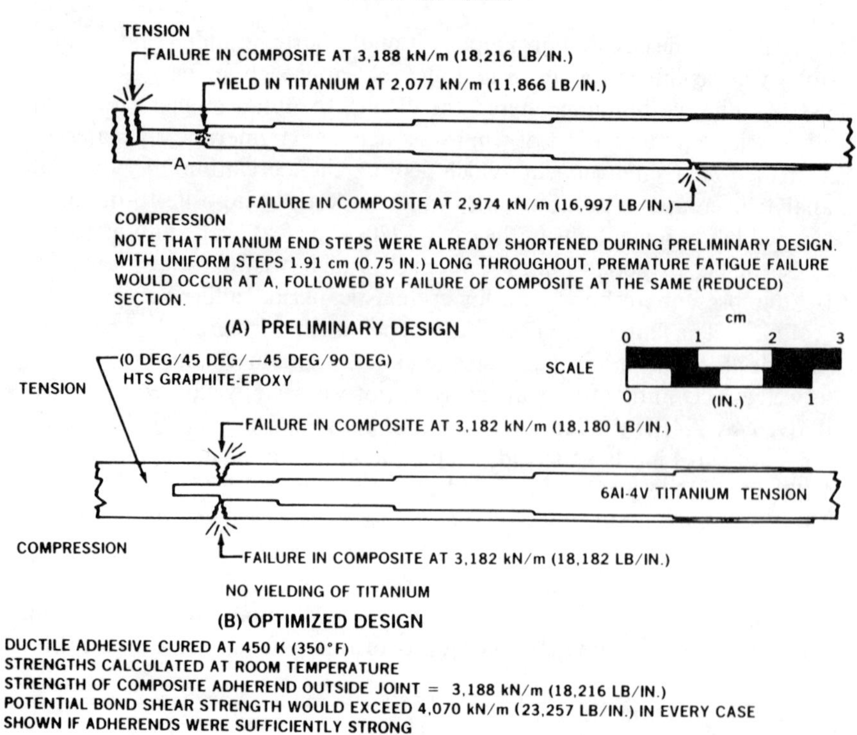

Fig. 20. Optimisation of details in stepped lap bonded joints.

the one on the left, even though that has been obtained by *removing* material rather than by adding reinforcement. The length and thickness of the end step on the titanium are very important, as shown in Fig. 20. The yielding of the titanium at the root of that step, shown in the preliminary design A in Fig. 20, has shown up as a design problem in two forms. That step would fail in fatigue if made too long, i.e. more than about $\frac{1}{4}$–$\frac{3}{8}$ in (6–9 mm), and result in a secondary failure because of the reduced effective thickness of the composite laminate.

An extreme case of poor detailing of that area is shown in Fig. 21, in which a 1 in (2·5 cm) end step was actually torn off under static load. That the solution to this problem is not the thickening of the end step should be evident from Fig. 22. In the load-introduction area of that specimen, a bundle of all 0° boron filaments 0·1 in (2·5 mm) thick had been terminated abruptly just beyond the end of the step plate in the foreground of the illustration. The application of load caused a

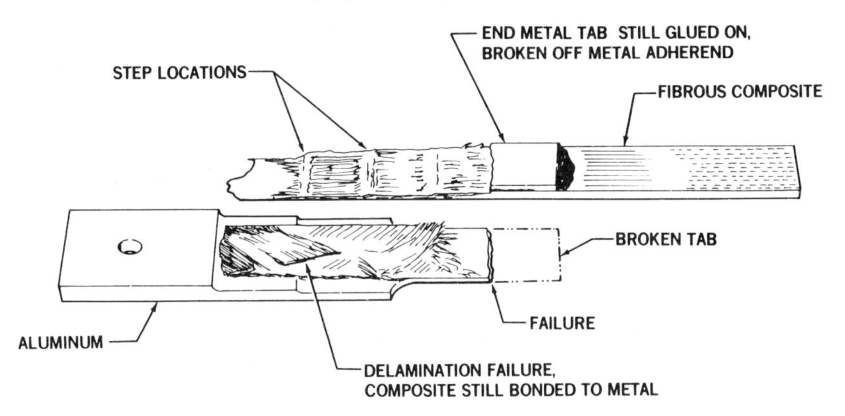

Fig. 21. Failure of stepped lap adhesive bonded joint.

delamination at each side of that bundle of strong stiff filaments because the resin matrix was not strong enough to transfer so much load in shear without failure.

A tip thickness of 0·030 in (0·75 mm) is more appropriate for the materials used on aircraft today and, even then, the core plies should not be cut off but be diverted around a low-modulus triangular wedge about 0·5 in (1·3 cm) long, as shown in Fig. 23 for the similar detail of a scarf joint with finite tip thickness. The last step on the other end of the joint

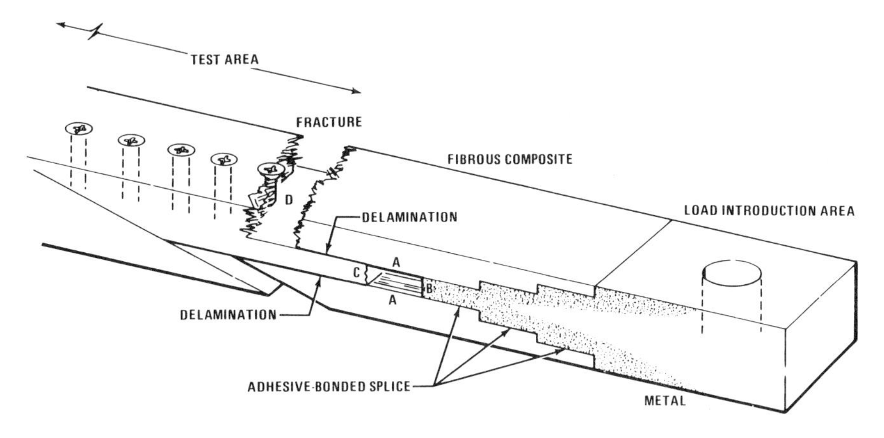

Fig. 22. Premature failure of stepped lap bonded joint by delamination. Initial failure is at A because the thickness B is excessive and the load in fibres C cannot be unloaded through the resin matrix. Final failure, at D, is by net-section tension on the top face and shear-out (not shown) on the lower face.

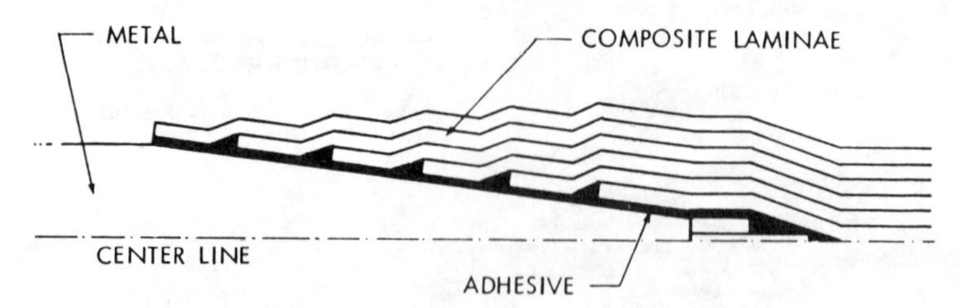

Fig. 23. Lay-up details for composite scarf joints.

must have thin outer adherends to prevent delamination due to induced peel stresses. The composite there should be no more than 0·030 in (0·75 mm) thick on each side of the step plate, for the same reasons that the ends of the splice plates are tapered in Fig. 13. That end step should also be limited to $\frac{3}{8}$–$\frac{1}{2}$ inch (9·4–12·7 mm) in length to prevent the thin steps being overloaded.

The remaining design variables are the length and number of steps. Of these two, the dominant effect is the number of steps. The length effect, for a given number of steps, is extremely slight, as can be seen from the three joint analyses in Fig. 24. Each joint had five steps and all thicknesses were the same at the corresponding stations. An increase in overlap by a

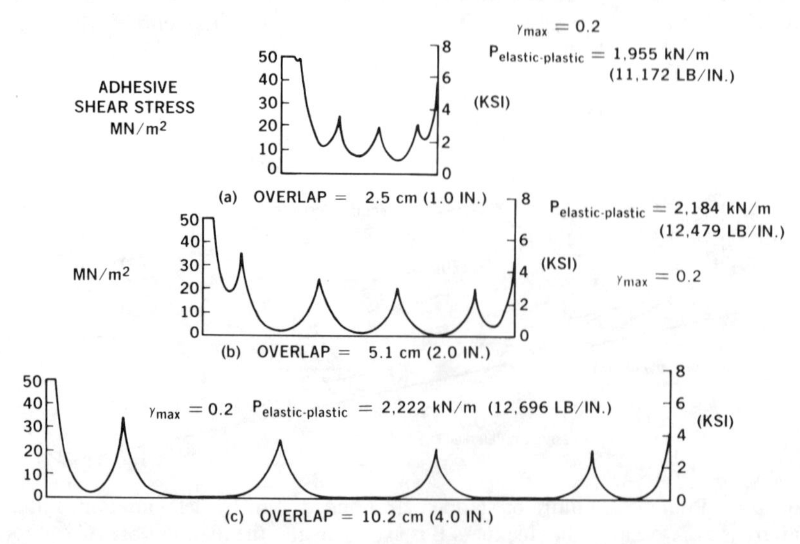

Fig. 24. Limited effect of overlap on strength of stepped lap bonded joints (all cross-sections are identical and all step lengths proportional).

factor of four is seen to increase the strength by only 14%. This highlights the large difference between the mechanism of load transfer in scarf joints (for which such an increase in area would have had a substantial effect on the joint strength) and stepped lap joints. The reason why the stepped lap joints are insensitive to the overall joint length, as the sum of a series of uniform steps, is that each of those steps is governed mathematically by precisely the same differential equation that applies to double lap joints. The insensitivity of the strength of those joints to overlap is shown in Fig. 11.

The strength increases associated with maximising the number of steps in the joint are substantial, as shown by a comparison between Figs 9, 8, and 6 of reference 19. The strength is maximised by having a separate step for each ply (0·005 in or 0·13 mm thick) and such a procedure may be appropriate for the highest load transfers between thick members. As a good starting point in design for typical composite laminates that are either quasi-isotropic or only slightly orthotropic, each step should be about 0·5 in (12·7 mm) long and the thickness increment should be in the range 0·02–0·03 in (0·5–0·76 mm) on each side of what is usually a titanium step plate. The exact increment should, of course, be a precise match for the corresponding increment of laminate thickness: for example, four layers of unidirectional tape at 0·005 in each would require 0·020 in steps, whilst two layers of satin weave cloth at 0·013 in each would require 0·026 in steps.

Just as with double lap joints, it is vital that at least one of the steps near the middle of the total bonded overlap contain a very lightly stressed deep elastic trough to prevent failure of the joint by creep. That step should probably be in the range 0·75–1·0 in (19–25 mm) long and its adequacy should be checked as part of the analysis. That is why program A4EI prints out not only the high adhesive stresses at the ends of each overlap but also the low stress at the middle of each step. That is taken to be indicative of the precise minimum, which is neither computed nor located.

7.7.3. Other considerations

It should be noted that the suggestions above are intended only to provide the starting point of the design process using the A4EG or A4EI computer programs. There is still usually room for improvement by minor 'tweaking' of the design once the internal stress distribution has been characterised. It is still desirable to establish that the potential shear strength of the adhesive would exceed by at least 50% the strength of the weaker

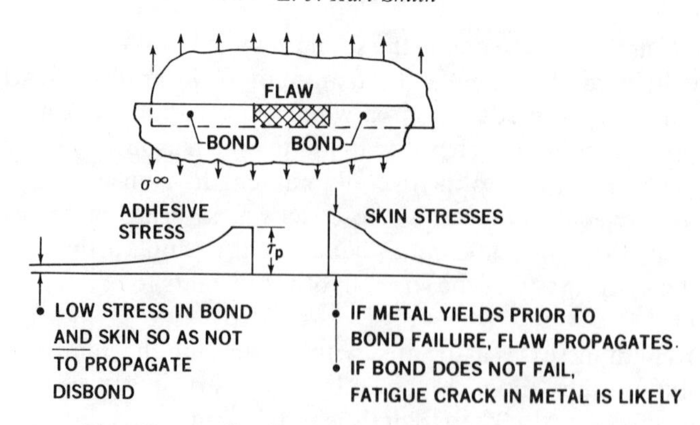

Fig. 25. Redistribution of load at flaws in bond.

adherend outside the joint. Otherwise the joint could become the weak link in the chain and suffer catastrophic disbonding as the consequence of only a minor local flaw, as explained in Fig. 25. For this safeguard it is acceptable to count on the entire length of the adhesive stress–strain curve to failure without any restriction on the peak strain developed under normal operating loads.

Practical experience with bonded stepped lap joints has taught that the layer of fibres or filaments immediately adjacent to the metal plate should never be perpendicular to the load direction. Otherwise, a much weaker joint results because of the tendency of such layers to split transversely and roll within themselves. This is easily understood by considering an axially loaded test coupon. The preferred fibre direction then would obviously be 0°, aligned along the specimen. A 90° ply would be quite unsuitable, with an intermediate performance coming from a +45° or −45° layer. Even with a large panel loaded primarily by in-plane shear, the same 0° direction is to be preferred because whichever 45° layer would be optimum for shear in one direction would be the worst possible for shear in the opposite direction. This consideration defines where the 0° plies are to be located within the stack-up, and therefore influences the basic laminate outside the joint as well.

As the load transfers increase to the level of tens of thousands of pounds per inch (several MN/m), it has already been found that the joint proportions are sufficiently critical to be of concern from the point of view of manufacturing tolerances. Experience in fabrication has already established that it is important not to overlap the end of the composite layers on to the next step or to stop them significantly short on their own steps. Otherwise, the laminate will be wrinkled within the joint area and will probably suffer internal delaminations during cyclic loading. Those, in turn, can so alter the internal load transfer as to cause a reduction in joint strength or life. Such problems become more acute as the incremental step thicknesses become larger. Indeed, for step thicknesses significantly greater than those recommended here, the author's analyses will not suffice. It would then be necessary to use an analysis which included a check on the possibility of delaminations within the composite adherends, as is done in ref. 21. A further important use for the analysis of stepped lap bonded joints is the establishment of the sensitivity of the joint strength to minor deviations from the nominal design. This task becomes progressively more important after the strength has been raised by repeated optimisation of the detailed dimensions.

The discussion above has been confined to symmetric stepped lap joints laid up on each side of a central step plate with the assembly co-cured and bonded together to ensure a good fit and the absence of warpage. A single-sided stepped lap bonded joint between dissimilar materials would warp so badly from a 350°F (177°C) cure as to probably break apart during the cool-down. If that problem were removed by the use of a room-temperature-cured adhesive, or the use of thermally compatible materials, a single bond surface stepped lap joint could be analysed as one side of a double bond joint. While there is some eccentricity in load path, that eccentricity is inevitably minor if the outermost steps have been restricted in thickness adequately to minimise any induced peel stresses.

7.8. SCARF JOINTS

It will seem strange to some that there is any need to have conducted research into the design of adhesively bonded scarf joints, particularly to those who still think of them as a means of transferring load uniformly across the bonded interface. Actually, that simplistic characterisation applies only to the transfer of load between two precisely identical members, both of which have perfect feather edges at their tips. The two dominant effects which cause real bonded scarf joints to deviate from that ideal are stiffness imbalance between the adherends and finite thicknesses at their ends. Adherend thermal mismatch between adherends of different materials is a further significant source of loss of structural efficiency.

Figure 26 shows how, for sufficiently long scarf lengths, the ratio of

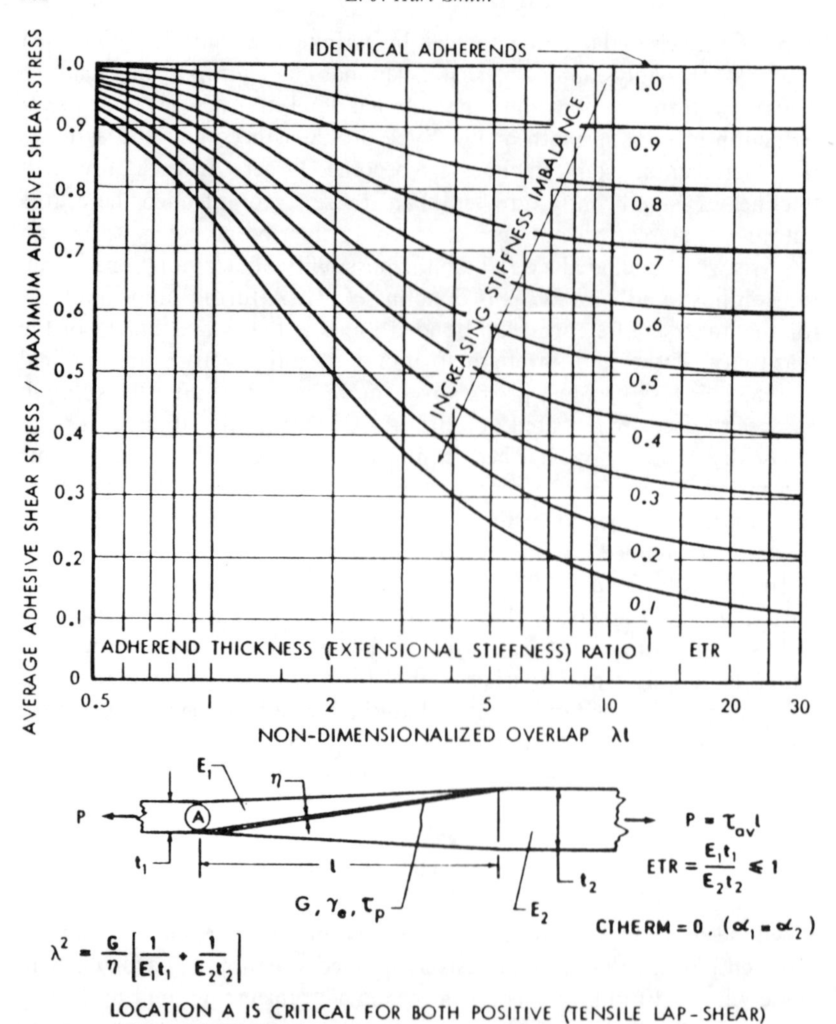

FIG. 26. Elastic strength of scarf bonded joints (effect of adherend stiffness imbalance).

average-to-peak adhesive shear strength tends asymptotically towards the lesser ratio of adherend stiffnesses. The critical location for the adhesive is usually at the point A shown. Most structural joints have scarf angles so low that the design is at or beyond the right-hand end of the diagram. Interestingly, even adherend thermal mismatch which weakens such joints (and has the maximum loss around $\lambda l = 2$) cannot overpower those asymptotic trends for large values of λl. Furthermore, adhesive ductility is effective in raising that ratio τ_{av}/τ_{max} only for the shorter overlaps also. Reference 17 contains a thorough derivation of the elastic–plastic transfer of load across idealised scarf joints. The design of such joints is simplified tremendously by the use of Fig. 26.

The practical difficulties of designing real scarf joints centre around the thickness of the tips of the adherends. Any attempt to actually create a feathered edge over a long length will inevitably result in weaknesses of the type shown in Fig. 27 due to bending and wrinkling of the metal adherends. The upper joint shown failed not in the thin basic section but across the tip of the internal doublers, where the nominal stress was only half as high, in repeated tests. The simpler design shown with external doublers was consistently stronger even though it had only half the bond area. If one accepts the need for a finite tip thickness, say 0·02–0·03 in (0·50–0·76 mm), one is then forced to analyse the joint by an approximation as a stepped lap joint in order to allow for the stress concentrations around the tip of the adherends. Indeed, the McDonnell Aircraft Company does not even keep the scarf joint analysis program in its active computer library. The use of the stepped lap joint programs to analyse scarf joints requires a fairly fine grid, even in those areas which are not critical. This can be seen from Fig. 28, which represents progressively more detailed models of the same scarf joint. The finite tip thickness was only 0·010 in (0·25 mm), yet there is still an appreciable stress concentration in the adhesive, even with the 32-step approximation. By assessing the entire output of the program instead of studying just the peak stresses, one can rapidly learn to make an accurate model with grids in which the fineness is varied as needed.

Whilst stepped lap joints and scarf joints have some very pronounced differences in their governing differential equations, one problem they share in common is the tendency for the thin tip of the stiffer member to fail in fatigue. With reference to Fig. 26, the load transfer is concentrated at the left-hand end of any unbalanced joint, so the build-up of load in the thicker member is not uniform. The load can, in fact, build up faster than the strength. Somewhat surprisingly, more failures of the adherends in Fig. 26 would occur at the tips of the stronger adherends rather than in the weaker adherend just outside the joint.

One potential problem with bonded scarf joints that is frequently overlooked is that the mathematically ideal form has a uniformly strained adhesive (ignoring the tip effects) and that, in consequence, the only way

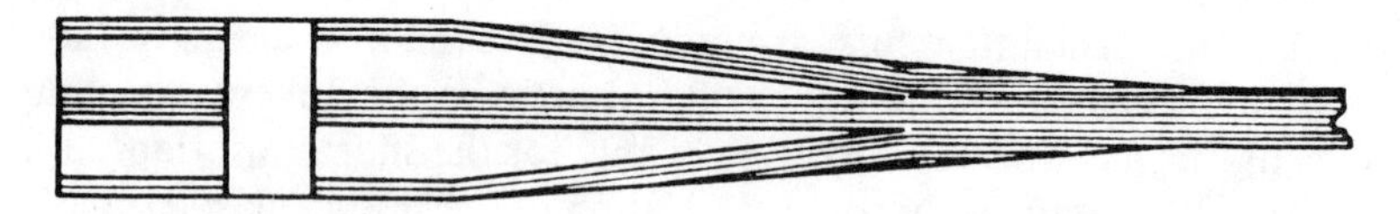

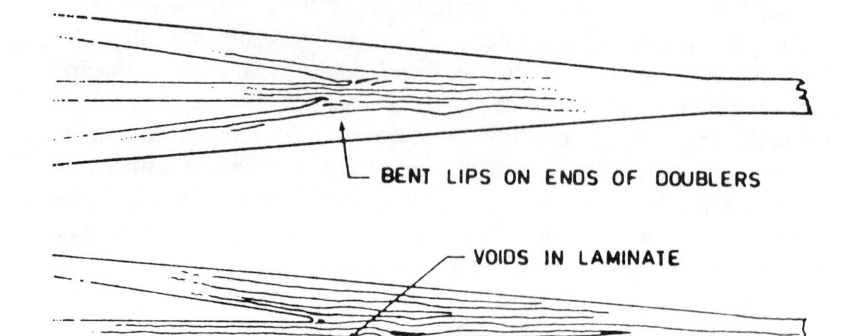

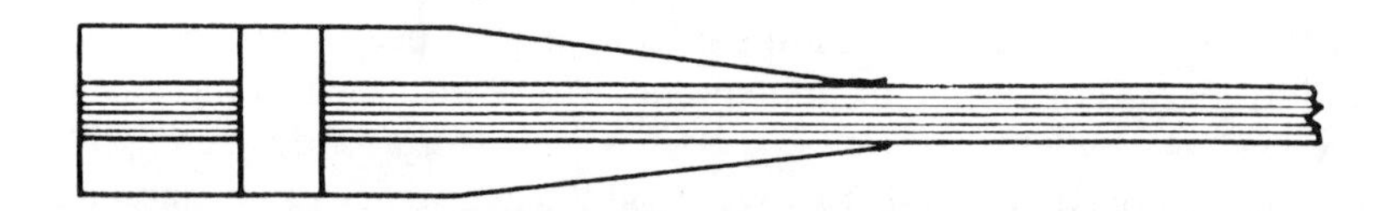

FIG. 27. Internal and external bonded doublers for composite laminates.

to restrict the minimum strain to prevent creep failures is to restrict the upper strain also. That leads to very long scarf joints with extremely small scarf angles, which is contrary to the concern expressed above about fatiguing the thin tips of scarfed adherends. A two-step scarf of graduated thickness build-up would seem to be an appropriate solution to this problem, as discussed in ref. 17.

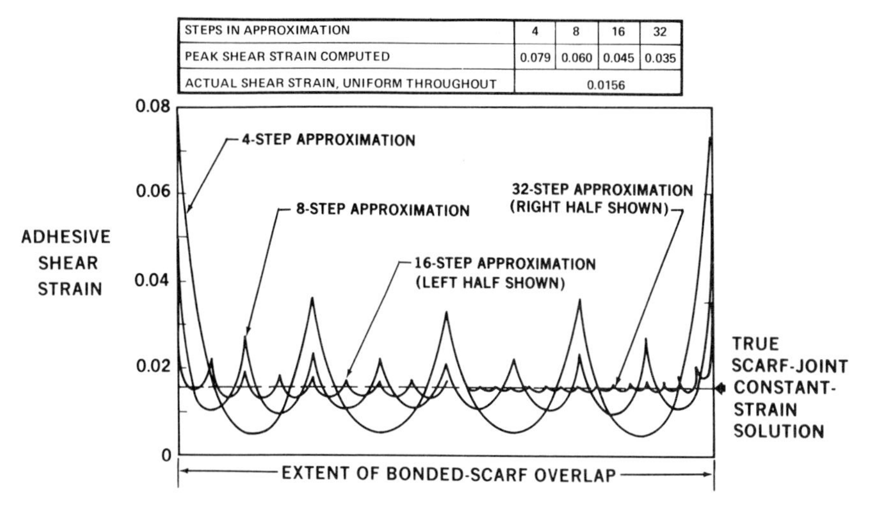

STEPS IN APPROXIMATION	4	8	16	32
PEAK SHEAR STRAIN COMPUTED	0.079	0.060	0.045	0.035
ACTUAL SHEAR STRAIN, UNIFORM THROUGHOUT	0.0156			

Fig. 28. Influence of precision of modelling on analysis of scarf joint. Best answers for a given total number of steps are achieved with a very fine grid near the ends and a coarser grid in the interior.

7.9. ADHESIVELY BONDED JOINTS OTHER THAN AT SPLICES

Whilst modern military aircraft with advanced composite structures contain many highly loaded adhesively bonded joints, most of the use of metal bonding in certain commercial transport aircraft, which is even more extensive, is in areas other than classical joints.

Both the British Aerospace (de Havilland) and Fokker companies have made extensive use of the bonding, rather than riveting, of stiffeners to skins with the prime objective of improving the static structural efficiency of panels loaded by compression or shear. The associated increases in fatigue lives and resistance to corrosion is like frosting the cake, since it would usually be the static ultimate design condition that established the weight of such structures. Whilst it is true that both manufacturers' choice of applications has been less limited as there is no involvement in making the largest of the transport aircraft, it is somewhat surprising that the applications of metal bonding in Europe have had such a different emphasis. The less widespread use of metal bonding in American aircraft structures has concentrated effort more on the local thickening of thin skins to permit countersinking and on the enhancement of the fatigue lives, in severe acoustic environments, of structures that are of minimum gauge and have considerable static strength margins. In all of these applications there has been no need for more than a cursory analysis of the strength of the adhesive bonds because of the minimal load transfer. For that reason, such reinforcements are usually made integral with the basic structure in fibrous composite construction.

The PABST programme[7] at Douglas identified a further area for the application of adhesive bonding in which good analysis was needed. This area of interest is the damage tolerance of adhesively bonded structures, such as stiffened panels, in which the bond loads are quite low provided that the metal elements are intact. However, if one element be broken by impact damage or due to fatigue cracking from one of the remaining rivet holes, the question arises as to whether a catastrophic widespread disbonding will follow or if such initial damage will remain confined. This topic is addressed in ref. 22, where it is shown that the likelihood of not containing the damage is remote in most of the structure that would be employed on fuselages or tail structures. This issue would become quite complex in fibrous composite construction because any initial damage would be more likely to be followed by delaminations within the basic material rather than by disbonding between the members.

7.10. FLAWS IN ADHESIVE BONDS

The variable-adhesive feature of the joint analysis program A4EI has permitted a comprehensive assessment of the effects in adhesives of flaws and porosity, or even the effects of moisture absorption or drying out, on the strength of bonded joints. References 2, 6 and 19 contain several examples depicting such effects. It was found quite generally that adhesive bonds have much more tolerance of flaws than is generally recognised. The critical location in a bonded joint is usually at one end of the overlap and flaws tend not to occur there; they form more often in the interior, where air bubbles can be trapped. Now, the interior of the joint is usually lightly loaded anyway and, even neglecting that effect, it was found that local flaws had no effect on the strength of bonded joints until they became so large as to shift the critical location away from the edge of the joint to a point adjacent to the flaw. Then, the load redistribution was so great that a substantial loss of strength would occur. This is shown particularly well in Figs 40–43 of ref. 6.

The history, and even current practice, of metal bonding is replete with instances of the structurally unnecessary repairs of flaws in bonded structures which have accomplished only a reduction in the service life of the structure by breaking the surface protection that had been afforded by the primer, anodise or etch. Reference 23 contains a more comprehensive discussion on the adhesively bonded repairs of metal structures.

A closely related topic is the use of adhesive bonding and mechanical fastening together to provide fail-safe characteristics. This topic is discussed thoroughly in ref. 19. The highlights are that, for thin and moderately loaded structures, adhesive bonding provides a fail-safe load path to protect the structure in the vicinity of the fastener holes. Conversely, thick, highly loaded, adhesively bonded structures need mechanical fasteners to protect the bond from unzipping catastrophically from any small initial damage. The vast differences between the behaviour of these combinations for thick and thin structures stems from the different relative strengths of the basic laminates and of the bonded joints, as shown in Fig. 4. The fail-safe load path changes at the point at which the adherends become stronger than the bond. The grossly dissimilar stiffnesses of the load paths through the bond and through the fasteners result in the two not acting together when both load paths are intact. The bond always transfers the load because it is so much stiffer. However, the combination is shown to be particularly useful once damage has been incurred to one or other of the alternative load paths.

7.11. REPAIR OF FIBROUS COMPOSITE STRUCTURES

The most vital step in the execution of an adhesively bonded repair to a fibrous composite structure is the thorough drying-out of the laminate to be repaired prior to hot bonding. Otherwise, even the small amount of water in the laminate will have a devastating effect on the strength of the adhesive or of the co-cured patch. This problem has been known for years (as explained in ref. 24) but has only recently attracted the attention it merited.[25] The difficulty of accomplishing this drying can be trivial for a thin skin on a control surface or fairing, which can be done with a small hot air gun, but could be horrendous for a thick component that may require a few days of gentle heating in an oven. The problem of removing large thick components from an aircraft and stripping them of heat-sensitive equipment, in conjunction with the difficulty that an adequate scarfing around any initial damage would leave little of the original structure remaining, has led to a recognition of the need for mechanical repairs in such cases. This, in turn, implies a need to allow for such repairs in the initial design, as with the AV-8B Harrier.

The other important step in the execution of adhesively bonded repairs, or of initial manufacture for that matter, is the thorough cleaning of the surfaces to be bonded. The need for mechanical abrasion, e.g. by grit blasting, is made very clear in ref. 26. Removal of peel plies is not a sufficient preparation for bonding. It is also important that the surface *not* be contaminated by 'cleaning solvent', as is often prescribed in official specifications. Scrubbing with an abrasive cleanser like 'Ajax' is preferable. Finally, the adequacy of the surface preparation should be verified by the standard water-break test,[27] just as for metal bonding.

Failure to adhere to these simple safeguards will nullify whatever effort had been expended in the analysis and design of bonded repairs. The problem is directly akin to the widespread corrosion problems that were encountered in America by the combination of the use of clad 7075 aluminium alloys with inadequate surface preparation and the first generation of 250°F (120°C)-cured epoxy adhesives which were too susceptible to the absorption of moisture.

7.12. CONCLUDING REMARKS

The purpose of this chapter has been to provide the designer with some of the finer details in the proportioning and analysis of various adhesively bonded joints, in a context in which the appropriate uses of each of the family become apparent.

The analysis methods discussed utilise an elastic–plastic shear model for the adhesive. The unsuitability for design, or even for adhesive selection, of the standard specifications of adhesive 'strength' is explained and a case is made for a more rational approach to the problem.

Only the foundations of the design process are presented here because the completion of the task often requires computer programs and, in any case, the sizes of the joint can vary with the fibrous reinforcement material, with the fibre pattern, and with the resin matrix. The simpler joints, of single-lap and double-lap configurations, can be designed accurately with no more than simple analytical formulae.

An extensive reference list contains the sources of more specific information related to the sizing and analysis of the various kinds of bonded joints, using the methods described here. While not all identified here, there are many production applications of adhesively bonded joints which have been designed by these methods. The various computer programs in the A4E... series, of which the early ones are freely available while the later ones can be obtained only on request to the USAF, have been used extensively throughout the aerospace industry for many years now.

ACKNOWLEDGEMENT

The origins of most of the material in this chapter are four US Government sponsored investigations, in which the author participated, at the Douglas Aircraft Company. The first is the NASA Langley Research Center contract NAS1–11234, 'Analysis and Design of Advanced Composite Bonded Joints', performed between 1971 and 1973. All of the material from that contract, as well as from the third and fourth contracts below, has been released by the US Government for unlimited distribution. The second contract was jointly managed and directed by the Air Force Flight Dynamics and Materials Laboratories, being the Primary Adhesively Bonded Structure Technology (PABST) program, F33615-75-C-3016, which occurred between 1975 and 1980. Some of those contract reports still have a restricted circulation, but all of the work in which the author's contributions have been documented have been released for unlimited distribution. Additional work was accomplished under the Air Force Flight Dynamics Laboratory contract F33615-79-C-3212, 'Design Methodology for Bonded–Bolted Composite Joints', during the period 1979 to 1981 (Volume II of that report, containing the computer codes, is restricted). Further information was developed under the Air Force Materials Laboratory contract F33615-80-C-5092, 'Adhesive Layer Thickness and Porosity Criteria for Bonded Joints', which extended from 1980 to 1982.

Also, it is probable that none of these developments would have occurred had not the original concept of an elastic–plastic model for adhesives in shear been sponsored under IRAD funding at the Douglas Aircraft Company, Long Beach, USA, throughout 1969 and 1970 to develop those ideas to a marketable level.

REFERENCES

1. Hart-Smith, L. J., Induced peel stresses in adhesive-bonded joints, Douglas Aircraft Company, McDonnell Douglas Corporation, Report MDC-J9422A, August 1982, USAF Contract Report AFWAL-TR-82-4172.

2. Hart-Smith, L. J., Further developments in the design and analysis of adhesive-bonded structural joints, Douglas Aircraft Company, McDonnell Douglas Corporation, Paper No. 6922, presented to *ASTM Conference on Joining of Composite Materials* (STP 749), Minneapolis, Minnesota, USA, April 1980.
3. Hart-Smith, L. J., Differences between adhesive behavior in test coupons and structural joints, Douglas Aircraft Company, McDonnell Douglas Corporation, Paper No. 7066, presented to *ASTM Adhesives Committee D-14 Meeting*, Phoenix, Arizona, USA, March 1981.
4. Hart-Smith, L. J., Design and analysis of adhesive-bonded joints, Douglas Aircraft Company, McDonnell Douglas Corporation, Paper No. 6059A, presented to *Air Force Conference on Fibrous Composites in Flight Vehicle Design*, Dayton, Ohio, USA, September 1972; *USAF Conference Proceedings*, AFFDL-TR-72-130.
5. Hart-Smith, L. J., Adhesive-bonded double-lap joints, Douglas Aircraft Company, NASA CR-112235, January 1973.
6. Hart-Smith, L. J., Effects of flaws and porosity on strength of adhesive-bonded joints, Douglas Aircraft Company, McDonnell Douglas Corporation, Report MDC-J4699, November 1981; USAF Contract Report AFWAL-TR-82-4172.
7. Hart-Smith, L. J., Adhesive bonding of aircraft primary structure, Douglas Aircraft Company, McDonnell Douglas Corporation, Paper No. 6979, presented to *SAE Aerospace Congress and Exposition*, Los Angeles, California, USA, October 1980.
8. Hart-Smith, L. J., Analysis and design of advanced composite bonded joints, Douglas Aircraft Company, NASA CR-2218, January 1973.
9. Volkersen, O., Die Nietkraftverteilung in Zugbeanspruchten Nietverbindungen mit konstanten Laschenquerschnitten, *Luftfahrtforschung*, **15**, 1938, 4–47.
10. Hart-Smith, L. J., Adhesive-bonded joints for composites—phenomenological considerations, Douglas Aircraft Company, McDonnell Douglas Corporation, presented to *Technology Conference Associates Conference on Advanced Composites Technology*, El Segundo, California, USA, March 1978.
11. Hart-Smith, L. J., Effects of adhesive layer edge thickness on strength of adhesive-bonded joints, Douglas Aircraft Company, McDonnell Douglas Corporation, Report MDC-J4675, May 1981; USAF Contract Report AFWAL-TR-82-4172.
12. Jones, R. and Callinan, R. J., Developments in the analysis and repair of cracked and uncracked structures, in: *Finite Element Methods in Engineering Proc. Third International Conference*, Sydney, Australia, 2–6 July, 1979, pp. 231–45.
13. Hart-Smith, L. J., Design and empirical analysis of bolted or riveted joints, Douglas Aircraft Company, McDonnell Douglas Corporation, Chapter 6 of this volume.
14. Goland, M. and Reissner, E., The stresses in cemented joints, *J. Appl. Mech.*, **11**, 1944, A17–A27.
15. Hart-Smith, L. J., Adhesive-bonded single-lap joints, Douglas Aircraft Company, NASA CR-112236, January 1973.

16. Corvelli, N. and Saleme, E., Analysis of bonded joints, Grumman Aerospace Corporation, Report No. ADR 02-01-70.1, July 1970. (Not available outside USA.)
17. Hart-Smith, L. J., Adhesive-bonded scarf and stepped-lap joints, Douglas Aircraft Company, NASA CR-112237, January 1973.
18. Smith, M. K., Hart-Smith, L. J. and Dietz, C. G., Interactive composite joint design, Douglas Aircraft Company, McDonnell Douglas Corporation, USAF Technical Report AFFDL-TR-78-38, April 1978.
19. Hart-Smith, L. J., Design methodology for bonded–bolted composite joints, Douglas Aircraft Company, McDonnell Douglas Corporation, USAF Technical Report AFWAL-TR-81-3154, February 1982.
20. Hart-Smith, L. J., Adhesive layer thickness and porosity criteria for bonded joints, Douglas Aircraft Company, McDonnell Douglas Corporation, USAF Technical Report AFWAL-TR-82-4172, December 1982.
21. Dickson, J. N., Hsu, T-M. and McKinney, J. M., *Development of an Understanding of the Fatigue Phenomena of Bonded and Bolted Joints in Advanced Filamentary Composite Materials*, Vol. I, Analysis Methods, June 1972, Lockheed Georgia Company, AFFDL-TR-72-64, Vol. I.
22. Hart-Smith, L. J., Adhesive bond stresses and strains at discontinuities and cracks in bonded structures, *Trans. ASME, J. Eng. Mat. Technol.*, **100**, 1978, 16–24.
23. Hart-Smith, L. J., Design and analysis of bonded repairs for metal aircraft structures, Douglas Aircraft Company, McDonnell Douglas Corporation, Paper No. 7089, presented to *International Workshop on Defense Applications of Advanced Repair Technology for Metal and Composite Structures*, Naval Research Laboratory, Washington, DC, USA, July 1981.
24. Hart-Smith, L. J., Design and analysis of bonded repairs for fibrous composite aircraft structures, Douglas Aircraft Company, McDonnell Douglas Corporation, Paper No. 7133, presented to *International Workshop on Defense Applications of Advanced Repair Technology for Metal and Composite Structures*, Naval Research Laboratory, Washington, DC, USA, July 1981.
25. Myhre, S. H., Labor, J. D. and Aker, S. C., Moisture problems in advanced composite structural repair, *Composites*, **13**(3), July 1982, 289–97.
26. Parker, B. M. and Waghorne, R. M., Surface pretreatment of carbon fibre-reinforced composites for adhesive bonding, *Composites*, **13**(3), July 1982, 280–8.
27. Hart-Smith, L. J., Ochsner, R. W. and Radecky, R. L., Surface preparation of fibrous composites for adhesive bonding or painting, *Douglas Service*, **41** (1st Quarter), 1984, 12–22.

Index